NUMERICAL SIMULATION RESEARCH PROGRESS

NUMERICAL SIMULATION RESEARCH PROGRESS

SIMONE P. COLOMBO
AND
CHRISTIAN L. RIZZO
EDITORS

Nova Science Publishers, Inc.
New York

For permission to use material from this book please contact us:
Telephone 631-231-7269; Fax 631-231-8175
Web Site: http://www.novapublishers.com

LIBRARY OF CONGRESS CATALOGING-IN-PUBLICATION DATA

Numerical simulation research progress / Simone P. Colombo and Christian L. Rizzo, editors.
p. cm.
ISBN 978-1-60456-783-0 (hardcover)
1. Numerical analysis--Simulation methods. 2. Numerical analysis--Research. 3. Quantitative research--Mathematical models. I. Colombo, Simone P. II. Rizzo, Christian L.
QA298.N86 2008
518--dc22
2008024212

Published by Nova Science Publishers, Inc. ✢ New York

CONTENTS

Preface		**vii**
Chapter 1	The Application of the Method of Characteristics for the Numerical Solution of Hyperbolic Differential Equations *Magdi Shoucri*	**1**
Chapter 2	Mixed Finite Difference-Spectral Numerical Approach for Kinetic and Fluid Description of Nonlinear Phenomena in Plasma Physics *Francesco Valentini, Marco Onofri and Leonardo Primavera*	**99**
Chapter 3	Numerical Simulations of the Nonlinear Solitary Waves *Jian-Qiang Sun, Meng-Zhao Qin and Hua Wei*	**141**
Chapter 4	Symmetry in Turbulence Simulation *Dina Razafindralandy, Aziz Hamdouni and Marx Chhay*	**161**
Chapter 5	The Shooting Method in Hydrothermal Optimal Control Problems *L. Bayón, J.M. Grau, M.M. Ruiz and P.M. Suárez*	**209**
Chapter 6	Exact N-Soliton Solutions of the Sharma-Tasso-Olver-Kadomtsev-Petviashvili (STO-KP) Equation *Abdul-Majid Wazwaz*	**249**
Chapter 7	Advances in Numerical Simulation of Granular Material *Mohammad Hadi Bordbar*	**257**
Index		**289**

PREFACE

Numerical simulation is the kind of simulation that uses numerical methods to quantitatively represent the evolution of a physical system. It pays much attention to the physical content of the simulation and emphasizes the goal that, from the numerical results of the simulation, knowledge of background processes and physical understanding of the simulation region can be obtained. In practice, numerical simulation uses the values that can best represent the real environment. The evolution of the system also strictly obeys the physical laws that govern the real physical processes in the simulation region. Then the result of such simulation can have a good representation of the real environment. From the result of such simulation we can safely draw proper conclusions and have a good understanding of the system. This new book presents leading research from around the world.

In Chapter 1, the method of characteristics for the numerical solution of hyperbolic type partial differential equations is applied to the kinetic equations of plasma physics, namely the Vlasov equations, for several selected problems especially pertinent to industrial plasmas. Eulerian Vlasov codes will be applied to study numerically a number of physical situations where plasma quasineutrality breaks down through boundary layers called plasma sheaths, which are either free or in contact with a wall. The plasma-wall transition problems are at the heart of an industrial revolution whose theme is the design of matter on the molecular scale. Examples will include a numerical solution for the problem of a classical sheath at a plasma-wall transition, a study of a capacitive discharge driven by a single frequency source, and a numerical study of the problem of ion extraction. The authors shall also present a numerical study of the sheath at a plasma-wall transition, in the presence of an external magnetic field, when the magnetic field is at grazing incidence with the wall. The problem of the formation of a charge separation and an electric field at a plasma edge in the presence of an external magnetic field will be studied numerically for the case of a cylindrical plasma column. The method of characteristics will be also applied to the relativistic Vlasov-Maxwell equations for the numerical simulation of two problems: the acceleration of particles in the wake field of a large amplitude laser pulse, and the interaction of a high intensity laser wave incident on an overdense plasma, when the frequency of the wave is below the plasma frequency. In all these problems, both electrons and ions will be treated using kinetic equations. The method of characteristics is quite general and can be applied to different equations of hyperbolic type in the field of mathematical physics. A rapid discussion will be presented for the application of this method to fluid type equations.

Several approaches are commonly used to study a system made of charged particles embedded in electromagnetic fields (commonly called a plasma): the kinetic one, where the dynamics of the system is studied through a statistical description in terms of the probability distribution function of finding a particle in a given volume of the phase space; the fluid one, which involves the solution of a system of partial differential equations similar to that describing neutral fluids.

Both in astrophysics and laboratory plasmas, one is often concerned with plasma equilibria, namely the instabilities which can develop in an ionized gas when an initial, inhomogeneous, equilibrium configuration is perturbed by fluctuations. In such problems, the nonlinear phase of the instability can be studied theoretically, in an effective way, only with the use of numerical simulations.

In such problems, the main numerical difficulty is provided by the fact that a very broad range of spatial and temporal scales are excited, requiring rather high resolutions both in space and time to approximate the numerical solution adequately. On the contrary, the geometries of the problem are generally rather simple (Cartesian, cylindrical or spherical), so that no need for special techniques for dealing with complicated or irregular domains, like finite element methods, are required. Furthermore, in the case of plasma instability problems, the solutions are rather regular, therefore no special techniques, like shock-capturing methods, or finite volume methods, are necessarily required.

The task of building equation solvers which implement the solution of a nonlinear set of partial differential equations with both inhomogeneous and periodic directions is often carried out with the use of mixed finite difference and spectral techniques. This permits a simple treatment of the boundary conditions by preserving a good numerical precision or, at least, the possibility to control effectively the numerical truncation errors.

In Chapter 2, the authors review the typical problems which are encountered when studying the effects of plasma instabilities by using either a kinetic or a fluid approach. The authors illustrate in detail some numerical codes which are currently used in the plasma physics domain. They further give examples of application of such codes to typical plasma problems and show how the numerical results obtained through the use of the mixed finite difference and spectral method give a description of physical phenomena which are in good agreement with the theoretical predictions.

The nonlinear solitary waves arise in fluid dynamics, wave phenomena in the atmosphere and the ocean, nonlinear optics, Bose-Einstein condensation, magnetohydrodynamics waves in warm plasma physics, etal. In Chapter 3, the symplectic and multi-symplectic methods in the structure preserving algorithms, which are well known methods in computation mathematical areas and have wide applications, are introduced. The soliton equations describing the evolution of the nonlinear solitary waves, such as the coupled nonlinear Schrodinger system, the Korteweg-de Veris equation, etal, were simulated by the symplectic and multi-symplectic methods. Numerical results showed that the symplectic and multi-symplectic methods can well simulate the evolution of the nonlinear solitary waves of the coupled nonlinear Schrodinger system, the Korteweg-de Veris equation, etal, moreover preserve the modulus square conserving property and the energy conserving property of the coupled nonlinear Schrodinger system, the Korteweg-de Veris equation, etal. the symplectic and multi-symplectic method in the structure preserving algorithms are advantage to simulating the behaviors of the nonlinear solitary waves.

Large-eddy simulation (LES) appears to be a good alternative between DNS (direct numerical simulation) and RANS (Reynolds-Averaged Navier-Stokes) in turbulence simulation. However, the modelisation of the effect of the small scales of the flow on the large ones, which is the cornerstone of this technique, remains a challenge in the domain. In Chapter 4, a recent and advanced method which aims to derive models, in isothermal and non-isothermal cases, respecting the physical properties and the mathematical structure of the equations is presented. This method is based on the symmetry group of the equations, which plays a fundamental role in the understanding of physical phenomena (existence of conservation laws, ...). A large part of the chapter is devoted to expose some interesting (classical and new) applications of the symmetry group theory. A discretisation method preserving the symmetry group is also presented. Such a discretisation method is necessary to avoid the violation of physical properties of the flow at the discrete scale. The method is grounded on the theory of moving frames which will be introduced in a didactical way and applied on Burgers'equation.

Chapter 5 presents an in-depth analysis of the behavior of the shooting method for the two-point boundary value problem that appears in the resolution of the short-term hydrothermal coordination problem when optimal control theory is used.

First the authors shall consider a problem without restrictions. The authors present a theorem that establishes sufficient conditions of existence of the extremals of the functional defined on the whole optimization interval. They analyze some additional hypotheses whose fulfilment guarantees that the extremals represent the strong relative minimum, and other hypotheses that ensure compatibility with any restriction on the admissible volume. They also present a result which makes it evident that the convexity of a functional is not necessary to guarantee that the extremals are solutions.

The authors then go on to study the problem arising in the case where the hydro- and thermal plants are subject to certain restrictions. Under these conditions, the nonemptiness of the set of admissible functions is sufficient to ensure the existence of a solution. They shall also study the conditions that guarantee non-existence of the boundary solutions. When we consider boundary solutions, the solution need not be "algebraically interior". To this end, the authors have developed the mathematical machinery to solve the problem in a very satisfactory way, from both the theoretical and algorithmic/ computational point of view.

Finally the authors illustrate the performance of our work with several numerical examples and the running of the proposed algorithms using Mathematica ©.

In Chapter 6 the authors derive a new completely integrable dispersive equation. The equation is obtained by extending the Sharma-Tasso-Olver (STO) equation using the extension sense of the Kadomtsev-Petviashvili (KP) equation. The newly derived Sharma-Tasso-Olver-Kadomtsev-Petviashvili (STO-KP) equation is studied by using the tanhcoth method to obtain kink solutions and periodic solutions. The powerful Hirota bilinear method is used to determine exact N-soliton solutions for this new integrable equation. The work highlights the power of the used methods and the structures of the obtained multiple-front solutions.

A granular flow is a mixture of grains and a fluid phase. Granular materials deform plastically like a solid under weak shear and they flow like a fluid under high shear. These materials exhibit other unusual behaviors, including pattern formation in the shaking of granular materials. The Discrete Element Method (DEM) and continuum type simulation are the most common methods used for the simulation of granular material in different flow regimes. In Chapter 7, some new improvements in the numerical simulation of this complex material are reviewed. In the first part, the collision between monosize and multisize spherical particles is reviewed and new activities in presenting a better contact force model for describing the collision between particles that has a main role in the accuracy of DEM models are presented. A detailed overview of a simplified model for collisions between particles of granular material is presented to be of use in molecular dynamic simulations. Detailed reported parameters of this new contact force model for two viscoelastic materials, glass and ice, are presented.

In the second part, the detailed overview of collective processes in gas-particle flows are used in developing a new set of hydrodynamic continuum equations to describe the deformation and flow of dense gas–particle mixtures. The constitutive equation used for the stress tensor provides an effective viscosity with a liquid-like character at low shear rates and a gaseous-like behavior at high shear rates. A review of the results of using this method in simulating a physical gas-particle system is done.

In the last part, a simplified discrete element method for the simulation of dense gas-particle flows is developed using the detailed overview of collective processes in gas-particle flows. The large eddy simulation technique is used for solving the governing equation of the continuous phase, and a Lagrangian method is used to predict the particle motion. In order to show the results of this simplified model, a simulation of a pentagonal prism containing glass balls with its base subjected to a sinusoidal vibration has been done.

In: Numerical Simulation Research Progress
Editors: Simone P. Colombo et al, pp. 1-98
ISBN: 978-1-60456-783-0

Chapter 1

THE APPLICATION OF THE METHOD OF CHARACTERISTICS FOR THE NUMERICAL SOLUTION OF HYPERBOLIC DIFFERENTIAL EQUATIONS

Magdi Shoucri
Institut de Recherche d'Hydro-Québec(IREQ), Varennes, Qué.,Canada J3X1S1

Abstract

The method of characteristics for the numerical solution of hyperbolic type partial differential equations is applied to the kinetic equations of plasma physics, namely the Vlasov equations, for several selected problems especially pertinent to industrial plasmas. Eulerian Vlasov codes will be applied to study numerically a number of physical situations where plasma quasineutrality breaks down through boundary layers called plasma sheaths, which are either free or in contact with a wall. The plasma-wall transition problems are at the heart of an industrial revolution whose theme is the design of matter on the molecular scale. Examples will include a numerical solution for the problem of a classical sheath at a plasma-wall transition, a study of a capacitive discharge driven by a single frequency source, and a numerical study of the problem of ion extraction. We shall also present a numerical study of the sheath at a plasma-wall transition, in the presence of an external magnetic field, when the magnetic field is at grazing incidence with the wall. The problem of the formation of a charge separation and an electric field at a plasma edge in the presence of an external magnetic field will be studied numerically for the case of a cylindrical plasma column. The method of characteristics will be also applied to the relativistic Vlasov-Maxwell equations for the numerical simulation of two problems: the acceleration of particles in the wake field of a large amplitude laser pulse, and the interaction of a high intensity laser wave incident on an overdense plasma, when the frequency of the wave is below the plasma frequency. In all these problems, both electrons and ions will be treated using kinetic equations. The method of characteristics is quite general and can be applied to different equations of hyperbolic type in the field of mathematical physics. A rapid discussion will be presented for the application of this method to fluid type equations.

1. Introduction

The method of characteristics for the numerical solution of differential equations of the hyperbolic type has been recently discussed [1], and some applications for the kinetic equations of plasmas have been presented, together with applications to fluid-type equations for the numerical solution of the shallow water equations and for the numerical solution of the equations of incompressible ideal magnetohydrodynamic (MHD) flows in plasmas. Numerical methods for hyperbolic equations are generally more complicated and difficult to develop compared to the numerical methods applied for parabolic or elliptic type partial differential equations. There have been important advances in the last few decades in the domain of the numerical solution of hyperbolic type partial differential equations using the method of characteristics, when applied to solve the initial value problem for general first order partial differential equations. The methods presented in [1] are Eulerian methods which use a computational mesh to discretize the equations on a fixed grid, and have been successfully applied to different important problems in plasma physics involving kinetic equations, such as laser-plasma interaction [2,3,4], the calculation of an electric field at a plasma edge [5], and to gyro-kinetic codes in plasma physics [6-12]. They present the great advantage of having a low noise level, which allows accurate results in the low density regions of the phase-space [13]. This is in contrast with the methods of particle-in-cell (PIC) for instance, frequently used in plasma physics, which have a level of numerical noise which decreases only as $1/\sqrt{N}$, where N is the number of particles in any particular computational cell. This noise problem becomes important if the physics of interest is in the low density regions of the phase-space. In fluids, the solution of the shallow water equations presented a very nice application of the method of characteristics [1,14-16]. In the applications presented in [1], the computation was usually done on a fixed grid, so no dynamical grid adjustment was necessary, and interpolation was restricted to the use of a cubic spline, so altogether the method remained relatively simple. The method of characteristics has been successfully applied in fluids to problems having shock wave solution [17]. Large Courant-Frederichs-Levy (CFL) computation parameter is possible, and therefore the time-step numerical limitation by large velocities can be removed, if the physics makes it possible.

Let us briefly review the numerical technique which describes the application of this method to a one-dimensional problem, in order to fix some ideas and notations (more details are presented in [1]). Consider the following simple hyperbolic type advection equation

$$\frac{\partial f}{\partial t}+c\frac{\partial f}{\partial x}=0. \tag{1.1}$$

If c is a constant, all the points on the solution profile will move at the same speed along the characteristics determined by the solution of $dx/dt=c$. Let us assume the initial condition $x(0)=x_0$. The solution of the characteristic equations gives the characteristic curves $x=x_0+ct$ (straight lines for the present case where c is a constant), where x_0 is the point where each curve intersects the x-axis at t=0 in the x-t plane. If at t=0 we have $f\equiv f(x_0)$, $x_0=x-ct$, then $f(x,t)=f(x-ct)$ (a simple translational motion of a wave), since the function $f(x,t)$ remains constant along the characteristics. This can be verified if we differentiate $f(x,t)$ along one of these curves to find the rate of change of f along the characteristic:

$$\frac{df(x(t),t)}{dt} = \frac{\partial f(x(t),t)}{\partial t} + \frac{dx}{dt}\frac{\partial f(x(t),t)}{\partial x} = \frac{\partial f}{\partial t} + c\frac{\partial f}{\partial x} = 0\,. \tag{1.2}$$

which verifies that f is constant along a characteristic curve. If c depends on x and t, the previous solution remains still approximately correct in a short time Δt, *i.e.* we can write $f(x,t+\Delta t) \approx f(x - c\Delta t, t)$. Different approximations are used in this case for the calculations of c and of $f(x - c\Delta t, t)$. These approximations involve interpolation using different polynomials, on a fixed Eulerian grid, from the known values of the function at the neighbouring grid points. These methods are usually called Euler-Lagrange methods, or semi-Lagrangian methods. Euler-Lagrange methods reflect better the fact that we are indeed dealing with a fixed Eulerian grid. In the results we present in this chapter, we use a simple cubic spline for the polynomial interpolation, as discussed in [1]. The cubic spline interpolation on a fixed Eulerian grid has shown to be accurate in many applications [13,18,19,20]. For the more general case where several dimensions are involved, the fractional step technique allows sometimes the reduction of the multi-dimensional equation to an equivalent set of one-dimensional equations [21]. For the shallow water equations for instance, this turned out to be very successful [14-16]. In plasma physics, since the early work reported in [22-30], the applications of the fractional step method associated with the method of characteristics to the kinetic equations of plasmas, especially to the Vlasov equation, is now widely used especially in problems of laser-plasma interaction and in gyro-kinetic codes (see the review works in [1,31]). Interpolation in several dimensions using a tensor product of cubic *B*-splines has been also successfully used recently [1,31-33], especially when the fractional step technique cannot be applied. The technique of multi-dimensional interpolation has long been extensively applied in the field of meteorology [34,35], where it is called the semi-Lagrangian method (although we prefer to call it the Euler-Lagrange method, since in this method we essentially use a fixed Eulerian grid, and use an iterative process to take care of the variation of the velocity along the characteristic curves). A more complete study on splines can be found in [36,37], and an important theoretical study on the method of characteristics can be found, for instance, in [38]. More applications to plasma physics problems can be found in [39-46].

In the present chapter, the numerical ideas outlined in this introduction, based essentially on the method of characteristics coupled to relatively simple cubic spline interpolations on fixed Eulerian grids as presented in [1], will be applied to selected problems in plasma physics. We study numerically a number of physical situations where plasma quasineutrality breaks down through boundary layers called plasma sheaths, which are either free or in contact with a wall or an electrode. Plasmas produced in the laboratories are generally in contact with a material surface, so understanding the physical processes in plasma-wall transitions and interactions is very important. In low pressure plasmas, plasma-wall interactions are at the heart of an industrial revolution whose theme is the design of matter on the molecular scale. Plasma etching is at the core of this revolution since the integrated circuit was invented in 1959. Treating a surface is generally the goal to achieve in these applications. So the control of the plasma properties of the processing systems in real time, during production, and the knowledge of the ion energy spectrum in front of the wall is fundamental for surface treatment. In probe measurements, a correct knowledge of the distribution functions at the plasma-wall transition is important in order to interpret correctly the results of the measurements (e.g. temperature, density, electric current). These problems generally involve the solution of kinetic equations of plasmas called the Vlasov equations. In section 2 we study the classical problem of a sheath formation at a plasma-wall transition. Macroscopic quantities (density, current, energy flux, etc…) will be calculated from the kinetic solution and shown to give very good agreement with available theoretical results. In section 3, we study the problem of capacitive coupling, which is important in many problems of surface treatment. Capacitive discharges are often driven by a single frequency source, or a combination of two frequency sources. We will restrict ourselves in the present chapter to the case of a single frequency source, and present solutions for the evolution of the phase-space structure of the distribution functions. In section 4, we study the problem of the modeling of ion extraction from a plasma. This problem is of great importance in many physical devices like ion accelerators, electrical discharges and in many related industrial applications. Again accurate solutions for the phase-space structure of the distribution functions will be presented. We study in section 5 the problem of a sheath in the presence of an external magnetic field, when the magnetic field is at grazing incidence with the wall. This problem is important to understand probe characteristics in the presence of magnetic fields. It is also of fundamental importance to optimize the heat load on the divertor or the limiter of a tokamak, and to study the effect of the scrape-off layers (the region where the plasma is directly in contact with the wall) on the tokamak wall. In this case the plasma can cause significant erosion to the material surface of a tokamak. We will show in this case that there exists a critical angle below which steady state oscillations exist, associated with an enhancement of the energy fluxes transmitted to the wall. In section 6, we apply the method of characteristics to the more complex cylindrical geometry to study the formation of an electric field at a plasma edge, in the presence of an external magnetic field along the axis of the cylinder. This basic problem is of great importance to understand the formation of an electric field at the edge of a plasma in tokamaks, in what is called the H-mode. The code in this case applies a numerical scheme based on a two-dimensional (2D) advection technique, of second order accuracy in time-step, where the value of the distribution function is advanced in time by interpolating in two dimensions along the characteristics (both in space and in velocity space), using a tensor product of cubic *B*-spline [32,33]. The simulation results show in this case that the electric field along the steep gradient is balanced exactly by the gradient of the pressure term. The problems in sections 5 and 6 are four dimensional in phase-space. Two problems in sections 7 and 8 will be studied by applying the method of characteristics for the numerical solution of the relativistic Vlasov-Maxwell equations. In section 7 we study the acceleration of particles in the wake field of large amplitude laser pulses. Plasma wake-field accelerators can double electron energy in a meter length, compared to full-scale accelerators such as the one at the Stanford Linear Acceleration Center (SLAC), which requires about two miles to achieve similar energy levels. In this case the numerical integration will take place by applying the method of characteristics in two dimensions, using for interpolation a tensor product of cubic *B*-spline [32]. This method allows to reach physically sensible answers even in very relativistic regimes. In section 8 we study the interaction of an intense laser wave whose frequency is below the plasma frequency, with the overdense plasma edge. If the laser wave is sufficiently intense to make electrons relativistic, the plasma frequency is modified by relativistic mass variation, and in the interaction between the wave and the plasma edge an electric field is formed at the edge due to the laser-produced electrons accumulating in front of the wave. This electric field accelerates ions. Laser-induced ion acceleration seems one of the most promising applications of laser pulses interacting with solid matter and high density targets. We will address all these problems by solving the appropriate kinetic equations for both electrons and ions. Our optic will be to present the details of the numerical methods and illustrate the wealth of physics which can be obtained from the kinetic equations, which is an invitation to continue future works to more complicated systems involving higher dimensions and more physics. Last but not least, a rapid discussion of recent applications of the method of characteristics to fluid-type equations will be presented in section 9. This problem has been already discussed in [1], and would require a chapter by itself. The conclusion in section 10 will mention rapidly some more recent applications of the method of characteristics to kinetic equations, especially in gyro-kinetic codes in tokamaks and in the field of semi-conductors.

2. The Collisionless Kinetic Sheath

The plasma-wall transition problem is a fundamental problem in several areas of physics where plasma is in contact with a wall. In this section, the solution of the collisionless kinetic sheath at a plasma-wall transition in the absence of an external magnetic field is studied numerically using an Eulerian Vlasov code for the two-dimensional phase-space problem (one space dimension and one velocity dimension). There is an abundant literature on this subject, where generally the electrons are treated using an adiabatic law (see for instance [47] for a review on this problem, and the more recent works in [48,49]). In the present work the Vlasov equations for both electrons and ions are solved directly in phase-space by a splitting or fractional step method previously reported in the literature [22-28], and integrated along their characteristics. Some of the basic macroscopic quantities associated with the sheath equilibrium will be calculated. We take advantage of the low numerical noise level of the Eulerian Vlasov code to calculate accurately the different parameters associated with the non-neutral plasma at a plasma-wall transition.

2.1. The Relevant Equations

The relevant equations to be solved are the Vlasov equations for electrons and ions, coupled with Poisson's equation. They are written in dimensionless form in the 2D phase-space x-v_x as follows:

$$\frac{\partial f_{e,i}}{\partial t} + v_x \cdot \frac{\partial f_{e,i}}{\partial x} + M_{e,i} E_x \cdot \frac{\partial f_{e,i}}{\partial v_x} = 0 . \tag{2.1}$$

The subscripts e and i refer to electrons and ions. $f_{e,i} = f_{e,i}(x, v_x, t)$ is the electron (ion) distribution function. The Vlasov equation in Eq. (2.1) are coupled to Poisson's equation for the potential :

$$\frac{\partial^2 \varphi}{\partial x^2} = -(n_i - n_e), \text{ where } n_{e,i} = \int_{-\infty}^{\infty} f_{e,i}\, dv_x , \tag{2.2}$$

n_e and n_i are the electrons and ions density respectively. The electric field is calculated from the equation:

$$E_x = -\frac{\partial \varphi}{\partial x} \tag{2.3}$$

Time is normalized to ω_{pi}^{-1}, where ω_{pi} is the ions plasma frequency, velocity is normalized to the acoustic velocity $c_s = (T_e / m_i)^{1/2}$, where T_e is the electron temperature and m_i is the ion mass. Space is normalized to the Debye length $\lambda_{De} = c_s / \omega_{pi}$. With this normalization, we have M_i =+1 in Eq. (2.1) for the ions, and $M_e = -m_i / m_e$ for the electrons. The condition for the sheath formation demands that the ions enter the sheath region with a high velocity which cannot be generated by thermal motion. This is called the Bohm criterion [50,51]. We assume ions are injected at the right boundary $x=L$ at $t=0$ with the sound speed $v_0 = \sqrt{1 + 1/t_{ei}}$, where $t_{ei} = T_e / T_i$, T_i is the ion temperature The ions distribution function at the right boundary then takes the form of a shifted Maxwellian:

$$f_i(x = L, v_x, t = 0) \sim Cg(v_x^2) \frac{e^{-t_{ei}(v_x - v_0)^2/2}}{\sqrt{2\pi / t_{ei}}} \tag{2.4}$$

C is a constant and $g(v_x^2)$ is a smoothing factor chosen to minimize the deformation of f_i at v_x=0 to allow for a smooth transition to f_i =0 for v_x>0, which we take in the following form [48,52]:

$$g(v_x^2) = (1 - e^{-2t_{ei}v_x^2}) \tag{2.5}$$

The electron distribution function $f_e(x=L, v_x, t=0)$ at $x=L$ is initially taken as a Maxwellian $f_e(x = L, v_x, t = 0) = N_e \exp(-m_e v_x^2 / (2m_i))$, normalized and truncated at a velocity V_c such that its density and its current are equal to the density and current obtained from Eq. (2.4). V_c is determined as follows. We assume that initially at the right boundary the injected ions have a density $n_i = \int_{-\infty}^{0} f_i(v_x)dv_x$. The initial density of the electrons at the right boundary should be equal to the density n_i :

$$n_i = \int_{-\infty}^{V_c} f_e(v_x)dv_x = N_e\sqrt{\frac{\pi m_i}{2m_e}}\left[erf\left(V_c / \sqrt{\frac{2m_i}{m_e}}\right)+1\right], \tag{2.6}$$

where $erf(y) = \frac{2}{\sqrt{\pi}}\int_0^y e^{-y^2}dy$ is the error function. Equation (2.6) gives a first relation between N_e and V_c. The ion current density at $x=L$, $J_i = \int_{-\infty}^{0} v_x f_i(v_x)dv_x$ (which is negative, the ions are injected at the right boundary in the negative direction towards the wall at the left), must be equal to the electron current density at $x=L$: $J_e = \int_{-\infty}^{V_c} v_x f_e(v_x)dv_x$. Substituting for f_e in this previous expression, we arrive after some straightforward algebra to the following transcendental equation for V_c :

$$\frac{J_i}{n_i} = -\sqrt{\frac{2m_i}{\pi m_e}}\frac{e^{-m_e V_c^2/2m_i}}{erf\left(V_c\sqrt{\frac{m_e}{2m_i}}\right)+1}, \tag{2.7}$$

which is solved numerically for V_c. We use deuterium ions, and an initial value for the ratio t_{ei} =20. With these parameters, the constant C in Eq.(2.4) is equal to 1 to allow for an initial ion density of 1 at $x=L$. This is not always so for lower values of t_{ei}, as presented in [52]. These distribution functions for ions and electrons are assumed to extend initially uniformly throughout the domain, so that the system is initially neutral. We assume in the present simulation that the ion and electron particles hitting a wall at $x=0$ are collected by a floating wall. Then:

$$\left.\frac{\partial E_x}{\partial t}\right|_{x=0} = -(J_{xi} - J_{xe})\Big|_{x=0} \text{ ; or } E_x\Big|_{x=0} = -\int_0^t (J_{xi} - J_{xe})\Big|_{x=0} dt \tag{2.8}$$

where the currents due to the particles hitting the wall are given by:

$$J_{xe,i}\Big|_{x=0} = \int_{-\infty}^{0} v_x f_{e,i}(x=0, v_x)dv_x \tag{2.9}$$

Integrating Eqs. (2.2) and (2.3) over the domain (0,L), we get:

$$E_x\Big|_{x=L} - E_x\Big|_{x=0} = \int_0^L (n_i - n_e)dx = \sigma \tag{2.10}$$

where σ is the total charge appearing in the system. We do not fix the distribution functions at the right boundary $x=L$. We rather assume that the plasma ions and electrons can circulate at the right boundary, *i.e.* that the plasma extends at the right boundary such that the point next to the last grid point is identical to the last grid point. With the present parameters and boundary conditions, the distribution functions at $x=L$ remained essentially the same as the initial values during the simulations. In the simulation results we present, the electric field at $x=0$ is calculated from Eq. (2.8). We found that the electric field at $x=L$ remained negligibly small. It follows from Eq. (2.10) that the charge accumulated on the plate calculated from Eq. (2.8) is practically equal and opposite in sign to the charge σ appearing in the system. This equality was verified to the third decimal during the present simulation. This has been also the case in the results which have been previously presented [52] for different values of t_{ei}. We take a small time step $\Delta t = 1.25\text{x}10^{-3}$ for the ions, and for the electrons this time step was smaller by a factor of 10, our main concern is to get the results as accurate as possible by following accurately the mobile electrons. The ratio of the electrons period of oscillation over the ions period of oscillation is of the order of 10^{-2} for a deuterium plasma. This means that with the present value of Δt we have about 10 time steps for the ions and 100 time steps for the electrons in each electrons plasma oscillation. The length of the system is L=40 Debye lengths, and 500 grid points were used in space, 200 grid points were used in velocity space for the ions, and 300 grid points in velocity space for the electrons. This generous number of grid points is to guarantee accuracy. The maximum and minimum velocities for the electrons are taken to be respectively $4./\sqrt{m_e/m_i}$ and $-4./\sqrt{m_e/m_i}$. For the ions, the maximum velocity is taken on the velocity grid to be $0.1\text{x}4./\sqrt{t_{ei}}$ (the ion distribution function is in fact 0 for v_x>0), and the minimum velocity is taken to be $-4.\text{x}4./\sqrt{t_{ei}}$, with t_{ei} =20 for the present simulation as we previously mentioned.

2.2. The Numerical Scheme

We use a method of fractional step [22-26] for the solution of Eq.(2.1). The numerical scheme is similar to what has been discussed in [1]. We discuss in details this scheme, which will be also used in sections 2 and 3. A method which has a precision of second order in time is obtained by splitting Eq.(2.1) as follows (we use f for either f_e or f_i):

Step1 - Solve $\dfrac{\partial f}{\partial t} + v_x \dfrac{\partial f}{\partial x} = 0$ for a step $\Delta t/2$ (2.11)

- Solve Poisson's equation for the electric field which we denote by E_x^*.

Step2 -Solve $\dfrac{\partial f}{\partial t} + M_{e,i} E_x^* \dfrac{\partial f}{\partial v_x} = 0$ for a step Δt (2.12)

Step3 -Solve $$\frac{\partial f}{\partial t} + v_x \frac{\partial f}{\partial x} = 0 \text{ for a step } \Delta t / 2 \quad (2.13)$$

In this 2D phase-space problem the shifts become fractional, i.e. each of the dimension of the phase-space is shifted separately along the corresponding characteristic. This splitting has the advantage that each of the x or v updates is a linear advection which is done by applying successively the following shifts to Eqs.(2.11-2.13):

$$f^a(x, v_x, t + \Delta t / 2) = f(x - v_x \Delta t / 2, v_x, t), \quad (2.14)$$

$$f^b(x, v_x, t + \Delta t) = f^a(x, v_x - M_{e,i} E_x^* \Delta t, t), \quad (2.15)$$

$$f(x, v_x, t + \Delta t / 2) = f^b(x - v_x \Delta t / 2, v_x, t), \quad (2.16)$$

That is, the shift is performed first in space a half time-step. Since v_x is an independent variable, the shift in Eq.(2.14) is done for each value of v_x. The shifted value is calculated using a cubic spline interpolation [1,36,37]. This is followed by the solution of Poisson`s equation for the calculation of the electric field $E_x^*$, and this electric field is used for the calculation of the shift in velocity space as indicated in Eq.(2.15), using again a cubic spline interpolation. Poisson`s equation in Eq.(2.2) is discretized in space as a tridiagonal matrix:

$$\varphi_{j-1} - 2\varphi_j + \varphi_{j+1} = -\Delta x^2 (n_{ij} - n_{ej}). \quad (2.17)$$

where $\Delta x = L / N_x$, the subscript j denotes the grid-point x_j. The ion and electron densities are calculated from Eq.(2.2). From Eq.(2.8) $E_x\big|_{x=0}$ is calculated (we have previously mentioned that $E_x\big|_{x=L}$ is negligible), which provides for the gradient of the potential at $x = 0$ from Eq.(2.3). We fix the potential at zero at the wall, then the tridiagonal matrix in Eq.(2.17) is solved. From φ we calculate E_x^* (Eq.(2.2)). We do verify that $E_x^*\big|_{x=L}$ is indeed negligible. Then the shift in Eq.(2.15) is calculated, and finally the second half of the spatial shift is repeated in Eq.(2.16).The overall precision of this numerical scheme is $O(\Delta t^2)$ [1,22].

In Eqs.(2.14-2.16) v_x is independent of x, and E_x is independent of v_x. Then a simple cubic spline interpolation, similar to what has been presented in [1] can be used to interpolate in Eq.(2.14-2.16). For instance, if we assume $0 < \Delta < 1$, with a uniform grid size normalized to 1 and Δ is a constant, then we use a Taylor expansion to calculate the shifted value $y_j(x_j + \Delta) = \tilde{y}_j$, with the function $y = f(x)$ and the notation $f(x_j) = f_j$:

$$y_j(x_j + \Delta) = \tilde{y}_j = f_j + p_j \Delta + \frac{1}{2} s_j \Delta^2 + g_j \Delta^3 \quad (2.18)$$

p_j, s_j and g_j are respectively the derivative, second derivative and third derivative of the function $f(x)$ at the grid point $j \equiv x_j$. We write that the function, its derivative and second derivative are continuous at every grid point, we get the following cubic spline relations on a uniform grid [36,37]:

$$p_{j-1} + 4p_j + p_{j+1} = 3(f_{j+1} - f_{j-1}) \tag{2.19}$$

$$s_{j-1} + 4s_j + s_{j+1} = 6(f_{j-1} - 2f_j + f_{j-1}) \tag{2.20}$$

$$g_{j-1} + 4g_j + g_{j+1} = -f_{j-1} + 3f_j - 3f_{j+1} + f_{j+2} \tag{2.21}$$

We can verify after substitution from Eqs.(2.19-2.21) in Eq.(2.18) the following relation:

$$\widetilde{y}_{j-1} + 4\widetilde{y}_j + \widetilde{y}_{j+1} = Af_{j-1} + Bf_j + Cf_{j+1} + Df_{j+2} \tag{2.22}$$

$$A = (1-\Delta)^3 \tag{2.23}$$

$$B = 4 - 3\Delta^2(1 + (1-\Delta)) \tag{2.24}$$

$$C = 4 - 3(1-\Delta)^2(1+\Delta) \tag{2.25}$$

$$D = \Delta^3 \tag{2.26}$$

We verify that for $\Delta = 0$ we have from Eq.(2.22):

$$\widetilde{y}_{j-1} + 4\widetilde{y}_j + \widetilde{y}_{j+1} = f_{j-1} + 4f_j + f_{j+1}$$

i.e.

$$\widetilde{y}_j = f_j$$

and for $\Delta = 1$ we verify that:

$$\widetilde{y}_{j-1} + 4\widetilde{y}_j + \widetilde{y}_{j+1} = f_j + 4f_{j+1} + f_{j+2}$$

i.e.

$$\widetilde{y}_j = f_{j+1}$$

as it should be.

The inversion of the tridiagonal matrix in Eq.(2.22) with appropriate boundary conditions determines $\widetilde{y}_j$. The calculation of $y_j(x_j - \Delta) = \widetilde{y}_j$ with $0 < \Delta < 1$ is done in a similar way and leads to the following relation:

$$\tilde{y}_{j-1} + 4\tilde{y}_j + \tilde{y}_{j+1} = Af_{j+1} + Bf_j + Cf_{j-1} + Df_{j-2} \tag{2.27}$$

These simple cubic spline relations to interpolate on a uniform grid will be used for the results presented in this section, and in the following sections 3, 4 and 5, unless otherwise stated.

2.3. Results

We have with the present parameters $v_0 = \sqrt{1.05}$ in Eq.(2.4), since t_{ei}=20. The distribution function for the ions at time t=30 is shown in Fig.(1). Equilibrium has already been reached. The upper plot in Fig.(1) shows the contours of the ion distribution function, accelerating towards the wall. The bottom curves in Fig.(1) show the distribution functions at different positions (from left to right) x=0, x=L/30, x=L/10, x=L/2, and x=L, taken from the upper plot. L=20 Debye length in the present simulation. For the plot at the bottom of Fig.(1), while the peak of the distributions remained essentially the same, we see the distribution functions cooling as they approach the wall. The upper plot in Fig.(2) shows the contour plot of the distribution function for the electrons at t=30, and the bottom curves give the distribution functions for the electrons at the same positions (from the lower to the higher curve) as for

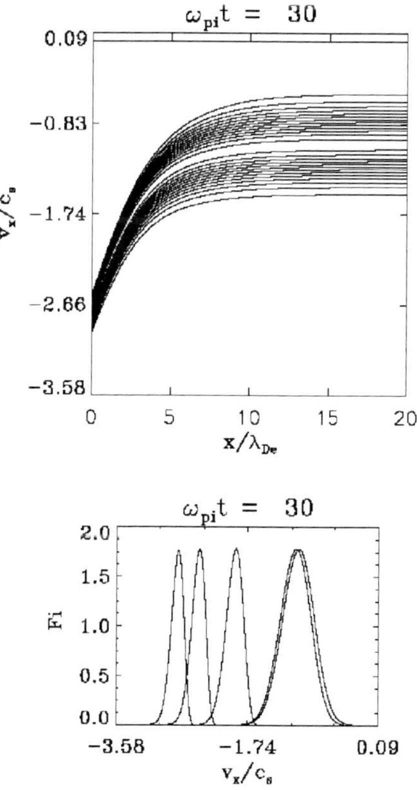

Figure 1. Distribution function for the ions.

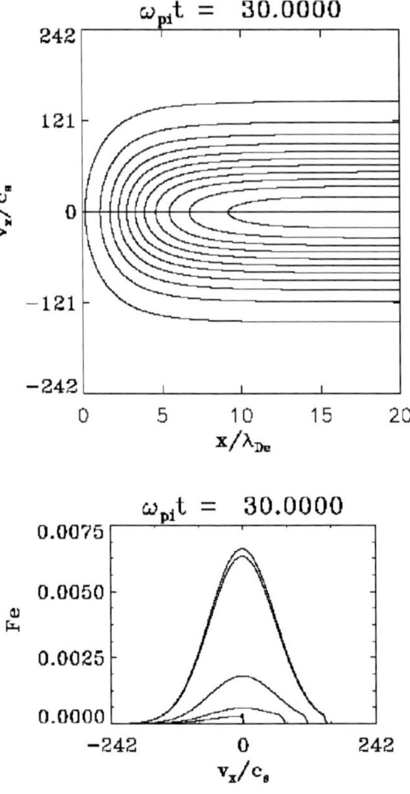

Figure 2. Distribution function for the electrons.

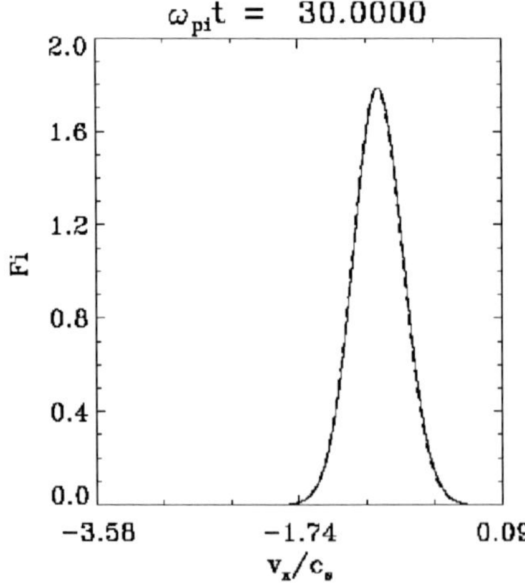

Figure 3. Distribution funtion for the ions at $x=L$.

the ions in Fig.(1). Fig.(3) shows the distribution function for the ions at the right boundary $x=L$, calculated from Eqs.(2.4) and (2.5) at $t=0$ (full curve), and calculated by the code at $t=30$ (dotted curve). We see in this case that at the right boundary the distribution function remains essentially unchanged, although it is not kept constant. Fig.(4) shows the plot of the potential profile, fixed to zero at the plate. The electron distribution functions in the bottom curves of

Fig.(2) show cuts at high velocities. These cuts appear at a cut-off velocity $v_c = \sqrt{2\Phi m_i / m_e}$, were Φ is the value of the potential at the position where the distribution is plotted. At *x*=0, *x*=*L*/30, *x*=*L*/10, *x*=*L*/2, and *x*=*L*, we can check from Fig.(4) that these values of the potential Φ are given respectively by 0, 0.748, 1.85, 3.10, and 3.15. These values correspond to cut-off velocities v_c in the corresponding distribution function respectively of 0, 74., 116.5., 151., and 152., which are the cut-off observed in the distribution curves in the plot at the bottom of Fig.(2) (we are using deuterium with $m_e / m_i = 1./(2.*1836.)$. Fig.(5) shows the plot of the electron density (full curve), the ion density (broken curve), and the dash-dot curve plots the quantity:

$$n_e(x) = n_e(L)\exp\left(\frac{\varphi(x) - \varphi(L)}{T_e(L)}\right) \tag{2.28}$$

$\varphi(x)$ is the potential shown in Fig.(4), and $\varphi(L)$, $T_e(L)$ and $n_e(L)$ are respectively the potential, the electron temperature and the density at the right boundary at *x*=*L*. We note the very good agreement between the electron density calculated from Eq.(2.28) (dash-dot curve in Fig.(5)) and the electron density calculated by the code from Eq.(2) (full curve in Fig.(5)). The temperature $T_e(x)$ is defined in our normalized units by:

$$T_e(x) = \frac{m_e}{m_i} \frac{\int_{-\infty}^{+\infty} (v_x - \langle v_x \rangle)^2 f_e(x, v_x) dv_x}{\int_{-\infty}^{+\infty} f_e(x, v_x) dv_x} \; ; \; \langle v_x \rangle = \frac{\int_{-\infty}^{+\infty} v_x f_e(x, v_x) dv_x}{\int_{-\infty}^{+\infty} f_e(x, v_x) dv_x} \tag{2.29}$$

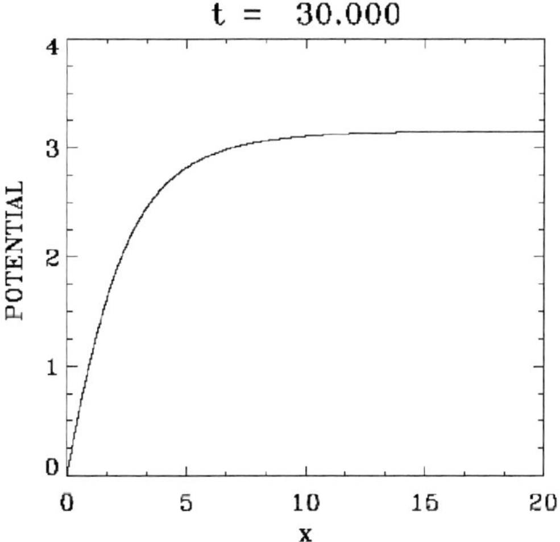

Figure 4. Potential profile.

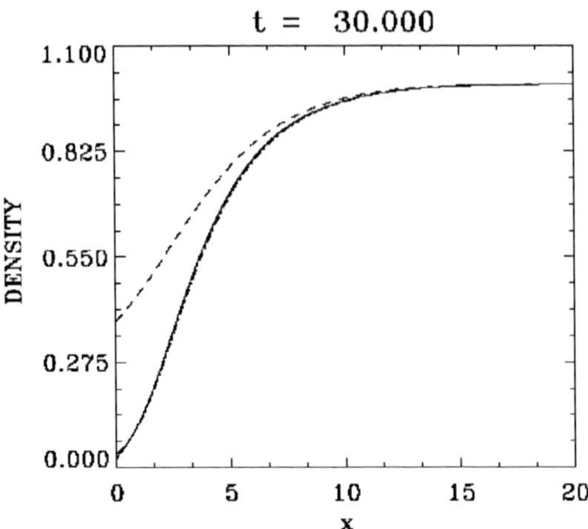

Figure 5. Ion (broken curve) and electron (full curve) densities. The dash-dot curve is the density from Eq.(2.28).

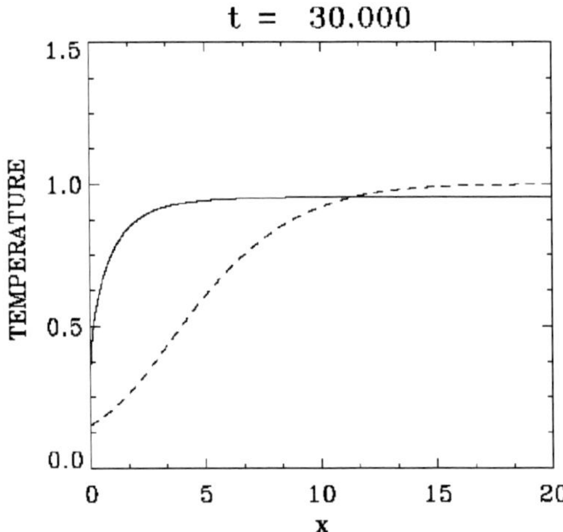

Figure 6. Temperature profile for the ions (multiplied by a factor of 20, broken curve) and for the electrons(full curve).

The electron temperature is plotted in Fig.(6) (full curve). We also show in Fig.(6) the ion temperature $T_i(x)$ (broken curve) defined in the same way as in Eq.(2.29), using the ion distribution function $f_i(x,v_x)$ (without the factor m_e/m_i in Eq.(2.29) in our normalized units). In Fig.(6) we are multiplying $T_i(x)$ by a factor of 20 to make the curve visible. We plot in Fig.(7) the product n_eT_e (full curve) and n_iT_i (dotted curve). Fig.(8) shows the electric field (full curve), and the charge ($n_i - n_e$, broken curve). The positive charge in front of the wall and the electric field are fading away as we penetrate from the wall at x=0 towards the plasma. The region where the charge density appears important is about 10 Debye lengths in front of the wall. Fig.(9) presents the electron current (full curve), the ion current (broken curve), and the total current ($J_{xi} - J_{xe}$, dotted curve), where $J_{xe,i}(x) = \int_{-\infty}^{+\infty} v_x f_{e,i}(x,v_x)dv_x$.

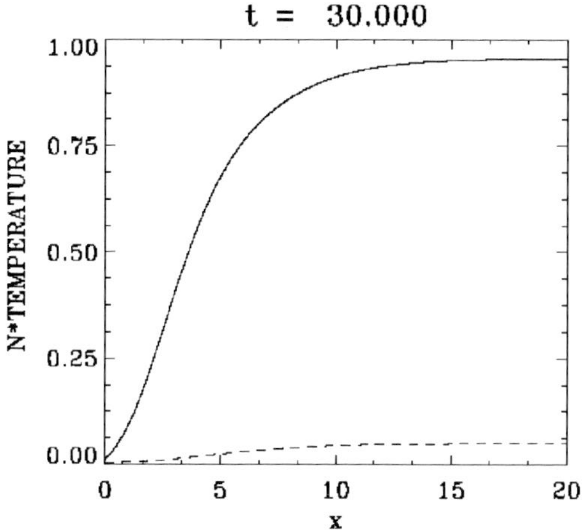

Figure 7. n_eT_e (full curve), n_iT_i (broken curve).

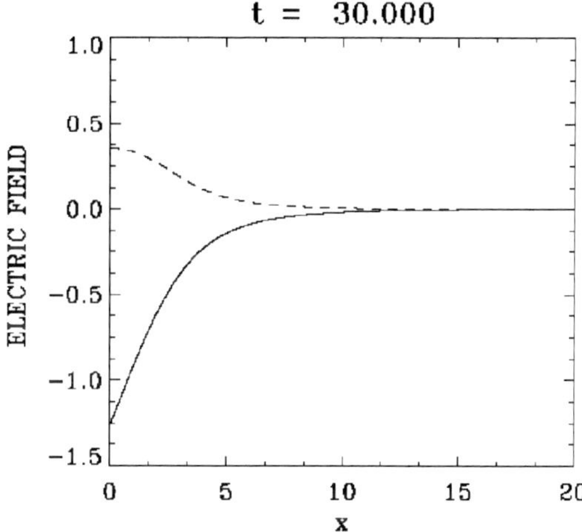

Figure 8. Electric field (full curve), and charge $n_i - n_e$ (broken curve).

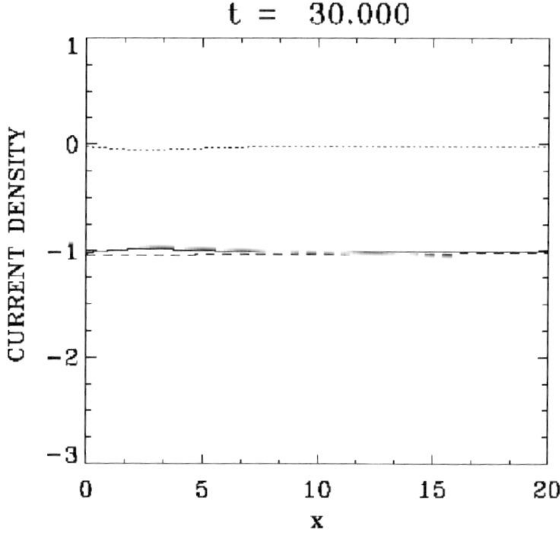

Figure 9. Current density J_{xe} (full curve), J_{xi} (broken curve), and $J_{xi} - J_{xe}$ (dotted curve).

We note the electron and ion currents are equal to -1 across the sheath, so that the total current remains zero. Note how in Figs.(5) and (8) the plasma quasineutrality breaks down in front of the floating wall, creating a positive charge so that the resulting electric field is maintaining the constant current throughout the sheath. Fig.(10) presents the sheath transmission factors for electrons and ions $\gamma_{e,i}$ defined in our normalized units as follows:

$$\gamma_e(x) = \frac{Q_e(x)}{T_e(L)\Gamma_e(L)} + 1; \; \gamma_i(x) = \frac{Q_i(x)}{T_i(L)\Gamma_i(L)} \tag{2.30}$$

referred to the values at the right boundary $x=L$. The electron and ion particle fluxes Γ_e and Γ_i are the currents $J_{xe,i}(L)$. The full curve in Fig.(10) gives γ_e as defined in Eq.(2.30), the broken curve gives γ_i, and the dash-dot curve gives the quantity $\gamma = 2. + \varphi(x)/T_e(L)$, which shows a very good agreement (the curves are essentially identical) with γ_e. The electron and ion energy fluxes Q_e and Q_i are given in our normalized units by :

$$Q_e(x) = \frac{1}{2}\frac{m_e}{m_i}\int_{-\infty}^{+\infty} v_x^3 f_e(x,v_x)dv_x; \; Q_i(x) = \frac{1}{2}\int_{-\infty}^{+\infty} v_x^3 f_i(x,v_x)dv_x \tag{2.31}$$

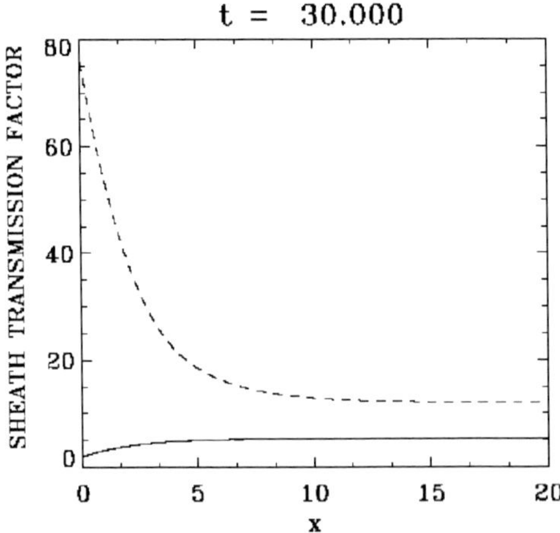

Figure 10. Sheath transmission factor for electrons (full curve) and ions (broken curve). The dash-dot curve is $\gamma = 2. + \varphi(x)/T_e(L)$.

Note that in all the previous integrals, infinity, the upper limit for the integrals, is automatically replaced by the cut-off velocity v_c at a given position x for the electrons (since f_e is zero for $v_x > v_c$) and by 0 for the ions (since f_i is 0 for $v_x > 0$). The electron and ion energy fluxes Q_e (full curve) and Q_i (broken curve) as defined in Eq.(2.31) are presented in Fig.(11). A comparison with the results at t_{ei} =1 and t_{ei} =10 in [52] shows little difference between the case t_{ei} =10 and t_{ei} =20. The difference is more pronounced for the case t_{ei} =1.

(Note that in the results in [52], Fig.(6) and Fig.(17) are incorrect, while Fig.(7) is for the temperatures T_e and T_i, and not for n_eT_e and n_iT_i ; the same thing apply to Fig.(18) in [52]).

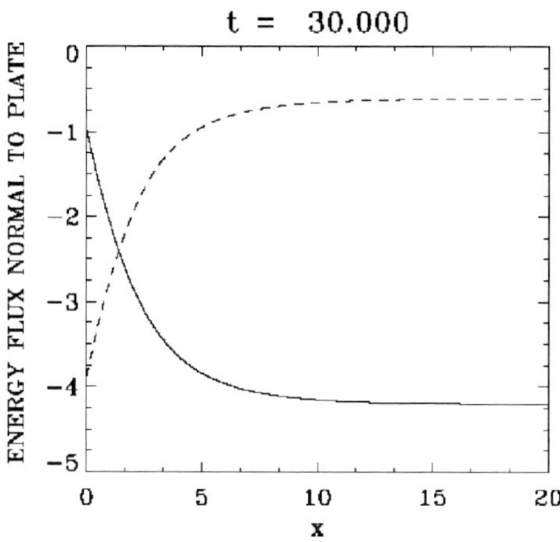

Figure 11. Q_e (full curve), Q_i (broken curve) (see Eq.(2.31)).

We have used an Eulerian Vlasov code to present a numerical solution for the problem of the kinetic sheath at a plasma-wall transition, where both electrons and ions have been treated using a kinetic Vlasov equation. The code provides an accurate representation of the distribution functions throughout the phase-space, especially in the low density region of the phase-space. Note especially the ion distribution function at x=0. Some of the basic macroscopic quantities associated with the sheath equilibrium have been calculated. Although for the present set of parameters the initial values presented in Eq.(2.4) at t=0 for instance remains essentially the same at x=L (see Fig. (3)), they deviate from this initial value in the other regions of the sheath, as indicated in Fig.(1) and Fig.(2), where plasma quasineutrality breaks down in such a way as to maintain a constant current across the sheath, as shown in Fig.(9). Many solutions for the sheath problem are usually presented by assuming an adiabatic law for the electrons [48,49]. In the present work we are providing a detailed description of the kinetic distribution functions associated with the sheath equilibrium, for both electrons and ions. This could be of significant importance for many problems of plasma-wall transition in nanotechnology, especially for the details of the distribution functions in front of the wall. More applications in the following sections will present further examples of the importance of calculating accurately the distribution functions for electrons and ions.

3. Study of the Phase-Space Dynamic in Capacitive Discharges

Capacitive discharges are often driven by a single frequency source, or a combination of two frequency sources. Different studies usually concentrate on the macroscopic quantities associated with the discharge (current, heating, control of the ion flux and ion energy impacting the electrode, etc...) [53]. In the present work, we use the method of characteristics applied to the Eulerian Vlasov code presented in section 2, to study the phase-space dynamic of the RF sheaths associated with capacitive discharges. The phase-space details of the distribution functions will be presented. We treat the plasma as completely collisionless. We consider a model where the plasma structure consists of a bulk plasma region where the density remains essentially constant and a sheath region adjacent to the wall. We simulate the sheath and its vicinity in the bulk plasma. The ions and electrons are described by the kinetic Vlasov equation written in Eq.(2.1), with the same normalization. The ions are injected at the right boundary of the bulk region of the simulation at the sound speed, to satisfy the Bohm criterion, while electrons have initially a distribution corresponding to a Maxwellian bulk plasma. The distribution functions for ions and electrons are initially determined as in section 2. The model is similar to what has been discussed in [54,55]. We take advantage of the very low numerical noise level of the Eulerian Vlasov code to calculate accurately the charge separation, the distribution functions and the different parameters associated with the non-neutral plasma in the entire domain where the charge evolves rapidly. We do not fix the distribution functions at the right boundary x=L. We rather assume that the plasma ions and electrons can circulate at the right boundary, *i.e.* that the plasma extends at the right boundary as in the model presented in [54,55], such that the point next to the last grid point is identical to the last grid point. The plasma parameters at the right boundary are left to evolve accordingly. We use singly ionized helium ions. A small time step $\Delta t = 1.25\text{x}10^{-4}$ was taken for the ions, and for the electrons this time step was smaller by a factor of 10, our main concern being to get the results as accurate as possible. The numerical scheme is the same fractional step technique presented in section 2. The calculation of the electric field is done through the solution of Eqs.(2.3). We assume that at the wall at x =0 we apply a potential $\phi|_{x=0} = V_{rf}\cos\omega t + V_{dc}$, where V_{rf} is the amplitude of the RF voltage, ω the angular frequency, V_{dc} is the self-bias voltage. To solve Poisson's equation, we divide the solution of the total potential ϕ into two parts:

$$\phi = \varphi + \psi \tag{3.1}$$

where φ and ψ satisfy the following equations:

$$\frac{\partial^2\varphi}{\partial x^2} = -(n_i - n_e)$$
$$\frac{\partial^2\psi}{\partial x^2} = 0, \tag{3.2}$$

The boundary conditions on ψ are chosen to be the same as those for ϕ. Then the boundary conditions for φ at the boundaries x =0 and x =L must be $\varphi|_{x=0} = \varphi|_{x=L} = 0$. We first solve Eq.(3.2) for ψ for the boundary conditions $\psi = 1$ at x =0, and $\psi = 0$ at x =L. In our case this solution is $\psi_0 = -x/L + 1$. Then the solution ψ is given at any time t by

$\psi = (V_{rf}\cos\omega t + V_{dc})\psi_0$. We then solve Eq.(3.2) for φ. And the total solution is then given by Eq.(3.1). Follow the solution of the electric field $E_x = -\dfrac{\partial\phi}{\partial x}$.

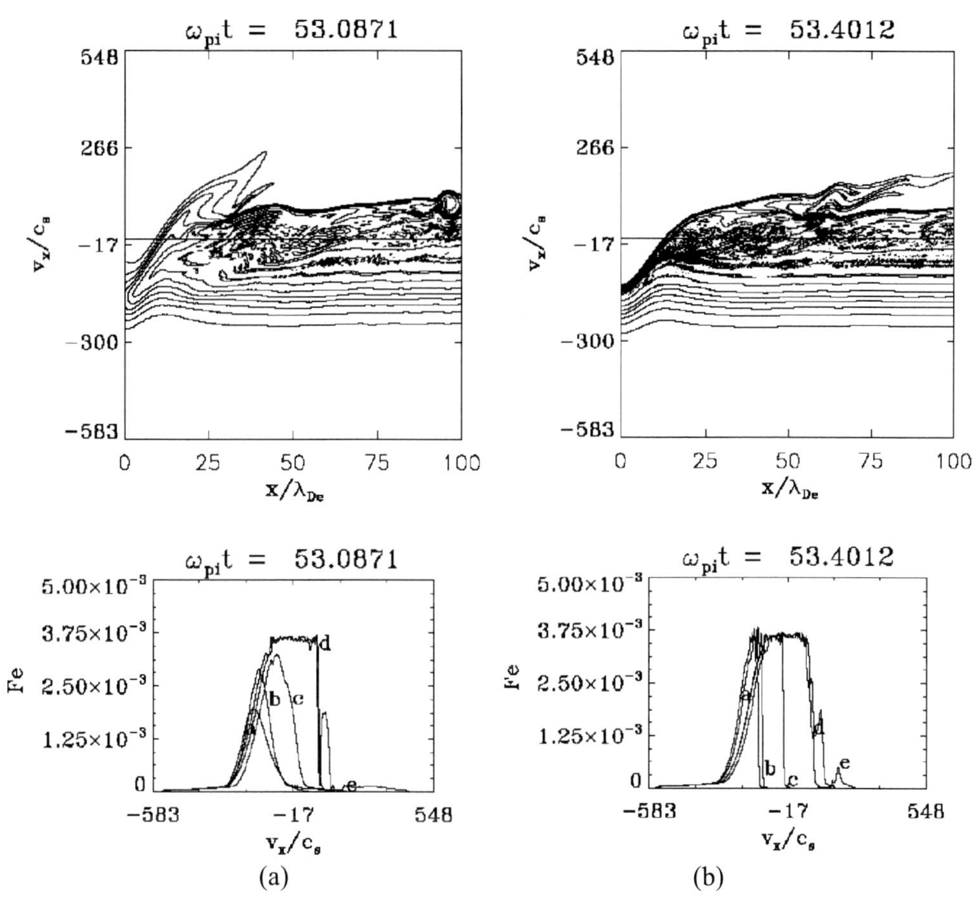

Figure 12. Distribution function for the electrons.

We consider the case the case $\omega = 10\omega_{pi}$. For the singly ionized helium plasma, we have $\omega_{pe} >> \omega >> \omega_{pi}$, all the RF current is essentially carried by the electrons. This is clearly seen in the curves of the current normal to plate to be presented in the following results. We also have $t_{ei} = T_e/T_i = 20$, $V_{rf} = -16$ and $V_{dc} = -10$. The maximum and minimum velocities for the electrons are taken to be respectively $6.4/\sqrt{m_e/m_i}$ and $-6.8/\sqrt{m_e/m_i}$. The maximum velocity for the helium ions is taken on the velocity grid to be $0.1\text{x}4./\sqrt{t_{ei}}$ (the ion distribution function is in fact 0 for $v_x>0$), and the minimum velocity is taken to be $-26./\sqrt{t_{ei}}$. The length of the system is L=100 Debye length, and 400 grid points were used in space, 240 grid points were used in velocity space for the ions,

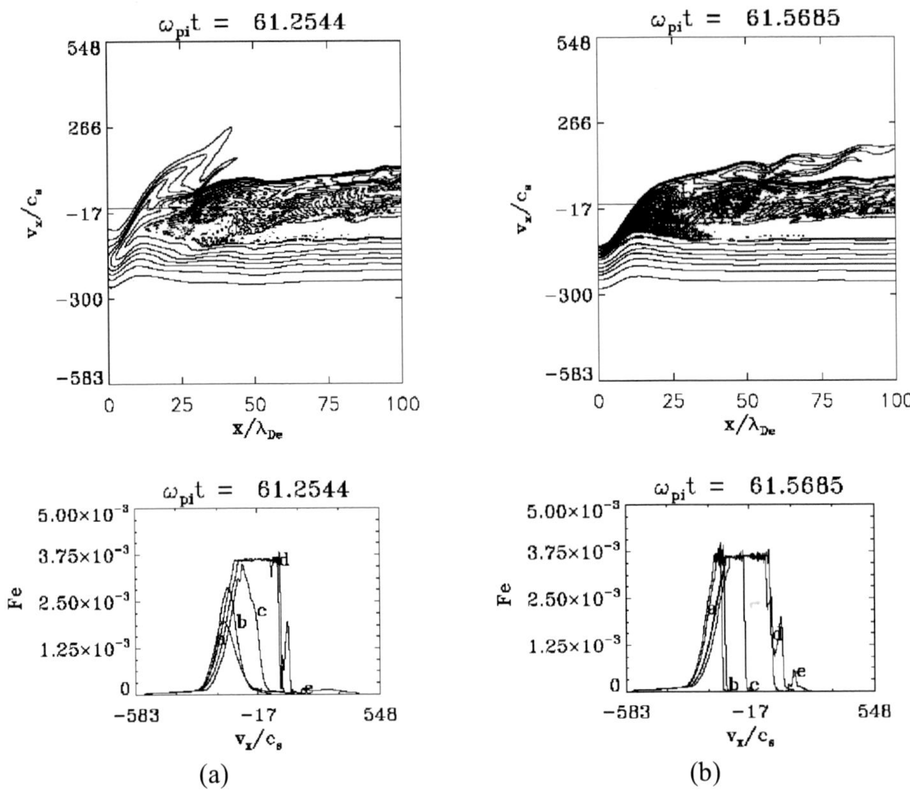

Figure 13. Distribution function for the electrons.

and 400 grid points in velocity space for the electrons. At steady state the curves repeat themselves accurately every RF period $T = 2\pi / \omega = 2\pi / 10$. Figs.(12a-b) show two plots of the electron distribution functions during a period, at t =53.087 and t =53.401. We show in Figs.(13a-b) two plots, at t =61.254 and t =61.568, of the electron distribution functions during a period at the same phase as in Figs.(12a-b). Inspite of the strong deformation of the electron distribution functions during a period, they repeat themselves identically at every period, with very minor differences which could be essentially numerical. During the oscillations, the middle parts of the electron distribution functions form a flat plateau. See also the steep profiles very nicely reproduced by the code. Similar results were obtained with different values of V_{rf} and V_{dc} .The lower curves in Figs.(12-13) are at the positions: a) x =0, b) x =L/30, c) x = L/10, d) x = L/2, e) x =L. Figs.(14) shows the ion distribution functions during a period at t =61.254 and t =61.568 (those obtained at t =53.087 and t =53.401 are identical). Note how the helium ions are accelerated towards the biased wall. Note also at x =0 how the distribution function contours are moving in velocity at t =61.254 and t =61.568. Again this small motion is repeating itself identically at every period. Fig.(15) shows the evolution of the potential during a period at: a) t=52.919, b) t=53.076, c) t=53.233, d) t=53.39, e) t=53.547.The same evolution for the potential is shown in Fig.(16) during a period at a) t=61.073, b) t=61.23, c) t=61.387, d) t=61.544, e) t=61.711. It is also interesting to note at the end of the period in Fig.(15) and(16) the curve e) repeating identically the curve a) at the beginning of the period. For the remaining results, we will show the curves during one period only, since they repeat themselves identically. The density in Fig.(17a) is given at a)

t=61.073, b) t=61.23, and in Fig.(17b) at a) t=61.387, b) t=61.544 and c) t=61.711, the curve c) at the end of the period repeating exactly the curve a) of Fig.(17a). The broken curve gives the ion density, showing little variation in time. Note how the electron and ion densities separate in front of the wall. During the oscillation, Figs.(17a-b) show clearly that in the sheath edge, the charge ($n_i - n_e$) of the sheath in front of the wall is oscillating between negative and positive values. Fig.(18) shows the electric field, plotted at the same time as in

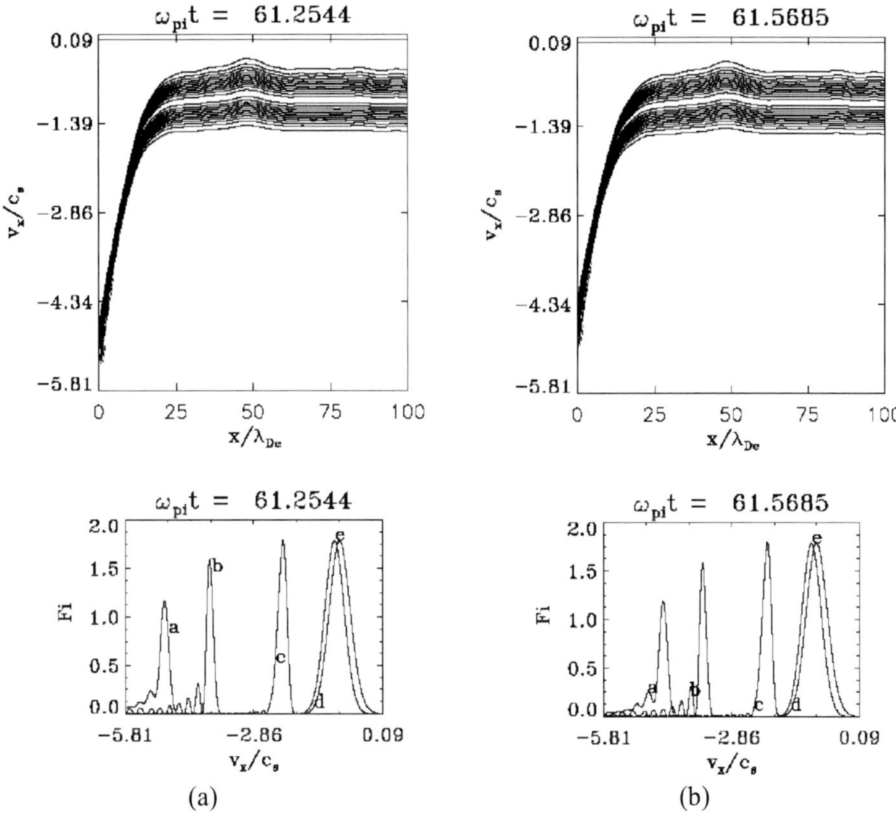

Figure 14. Distribution function for the ions.

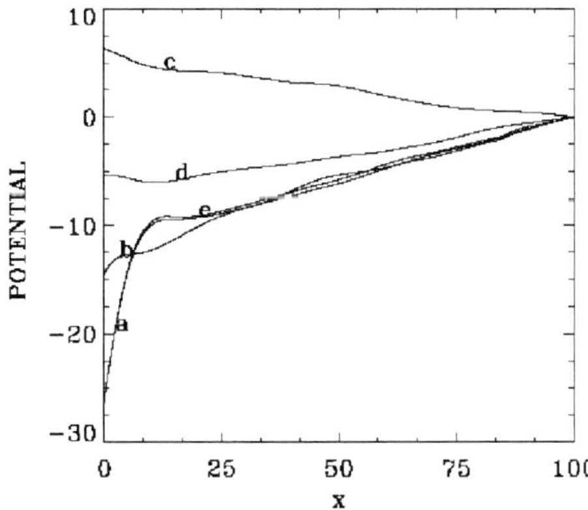

Figure 15. Potential at a) t=52.919, b) t=53.076, c) t=53.233, d) t=53.39, e) t=53.547.

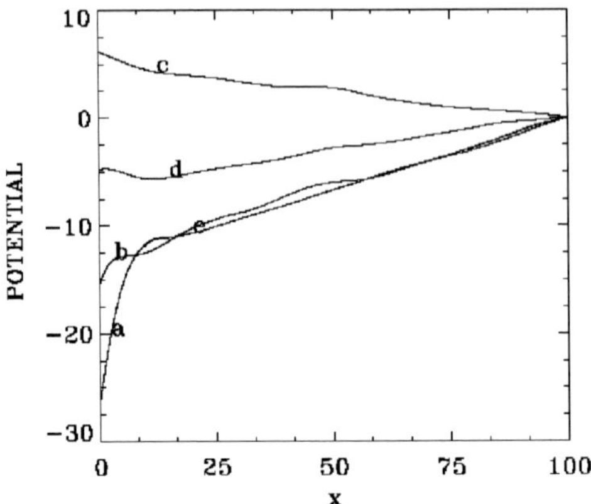

Figure 16. Potential at a) t=61.073, b) t=61.23, c) t=61.387, d) t=61.544, e) t=61.711.

Fig.(16). Note how the electric field can reach positive values in front of the wall during the period of oscillations. The strong negative electric field in Fig.(18) at a) t=61.083 is pushing

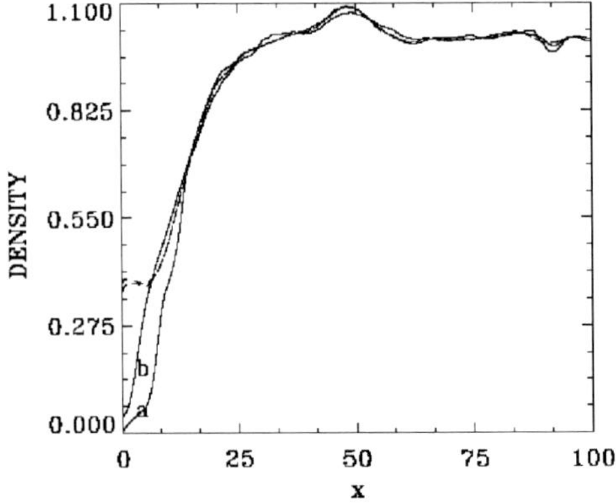

Figure 17a. Density profiles at a) t=61.073, b) t=61.23 (full curves for electrons, broken curves for ions).

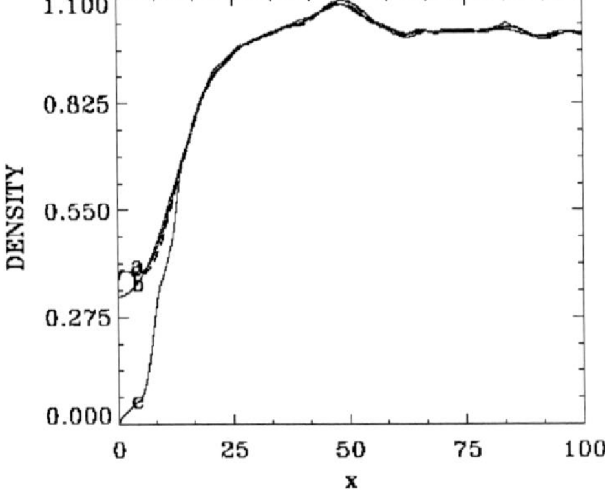

Figure 17b. Density profiles at a) t=61.387, b) t=61.544, c) t=61.711 (full curves for electrons, broken curves for ions).

back the electrons, giving the density curve at a) in Fig.(17a) or c) in Fig.(17c). Fig.(19) presents the current normal to the plate (essentially carried by the electrons, J_{xe} full curves, and broken curves J_{xi} for the ions). Fig.(20) is the energy flux normal to the plate, calculated from Eq.(2.31), and Fig.(21) is the temperature (full curves electrons, broken curves ions), calculated from Eq.(2.29), calculated at the same time as in Fig.(16). Note again how the curve e) at the end of the period repeats exactly the curve a) at the beginning of the period. The strong peak in Fig.(21) for the curves a) and e) can be artificial, due to the fact that we are dividing by n_e in Eq.(2.29), and n_e close to x=0 is very small.

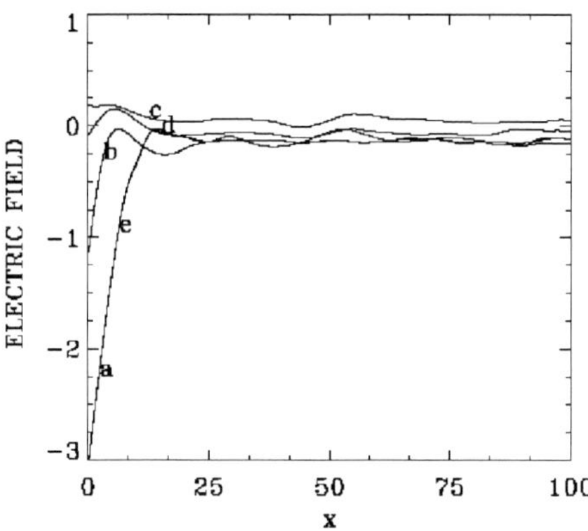

Figure 18. Electric field (plotted at the same time as in Figure (16)).

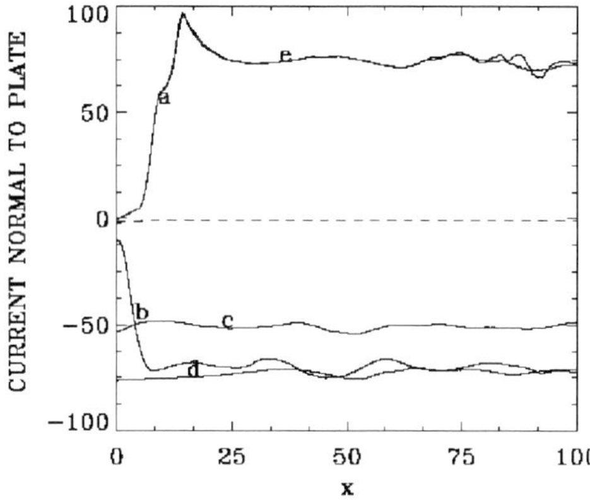

Figure 19. Current density normal to the wall (plotted at the same time as in Figure (16), full curves for electrons, broken curves for ions).

The model we have presented is similar to what has been discussed in [55] using PIC codes. We have, however, presented a complete solution for the distribution functions, showing the details of the plasma dynamics in the phase-space, especially for the strongly distorted, rapidly varying electron distribution functions. The results presented in this section for capacitively coupled plasmas are consistent. The exact repetition of the solution at every period allows a good control of important parameters, like the current for instance, by adjusting V_{rf} and V_{dc}. Note how the code is handling very nicely the steep gradients in the electron distribution functions. The method can be extended without difficulty to the case of dual-frequency excitations and multiple frequency capacitive discharges [53,54], by a simple adjustment of V_{rf} and V_{dc} in the boundary conditions on the excitation of the potential in Eq.(3.2). Note that higher potentials will necessitate using a longer system, since in this case the deformation penetrates deeper in the plasma. More realistic distribution functions for capacitively coupled plasmas require the inclusion of realistic elastic and inelastic collisions (excitation of electronic states and ionizations). The work in [45] contains more complete physics and gives a good indication and guidance on how to handle these problems.

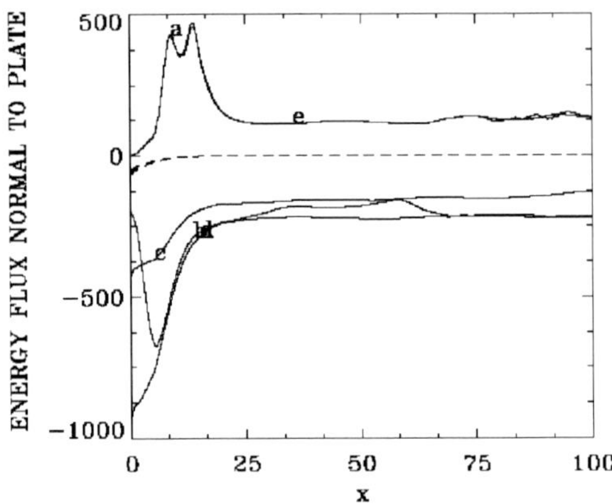

Figure 20. Energy flux normal to the wall. Q_e for the electrons (full curve), Q_i for the ions (broken curves) (same times as in Figure 16).

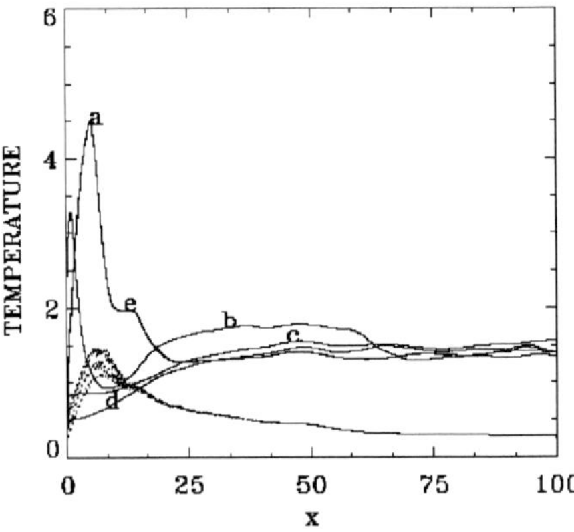

Figure 21. Temperature profiles for the electrons (full curves) and ions (multiplied by a factor of 5, broken curves) (same times as in Figure 16).

4. A One-Dimensional Ion Extraction Model

The modeling of ion extraction from a plasma is of great importance in many physical devices like ion accelerators, electrical discharges and in many related industrial applications. In the present section, we consider the problem of the one-dimensional model of ion extraction from a neutral plasma located between two electrodes. The neutral plasma is initially placed behind an anode held to a zero electric potential, and has an aperture such that the ions are attracted in the inter-electrode space by facing a highly negative cathode. The electrons are attracted by the ions which move towards the cathode, but on the other hand are repulsed by the negative cathode. As a result electrons and ions are penetrating from the anode boundary, and form a plasma zone where quasi-neutrality holds, and are penetrating to a sheath region from where the ion beam is extracted or emitted [56]. In the present simulation, the cathode and anode are located respectively at x =0 and $x = L$. Many theories of ion extraction [56] have treated this problem using an adiabatic law for the electrons. In the present section, a full kinetic equation is used for both the electrons and the ions. The electric potential is held at a constant potential V_0 at the cathode at x =0, and to zero at the anode at $x = L$, where the electrons have a Maxwellian distribution:

$$f_e(x = L, v_x) = \frac{e^{-m_e v_x^2/(2m_i)}}{\sqrt{2\pi m_i / m_e}} \tag{4.1}$$

Density is normalized to 1. We use the same normalization as in section (2.1). We assume the ions, which are accelerated by the electric field, are injected initially at the anode at the right boundary into the inter-electrode region with a velocity v_0 equal to the sound velocity $v_0 = \sqrt{1 + 1/t_{ei}}$:

$$f_i(x = L, v_x, t = 0) = \frac{e^{-t_{ei}(v_x - v_0)^2/2}}{\sqrt{2\pi / t_{ei}}} \tag{4.2}$$

This boundary condition on the ions at $x = L$ is not fixed, but left free to evolve, *i.e.* we assume the plasma extends beyond the anode, and that the point next to the last grid point at $x = L$, is identical to the last grid point. In the present calculation t_{ei} =1, and we consider deuterium ions with $m_e / m_i = 0.5/1836$. We take the length of the system L = 40 (in terms

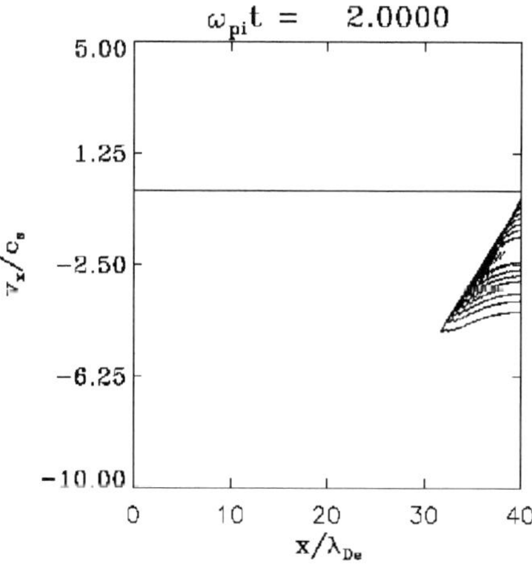

Figure 22. Ion distribution function.

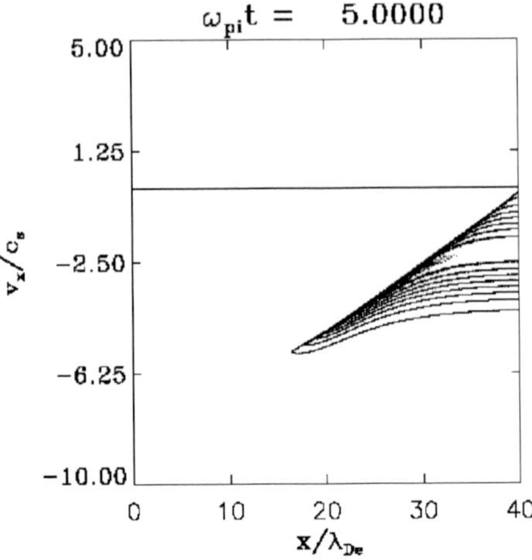

Figure 23. Ion distribution function.

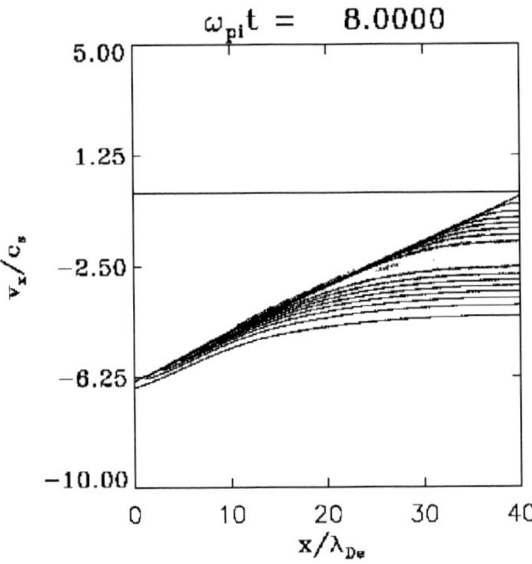

Figure 24. Ion distribution function.

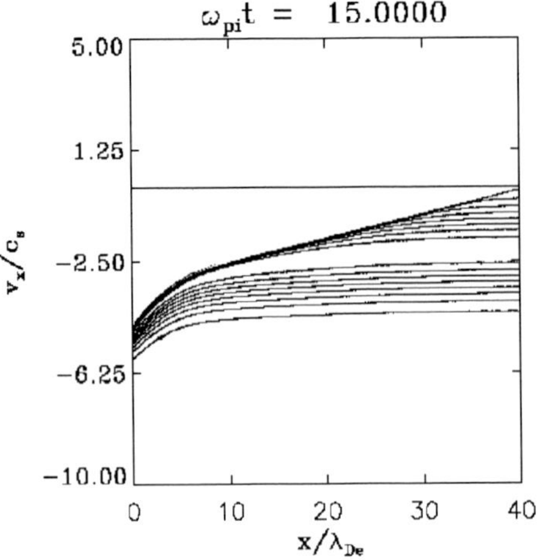

Figure 25. Ion distribution function.

of the Debye length), so the parameter $\varepsilon = 1/L$ is very small. The potential at the cathode at x =0 is set to V_0=-7 (normalized to T_e/e), so that $\eta = 1/|V_0|$ is also a small parameter. The ion injection energy is $\frac{1}{2}v_0^2$ =1 (v_0 is normalized to the acoustic speed $c_s = \sqrt{T_e/m_i}$).

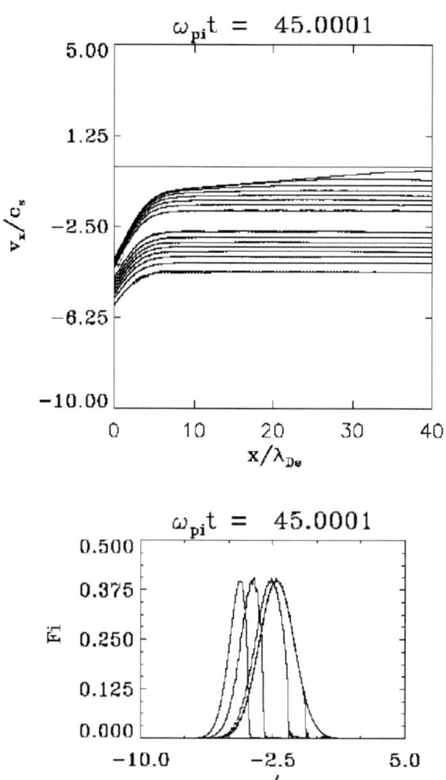

Figure 26. Ion distribution function.

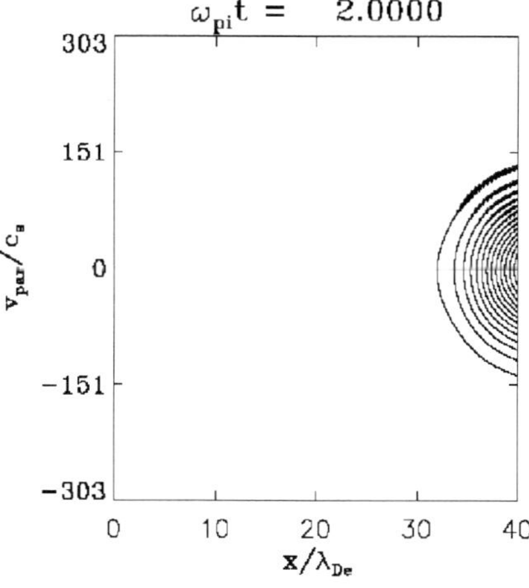

Figure 27. Electron distribution function (v_{par} is v_x).

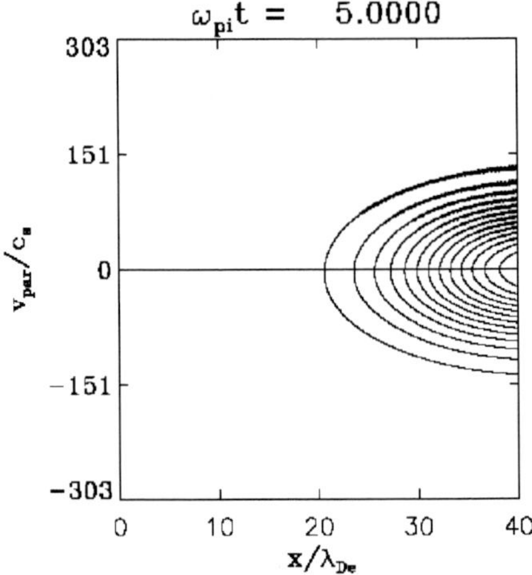

Figure 28. Electron distribution function (v_{par} is v_x).

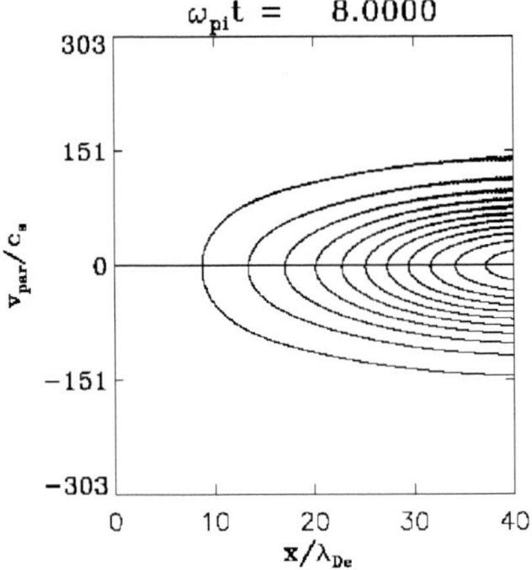

Figure 29. Electron distribution function (v_{par} is v_x).

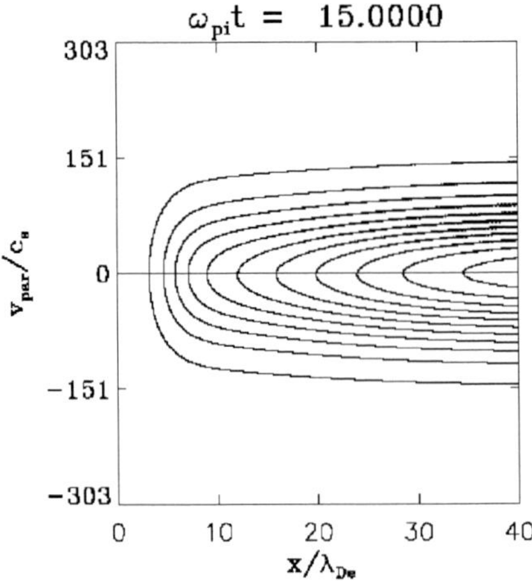

Figure 30. Electron distribution function (v_{par} is v_x).

Figs(22-25) show different phase-space shots of the penetration of the ions in the inter-electrode space. In Fig.(26), we show the ion distribution function at t =45, where a steady state has long been reached. The lower curves in Fig.(26) show the cuts in the upper contour of the distribution function at (from right to left) a) x =0, b) x =L/30; c) x =L/10, d) x =L/2, and e) x =L. Note the steep gradient of the distribution functions at x=0 and x=L/30, and even x=L/2. Figs.(27-30) show the contour plot of the penetrating electrons, at the same time as in Figs.(22-25). Fig.(31) shows the electron distribution function at t =45, when a steady state

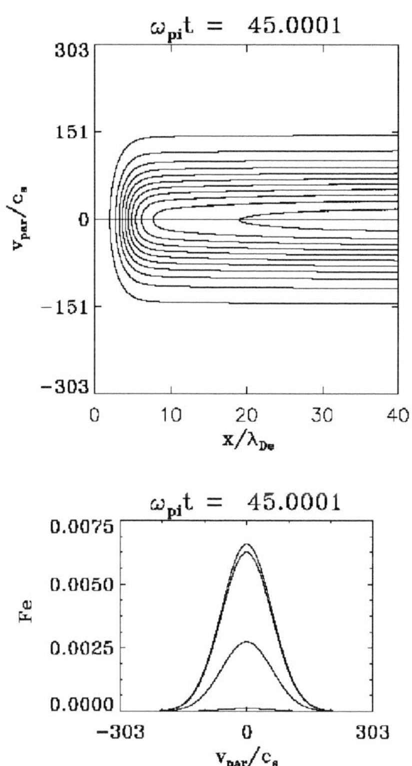

Figure 31. Electron distribution function (v_{par} is v_x).

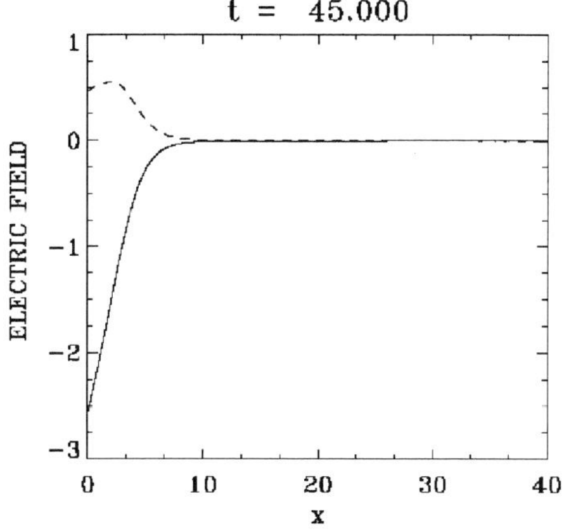

Figure 32. Electric field (full curve) and charge (broken curve).

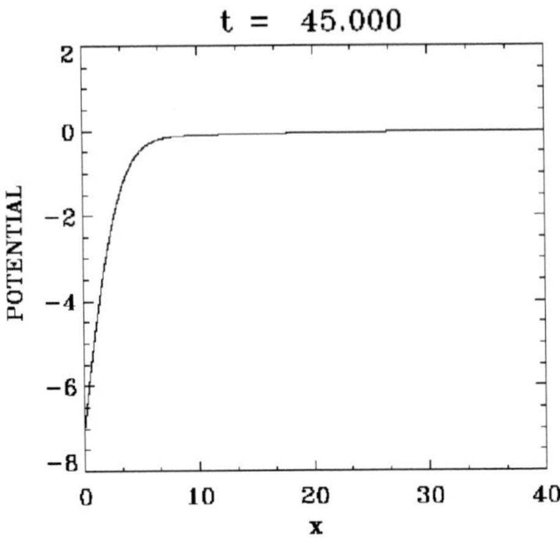

Figure 33. Potential.

has been reached. The lower curves in Fig.(31) show the cuts in the upper contour plots at, from bottom to top, the same positions as in Fig.(26). The electrons are penetrating in the inter-electrode region in such a way as to maintain a neutral zone (the plasma) so that the potential towards the right boundary is constant (zero), and $\partial\varphi / \partial x = 0$, so that the electric field is also zero (see Figs.(32,33)). The dotted curve in Fig.(32) is the charge ($n_i - n_e$), essentially zero towards the right boundary. There is a point reached where the potential deviates slowly from $\partial\varphi / \partial x = 0$. At this point, the ions are extracted following the Child-Langmuir law [57]. Fig.(34) shows how the electron and ion density separates, creating the charge and electric field as in Fig.(32), so that the ion current is essentially maintained constant (dotted curve in Fig.(35)). The electrons, which remained essentially Maxwellian, give zero current. Fig.(36) shows the electron and ion temperature, calculated from Eq.(2.29). The electrons appear to maintain their Maxwellian distribution until very close to the x =0 electrode. Note the cooling appearing in the ion distribution (broken curve in Fig.(36)). The ion distribution evolves in such a way that the total energy of an ion at each position is conserved $\frac{1}{2}v_i^2 + \varphi(x) = \frac{1}{2}v_{0i}^2$, where v_{0i} is the entry velocity of an ion at x =L.

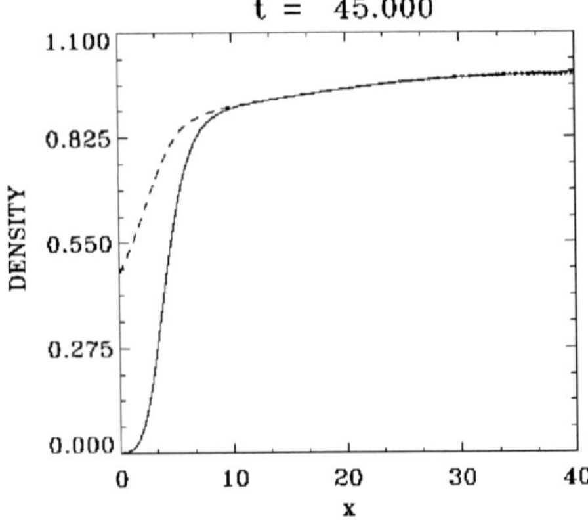

Figure 34. Electron density (full curve) and ion density (broken curve).

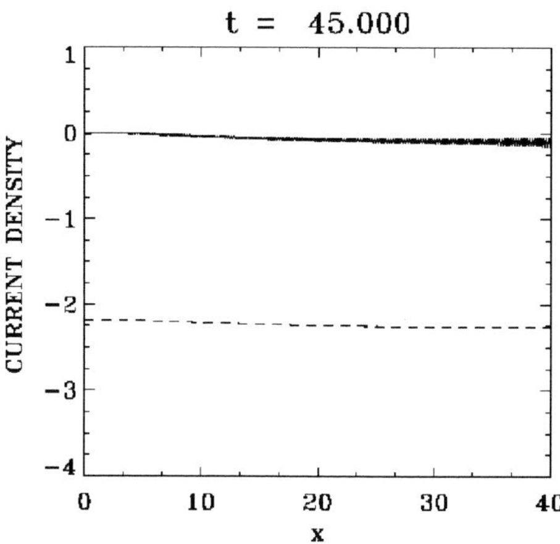

Figure 35. Electron current density (full curve) and ion current censity (broken curve).

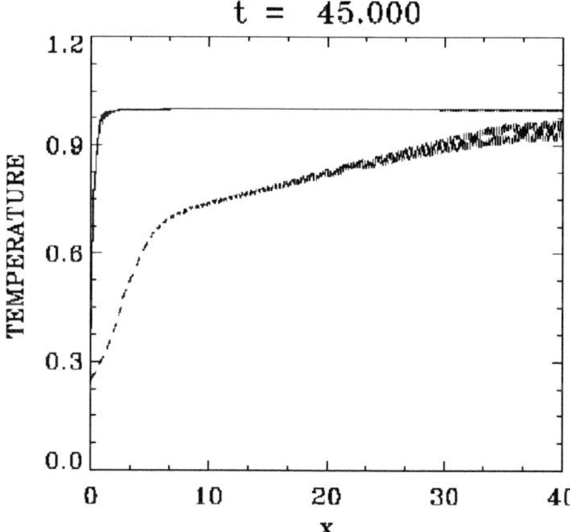

Figure 36. Electron temperature (full curve), and ion temperature (broken curve).

For instance at the center of the ion distribution at x =L, we have the injected velocity $v_{0i} = \sqrt{1+1/t_{ei}} = 1.414$. This central point of the distribution will arrive at x =0, where $\varphi = -7$, with a velocity equal to 4, which is what is observed in Fig.(26). An ion having at the injection point at x =L a velocity close to 0, will reach the position x =0 where $\varphi = -7$, with a velocity equal to $\sqrt{14}$ =3.74, which is again what is observed in Fig.(26).

We have seen in this section for the ion extraction problem how a space-charge layer is generated at the cathode and how the charge and potential adjust themselves so that a constant value of the ion current, extracted from a neutral plasma, is maintained. Note again how the different profiles and the details of the distribution functions in the phase-space are given without numerical noise with the Eulerian Vlasov code.

5. Oscillations of the Collisionless Sheath at Grazing Incidence of the Magnetic Field

In this section, we apply the method of characteristics for the numerical solution of the kinetic equations of plasmas, to study the problem of a collisionless sheath for a plasma-wall transition in the presence of an oblique external magnetic field, which is at grazing incidence to the wall. This problem is important to understand probe characteristics in the presence of an external magnetic field, and to determine the heat load on the divertor and on the limiter in many plasma-divertor and plasma-limiter transition in tokamaks. A similar problem was studied for a hydrogen plasma in [58], and it was shown that at small angles of incidence of the magnetic field to the wall, the electrons which are frozen along the magnetic field lines have a tendency to reduce the transport of the ions (essentially the ion current and ion heat flux) to the wall across the magnetic field, resulting in what can be qualified as a plasma detachment. We are interested here in the case of a large ratio of the ion gyroradius to the Debye length. In this case there is another mechanism than transport which brings the ions into contact with a wall at grazing incidence of the magnetic field, namely the ions gyrating around the magnetic field B can be scraped-off by the wall [59]. Below a critical angle, the electrons moving parallel to B can no longer follow the ions gyrating perpendicular to B. The electrons, frozen by the magnetic field, are running along the magnetic field in their attempt to catch the ions gyrating across the magnetic field. The electrons then determine the characteristic times for information propagation. This results in low frequency oscillations appearing in the sheath, which are not related to edge turbulence. The charge separation at a sheath-wall transition is sensitive to the effect of the finite ion gyroradius, or more precisely to the ratio of the ion gyroradius to the Debye length. It is of course of interest to study how this parameter would affect the physics of the sheath, not only at the sheath-wall transition, but in the entire domain from the entrance at the sheath edge, to the sheath-wall transition. In the absence of an external magnetic field, it is well known that the condition for the sheath formation demands that the ions enter the sheath region accelerated in a presheath region to the sound velocity. This is known as the Bohm criterion [50,51] (see section 2 of this chapter). In the present case where an external magnetic field is present, and in the case of a large angle of incidence, there is a magnetic presheath which direct the ion flow along the magnetic field toward the wall [49]. The ions are penetrating this magnetic presheath with a velocity along the magnetic field equal to the sound velocity, and acquire a velocity perpendicular to the magnetic field as they approach the wall. However, as recently discussed in [59], the plasma-wall transition cannot always be clearly divided into separate regions, especially at grazing angles of incidence of the magnetic field, and when $\rho_i >> \lambda_{De}$, where ρ_i is the ion gyro-radius and λ_{De} is the Debye length, and in the presence of a high magnetic field such that $\omega_{ci} / \omega_{pi} \approx 0.1$, where ω_{ci} is the ion cyclotron frequency and ω_{pi} is the ion plasma frequency. So in the present section we will simply refer to the domain where we study the physics of the sheath as the sheath domain, since the motion along the magnetic field and across the magnetic field are coupled, and in our case of grazing incidence of the magnetic field, there is no clear cut point where the motion deviates from a motion along the magnetic field to a motion having a component perpendicular to the magnetic field. Many papers have studied the sheath problem in a magnetized plasma using an adiabatic law for the electrons (see for instance [48] and the recent work in [49]). It is of course of interest to use a kinetic description to study how the electrons, frozen by the magnetic field lines, are following the gyrating ions penetrating the magnetic sheath edge at the sound speed. We use an Eulerian Vlasov code in one spatial dimension in which the electrons, assumed to move only along the magnetic field lines, are described by a parallel-*B* kinetic equation, whereas the ions are described by a kinetic equation in one special dimension and in the full velocity space, which integrates exactly the ion orbits in the four dimensional phase-space considered. A deuterium plasma is considered in the present study.

5.1. The Kinetic Model for the Magnetized Sheath

The pertinent equations are those presented in Ref [59]. We consider a one-dimensional (1D) slab geometry in which the inhomogeneous direction is in the direction x normal to the wall. The y and z directions are assumed homogeneous. The constant magnetic field is located in the (x,y)-plane, and makes an angle α with the y axis. The magnetized electrons are restricted to move with a velocity $v_{\|}$ along the magnetic field and are described using a kinetic equation in the direction along the magnetic field, with a distribution function $f_e(x,v_{\|},t)$ obeying the Vlasov equation:

$$\frac{\partial f_e}{\partial t}+v_{\|}\sin\alpha\frac{\partial f_e}{\partial x}-\frac{m_i}{m_e}E_x\sin\alpha\frac{\partial f_e}{\partial v_{\|}}=0 \tag{5.1}$$

Time is normalized to the ion plasma frequency ω_{pi}^{-1}, velocity is normalized to the acoustic velocity $c_s=\sqrt{T_e/m_i}$, and length is normalized to Debye length $\lambda_D=c_s/\omega_{pi}$. The potential is normalized to T_e/e, and the density is normalized to the peak initial central density n_o. The ions are treated with a kinetic Vlasov equation in one spatial dimension, which is written for the distribution function $f_i(x, v_{x,} v_y, v_z, t)$ as:

$$\frac{\partial f_i}{\partial t}+v_x\frac{\partial f_i}{\partial x}+(E_x-v_z\omega_{ci}\cos\alpha)\frac{\partial f_i}{\partial v_x}+v_z\omega_{ci}\sin\alpha\frac{\partial f_i}{\partial v_y}+\omega_{ci}(v_x\cos\alpha-v_y\sin\alpha)\frac{\partial f_i}{\partial v_z}=0 \tag{5.2}$$

The potential and the electric field are calculated from the Poisson equation given in Eq.(2.2), where:

$$n_i(x)=\int f_i(x,\vec{v})d\vec{v}\,;\ n_e(x)-\int f_e(x,v_{\|})dv_{\|} \tag{5.3}$$

We assume a completely absorbing wall at x=0. We consider that the right boundary at x=L is the sheath edge at the sheath domain entrance, where ions are penetrating at the ion sound velocity. At x=L, the in-streaming ions have initially a parallel to B shifted Maxwellian distribution of the form [48]:

$$f_i(L,\vec{v}) \sim \frac{g(v_{\|})}{(2.\pi / t_{ei})^{3/2}} e^{-t_{ei}(v_{\|}-v_0)^2/2.} e^{-t_{ei}(v_{\perp}^2+v_z^2)/2.}; \tag{5.4}$$

with $t_{ei} = T_e / T_i$. $v_{\|}$ is the velocity component parallel to the magnetic field and $v_{\perp}$ the velocity component normal to the magnetic field. The shift v_0 in Eq.(5.4) is chosen to be the negative sound velocity $\sqrt{1+1/t_{ei}}$. The smoothing factor $g(v_{\|}) = (1-\exp(-2t_{ei}v_{\|}^2))$ is chosen to provide a smooth transition to f_i =0 for $v_{\|} > 0$. The electron distribution function $f_e(x=L, v_{\|}, t=0)$ at x=L is initially taken as a Maxwellian

$$f_e(x=L, v_{\|}, t=0) = N_e \exp(-m_e v_{\|}^2 /(2m_i)) \tag{5.5}$$

normalized and truncated at a velocity V_c such that its density and its current are equal to the density and current obtained from Eq.(5.4). V_c is determined by generalizing the method presented in section 2 with one velocity dimension. We outline here the important steps. We assume that initially at the right boundary the injected ions have a density $n_i(x=L) = \int_{-\infty}^{0} f_i(\vec{v})d\vec{v}$. The initial density of the electrons at the right boundary should be equal to the density n_i:

$$n_i = \int_{-\infty}^{V_c} f_e(v_{\|})dv_{\|} = N_e \sqrt{\frac{\pi m_i}{2m_e}} \left[erf\left(V_c / \sqrt{\frac{2m_i}{m_e}} \right) + 1 \right], \tag{5.6}$$

where $erf(y) = \frac{2}{\sqrt{\pi}} \int_0^y e^{-y^2} dy$ is the error function. Equation (5.6) gives a first relation between N_e and V_c. The ion current density at x=L penetrating the plasma in the direction parallel to B is given by: $J_i = \int_{-\infty}^{0} v_{\|} f_i(\vec{v})d\vec{v}$ (which is negative, the ions are injected at the right boundary towards the left in the plasma), and must be equal to the electron current density at x=L: $J_e = \int_{-\infty}^{V_c} v_{\|} f_e(v_{\|})dv_{\|}$. Substituting for f_e from Eq.(5.5) we arrive after some straightforward algebra at the following transcendental equation for V_c:

$$\frac{J_i}{n_i} = -\sqrt{\frac{2m_i}{\pi m_e}} \frac{e^{-m_e V_c^2/2m_i}}{erf\left(V_c \sqrt{\frac{m_e}{2m_i}} \right) + 1}, \tag{5.7}$$

which is solved numerically for V_c. This allows the electron density and current along the magnetic field to be equal to the ion density and current along the magnetic field at the sheath entrance $x=L$ (see Ref.[48]). The initial condition is taken to be spatially uniform with $f_{i,e}(x,\vec{v}) = f_{i,e}(L,\vec{v})$ throughout the domain. $m_e/m_i = 0.5/1836$ for deuterium. We assume that at the sheath entrance at the right boundary at $x = L$ the plasma extends to an identical plasma, so that the point next to the last grid point is identical to the last grid point. At the left boundary, particles hitting the floating wall at $x=0$ are lost from the system and collected through the current delivered at the wall:

$$\frac{\partial E_x}{\partial t}|_{x=0} = -J_x|_{x=0} = -(J_{xi} - J_{xe})|_{x=0}; \tag{5.8}$$

from which:

$$E_x|_{x=0} = -\int_0^t J_x|_{x=0}\, dt \equiv -\frac{\partial \phi}{\partial x}|_{x=0} \tag{5.9}$$

where

$$J_{xi} = \int d\vec{v} v_x f_i(x,\vec{v}); \; J_{e||} = \int dv_{||} v_{||} f_e(x, v_{||})$$

and $J_{xe} = J_{e||}\sin\alpha$ .We integrate over the domain the equation $\partial E_x/\partial x = (n_i - n_e)$, we get:

$$E_x\Big|_{x=L} - E_x\Big|_{x=0} = \int_0^L (n_i - n_e)dx = \sigma \tag{5.10}$$

The boundary condition on E_x in Eq.(5.9) is used to solve for the potential in Eqs.(2.2-2.3). The resulting electric field at $x = L$ is related to the electric field at $x = 0$, and to the charge σ appearing in the system, through Eq.(5.10). We verify this equation is satisfied at every time-step. Equations (5.1)-(5.3) are solved using a method of characteristics coupled to a fractional step [1,27-29] (to be discussed in details in section 5.2). We run the code and let the initially neutral plasma evolve to a steady state. The motion of the ions is advanced with a time-step $\Delta t = 0.04$, and the motion of the electrons is advanced with $\Delta t = 0.005$,(ions are advanced by one time-step for every eight time-step for the electrons). We use 200 grid points in space in a domain L = 100 Debye lengths, and 80 grid points in each velocity space direction for the ions, with velocity extrema for the ions equal to ± 4 ion thermal velocities ($\pm 4\sqrt{T_i/T_e}$ the acoustic velocity c_s), and for the electrons the velocity extrema vary between + 0.8x4 electron thermal velocities to -1.2x4 electron thermal velocity ($0.8 * 4\sqrt{m_e/m_i}$ and $-1.2 * 4\sqrt{m_e/m_i}$ in terms of acoustic speed, in order to center the grid in the region where the electron distribution is dominant), with 250 grid points in space. For an arbitrary plasma profile, we have the following relations for the temperature:

$$T_{ix,y,z}(x)=\frac{1}{n_i}\int d\vec{\upsilon}(v_{x,,y,z}-<v_{x,y,z}>)^2 f_i(x,\vec{v}) \qquad (5.11)$$

where
$$<v_{x,y,z}>=\frac{1}{n_i}\int d\vec{v}v_{x,y,z}f_i(x,\vec{v}) \qquad (5.12)$$

5.2. The Numerical Scheme

The numerical scheme used to solve Eqs.(5.1-5.2) applies a method of fractional steps which has been discussed in the literature coupled to an integration along the characteristics [1,27-29]. A method which has a precision of second order in time is obtained by splitting Eqs.(5.1-5.2) following the steps:

$$\frac{H}{2}\frac{V_x}{2}\frac{V_y}{2}V_z\frac{V_y}{2}\frac{V_x}{2}\frac{H}{2} \qquad (5.13)$$

Where the operators H, V_x, V_y, V_z represent the shift operators for the distribution function along the characteristics of Δt, respectively in the directions x, v_x, v_y, and v_z. The shift are calculated using cubic spline interpolation, as discussed for instance in section 2. This translates into the following steps, starting from the time $t=n\Delta t$, to move to the time $t=(n+1)\Delta t$:

Step1 -Solve
$$\frac{\partial f_e}{\partial t}+v_{\|}\sin\alpha\frac{\partial f_e}{\partial x}=0 \text{ for a step } \Delta t/2 \qquad (5.14)$$

Solve
$$\frac{\partial f_i}{\partial t}+v_x\frac{\partial f_i}{\partial x}=0 \text{ for a step } \Delta t/2 \qquad (5.15)$$

The solutions are written $f_{e1}(x,v_{\|})=f_e^n(x-v_{\|}\sin\alpha\Delta t/2,v_{\|})$

$$f_{i1}(x,v_x,v_y,v_z)=f_i^n(x-v_x\Delta t/2,v_x,v_y,v_z)$$

Solve Poisson’s equation for the electric field which we denote by E_x. Poisson`s equation is discretized in space as a tridiagonal matrix as indicated in Eq.(2.17).

Step2 -Solve
$$\frac{\partial f_e}{\partial t}-\frac{m_i}{m_e}\sin\alpha E_x\frac{\partial f_e}{\partial v_{\|}}=0 \text{ for a step } \Delta t/2 \qquad (5.16)$$

Solve $$\frac{\partial f_i}{\partial t}+(E_x - v_z\omega_{ci}\cos\alpha)\frac{\partial f_i}{\partial v_x}=0 \text{ for a step } \Delta t/2 \tag{5.17}$$

The solutions are written $f_{e2}(x,v_{\|}) = f_{e1}(x,v_{\|}+\frac{m_i}{m_e}\sin\alpha E_x\Delta t/2)$

$$f_{i2}(x,v_x,v_y,v_z) = f_{i1}(x,v_x-(E_x - v_z\omega_{ci}\cos\alpha)\Delta t/2, v_y, v_z)$$

Step3 - Solve $$\frac{\partial f_i}{\partial t}+v_z\omega_{ci}\sin\alpha\frac{\partial f_i}{\partial v_y}=0 \text{ for a step } \Delta t/2 \tag{5.18}$$

The solution is written $f_{i3}(x,v_x,v_y,v_z) = f_{i2}(x,v_x,v_y - v_z\omega_{ci}\sin\alpha\Delta t/2, v_z)$

Step4 -Solve $$\frac{\partial f_i}{\partial t}+\omega_{ci}(v_x\cos\alpha - v_y\sin\alpha)\frac{\partial f_i}{\partial v_z}=0 \text{ for a step } \Delta t \tag{5.19}$$

The solution is written

$$f_{i4}(x,v_x,v_y,v_z) = f_{i3}(x,v_x,v_y,v_z - \omega_{ci}(v_x\cos\alpha - v_y\sin\alpha)\Delta t)$$

Step5 Repeat **Step3** to calculate $f_{i5}(x,v_x,v_y,v_z)$ from f_{i4}.

Step6 Repeat **Step2** to calculate $f_{i6}(x,v_x,v_y,v_z)$ from f_{i5}.

Step7 Repeat **Step1** to calculate $f_i^{n+1}(x,v_x,v_y,v_z)$ from f_{i6}.

One can verify for the ions (the same procedure can be applied for the electrons, see [1]), that by following the previous sequence, at time $t=(n+1)\Delta t$ the ion distribution function can be written :

$$f_i^{n+1}(x,\vec{v}) = f_i\ (x^*, v_x^*, v_y^*, v_z^*, t=n\Delta t) \tag{5.20}$$

With:

$$x^* = x - v_x\Delta t + \frac{1}{2}E_x^2\Delta t^2 - v_z\omega_{ci}\cos\alpha\frac{\Delta t^2}{2} \tag{5.21}$$

$$v_x^* = v_x - E_x\Delta t + v_z\omega_{ci}\cos\alpha\Delta t - v_z\omega_{ci}^2\cos^2\alpha\frac{\Delta t^2}{2} + v_y\omega_{ci}^2\sin\alpha\cos\alpha\frac{\Delta t^2}{2} \quad (5.22)$$

$$v_y^* = v_y - v_z\omega_{ci}\sin\alpha\Delta t + v_x\omega_{ci}^2\sin\alpha\cos\alpha\frac{\Delta t^2}{2} - v_y\omega_{ci}^2\sin^2\alpha\frac{\Delta t^2}{2} \quad (5.23)$$

$$v_z^* = v_z - v_x\omega_{ci}\cos\alpha\Delta t + v_y\omega_{ci}\sin\alpha\Delta t + \omega_{ci}\cos\alpha E_x\frac{\Delta t^2}{2} \quad (5.24)$$

The electric field E_x is calculated at the following position and time::

$$E_x \equiv E_x(x - v_x\frac{\Delta t}{2}, t = n\Delta t + \frac{\Delta t}{2}) \quad (5.25)$$

From Eq.(5.2), the equations of the characteristics (which describe the ions motion) are given by:

$$\frac{dx}{dt} = v_x \quad (5.26)$$

$$\frac{dv_x}{dt} = E_x - v_z\omega_{ci}\cos\alpha \quad (5.27)$$

$$\frac{dv_y}{dt} = v_z\omega_{ci}\sin\alpha \quad (5.28)$$

$$\frac{dv_z}{dt} = v_x\omega_{ci}\cos\alpha - v_y\omega_{ci}\sin\alpha \quad (5.29)$$

By integrating Eqs.(5.26-5.29) from $n\Delta t$ to $(n+1)\Delta t$, we get the following set of coupled equations:

$$x^n = x^{n+1} - v_x^{n+1}\frac{\Delta t}{2} - v_x^n\frac{\Delta t}{2} \quad (5.30)$$

$$v_x^n = v_x^{n+1} - E_x^{n+1/2}\Delta t + v_z^{n+1}\omega_{ci}\cos\alpha\frac{\Delta t}{2} + v_z^n\omega_{ci}\cos\alpha\frac{\Delta t}{2} \quad (5.31)$$

$$v_y^n = v_y^{n+1} - v_z^{n+1}\omega_{ci}\sin\alpha\frac{\Delta t}{2} - v_z^n\omega_{ci}\sin\alpha\frac{\Delta t}{2} \quad (5.32)$$

$$\begin{aligned} v_z^n &= v_z^{n+1} - v_x^{n+1}\omega_{ci}\cos\alpha\frac{\Delta t}{2} - v_x^n\omega_{ci}\cos\alpha\frac{\Delta t}{2} \\ &+ v_y^{n+1}\omega_{ci}\sin\alpha\frac{\Delta t}{2} + v_y^n\omega_{ci}\sin\alpha\frac{\Delta t}{2} \end{aligned} \quad (5.33)$$

It is straightforward to calculate the following solution for the set of Eqs.(5.30-5.33) :

$$x^n = x^{n+1} - v_x^{n+1}\Delta t + E_x^{n+1/2}\frac{\Delta t^2}{2} - v_z^{n+1}\omega_{ci}\cos\alpha\frac{\Delta t^2}{2} + O(\Delta t^3) \quad (5.34)$$

$$\begin{aligned} v_x^n &= v_x^{n+1} - E_x^{n+1/2}\Delta t + v_z^{n+1}\omega_{ci}\cos\alpha\Delta t - v_x^{n+1}\omega_{ci}^2\cos^2\alpha\frac{\Delta t^2}{2} \\ &+ v_y^{n+1}\omega_{ci}^2\sin\alpha\cos\alpha\frac{\Delta t^2}{2} + O(\Delta t^3) \end{aligned} \quad (5.35)$$

$$\begin{aligned} v_y^n &= v_y^{n+1} - v_z^{n+1}\omega_{ci}\sin\alpha\Delta t - v_y^{n+1}\omega_{ci}\sin^2\alpha\frac{\Delta t^2}{2} \\ &+ v_x^{n+1}\omega_{ci}\sin\alpha\cos\alpha\frac{\Delta t^2}{2} + O(\Delta t^3) \end{aligned} \quad (5.36)$$

$$\begin{aligned} v_z^n &= v_z^{n+1} - v_x^{n+1}\omega_{ci}\cos\alpha\Delta t + v_y^{n+1}\omega_{ci}\sin\alpha\Delta t \\ &+ E_x^{n+1/2}\omega_{ci}\cos\alpha\frac{\Delta t^2}{2} - v_z^{n+1}\omega_{ci}^2\frac{\Delta t^2}{2} + O(\Delta t^3) \end{aligned} \quad (5.37)$$

By comparing Eqs.(5.21-5.24) and Eqs.(5.34-5.37), we see that the fractional step scheme we are using integrates the distribution function for the ions along the characteristics correctly to second order in Δt [1,27-29]. Following the same steps as before, a similar result can be obtained for advancing by one time-step the distribution function of the electrons [1].

5.3. Results

The ratio of the ion gyroradius to the Debye length is given by:

$$\frac{\rho_i}{\lambda_{De}} = \sqrt{\frac{2T_i}{T_e}}\frac{1}{\omega_{ci}/\omega_{pi}} \quad (5.38)$$

If we assume initially the ions distribution with $T_{ix} = T_{iz} = T_i$, then the factor $2T_i$ in the calculation of the gyroradius in Eq.(5.38) takes into account that the perpendicular temperature of the ions $< \upsilon_x^2 > + < \upsilon_z^2 > = 2T_i / m_i$, assuming an initially uniform plasma. The results we present are obtained for an angle of incidence of the magnetic $\alpha = 0.75^0$. The total length L of the system is taken to be 100 Debye lengths. We consider the case where $T_i / T_e = 2$ and $\omega_{ci} / \omega_{pi} = 0.1$. In this case from Eq.(5.38) $\rho_i / \lambda_{De} = 20$ and the length of the system is about 5 gyroradii, and in Eq.(5.4) $v_0 = \sqrt{3} = 1.73$. There exists a critical angle α_c where the transport of the ions perpendicular to the magnetic field is equal to the transport of the electrons along to the magnetic field (the *x*-components of the ion and electron velocities are equal to each other at this critical angle). Ions need half a gyro-period τ_i to travel twice the gyroradius ρ_i. In the time $\tau_i / 2$, electrons will travel along the magnetic field a distance $\sqrt{T_e / m_e}\tau_i / 2$. Hence we have the following relation for the critical angle [59]:

$$\tan \alpha_c = \frac{2\rho_i}{\sqrt{T_e / m_e}\tau_i / 2} = \frac{2\sqrt{2T_i / m_i}\tau_i / (2 * \pi)}{\sqrt{T_e / m_e}\tau_i / 2} = 0.9\frac{\sqrt{T_i m_e}}{\sqrt{T_e m_i}}$$

At this critical angle, the electrons moving along the magnetic field can still adjust to the gyration of the ions perpendicular to the magnetic field in their attempt to compensate for a charge separation. For the present set of parameters, this critical angle is $\tan \alpha_c = 0.9\,(T_i m_e / T_e m_i)^{1/2} = 0.9 / 42.85$ or equivalently $\alpha_c = 1.2^{\circ}$. There are other effects like gyro-cooling [48] which reduces T_i and ρ_i in front of the plate, and polarization drift (towards the wall) from the accelerating electric field, which possibly do not exactly compensate, so that the factor 0.9 might be approximate. For the present calculation we have taken $\alpha = 0.75^{\circ}$, therefore $\alpha < \alpha_c$. Fig.(37) shows the potential profile during a half period of the evolution of the oscillation (at *t*=250,*t*=275 and *t*=300). The period of this oscillation for the present set of parameters is about 100. The electric field at the sheath entrance (the slope of the potential at *x*=*L*) is oscillating between positive and negative values. We present in Fig.(38) the electric field profiles at *t*=250 and *t*=300. In Fig.(39) we present the electric field at the intermediate time *t*=275 (full curve), at a time where the slope of the potential at the sheath entrance is close to zero (see Fig.(37)). The dotted curve in Fig.(39) plots the quantity $\nabla P_i / n_i$, where the pressure $P_i = 0.5(n_i T_{ix} + n_i T_{iz})$. The dash-dot curve in Fig.(39) represents the Lorentz force in the *x* direction for the ions, which in our units is written $0.1 < \upsilon_z > / \cos\alpha$. The broken curve in Fig.(39) represents the combined force term $\nabla P_i / n_i + 0.1 < \upsilon_z > / \cos\alpha$. The broken curve appears to follow the electric field curve, deviating only in front of the plate. This deviation is due to the fact that n_i is small in front of the plate, so the division by n_i might not be very accurate. Fig.(40) shows the density profiles (full curve electrons, broken curve ions), at a) *t*=250 and at b) *t*=300. The oscillation

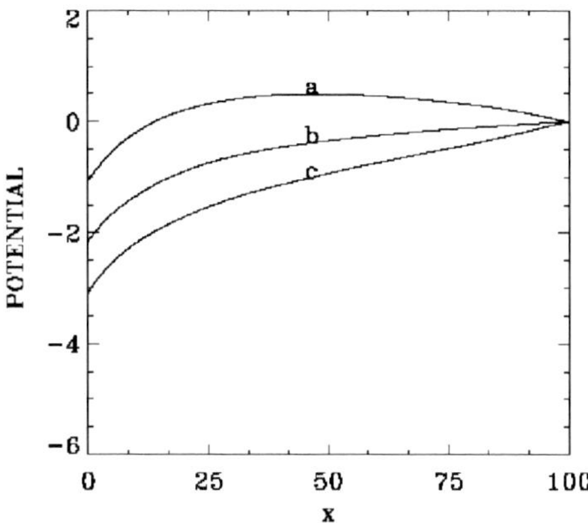

Figure 37. Potential profile at a) t=250, b) t=275, c) t=300.

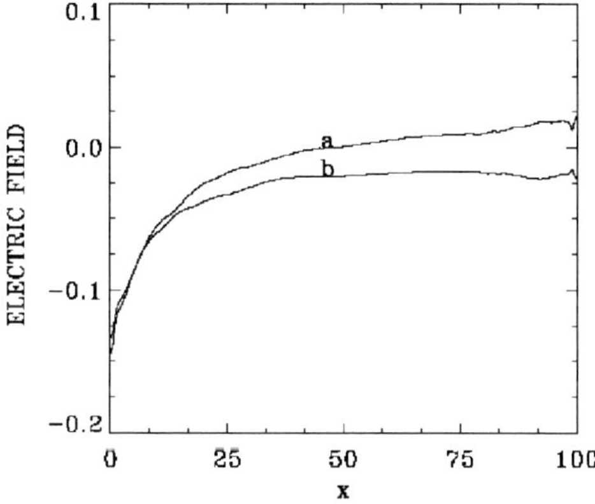

Figure 38. Electric field profile at a) t=250, b) t=300.

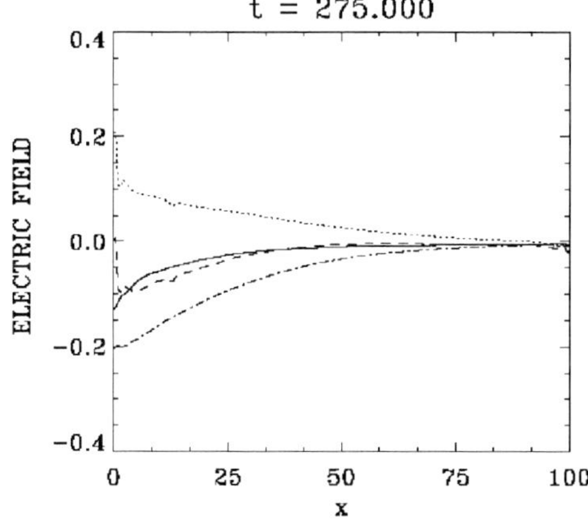

Figure 39. Electric field at t=275 (full curve), $\nabla P_i / n_i$ (dotted curve), $0.1 < \upsilon_z > / \cos\alpha$ (dash-dot curve), $\nabla P_i / n_i + 0.1 < \upsilon_z > / \cos\alpha$.

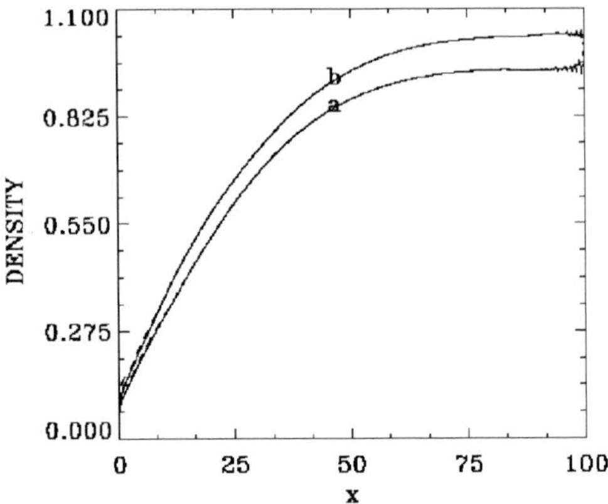

Figure 40. Density profiles (full curves electrons, broken curves ions) at a) t=250, b) t=300 (broken curve).

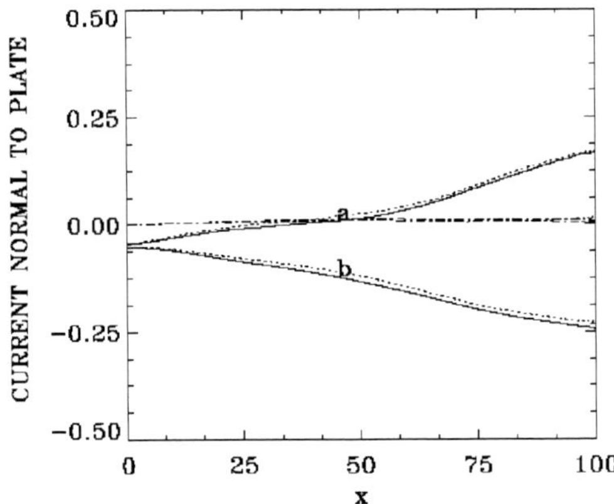

Figure 41. Current normal to plate (full curves electrons, broken curves ions) at a) t=225, b) t=300 (dash-dot curve total current, essentially zero).

of the density curves reflects the fact that particles are moving in and out during an oscillation period, maintaining a total current of zero normal to the plate in the sheath domain. This is observed in Fig.(41), where the currents normal to the plate are presented (full curve $J_{e\|}\sin\alpha$ for the electrons, dotted curve J_{xi} for the ions, and the dash-dot curve for the difference $(J_{xi} - J_{e\|}\sin\alpha)$ which is essentially zero). These curves are taken at t=225, and at t=275. Since the total current normal to the plate is zero, then Eq.(5.9) demands that $E_x\big|_{x=0}$ be constant at the wall, as observed in Fig.(38). Since on the other hand $E_x\big|_{x=L} - E_x\big|_{x=0} = \sigma$ (see Eq.(5.10)), therefore the variation of the total charge σ in the sheath domain has to be followed by the same variation of the electric field $E_x\big|_{x=L}$ at $x = L$. This is exactly what is observed in Fig.(38). Equation.(5.10) is verified at every time-step by the numerical code. Fig.(42) represents the charge (n_i-n_e), at t=250 and t=300. The two curves are close in the figure, as we can see also from Fig.(40) that the electron and ion densities remain close. But due to the large ratio of ρ_i / λ_{De} a small charge separation translate into an effective large difference in the corresponding potential and electric field. There is a positive charge in front of the plate over a length of about 20 Debye lengths, i.e. over about one gyroradius. There is a tendency for the positive charge in front of the plate to decrease for small angles α , the electrons running along the magnetic field can reach closer to the wall.

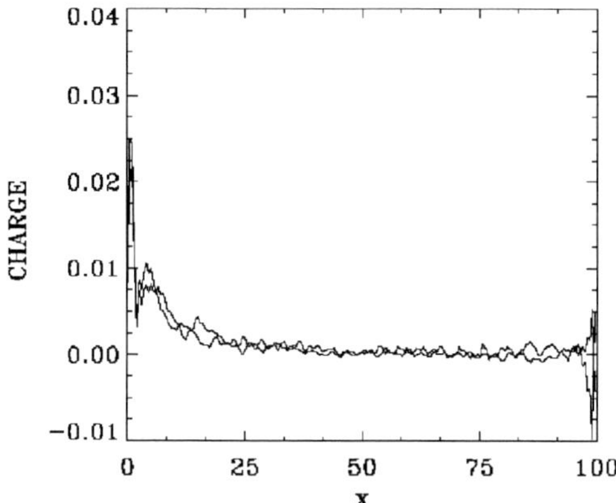

Figure 42. Charge ($n_i - n_e$) at t=250 and t=300 (the curves are very close).

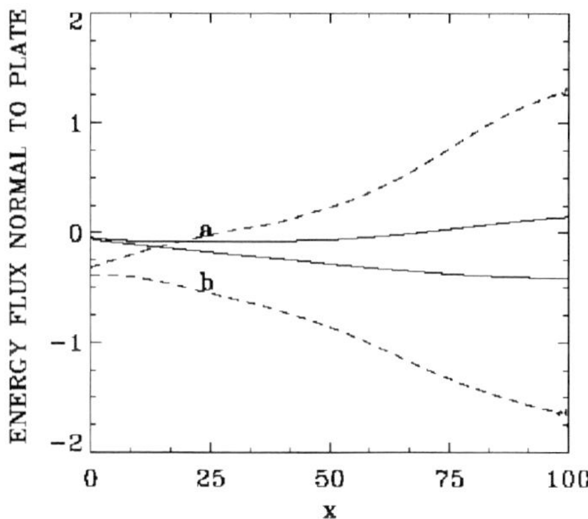

Figure 43. Energy flux (full curves electrons, broken curves ions) at a) t=225, b) t=275.

The energy fluxes $Q_i(x)$ and $Q_e(x)$ normal to the plate for the ions and electrons are given in Fig.(43), defined by:

$$Q_i(x) = \frac{1}{2} \int d\vec{v} v_x v^2 f_i(x, \vec{v}) \ ; \ Q_e(x) = \frac{m_e}{m_i} \frac{1}{2} \sin\alpha \int dv_{\parallel} v_{\parallel}^3 f_e(x, v_{\parallel}) \qquad (5.39)$$

(full curves for the electrons, broken curves for the ions, at a) t=225, and b) t=275). Usually, at such a small grazing incidence of the magnetic field, the frozen electrons along the magnetic field can prevent the transport of the ions across the magnetic field, causing a plasma detachment (see for instance, the results in [58]). However, we note during the oscillations of the ion energy flux (the broken curve in Fig.(43)), the ions are maintaining a negative contribution at the plate (in contrast, for instance, to what we see for $\alpha = 0.75^0$ in [58]), reflecting the fact that a population of ions are being scraped-off in front of the plate. A similar result showing an increase in the energy flux below a critical angle was also reported in [60], however the ratio of the ions gyroradius to the Debye length in [60] was 200. The results in [60] show very clearly the existence of a critical angle at grazing incidence of the magnetic field, below which the energy flux to the wall increases. The present results agree with the results in [60] that below a critical angle, the energy flux to the wall increases.

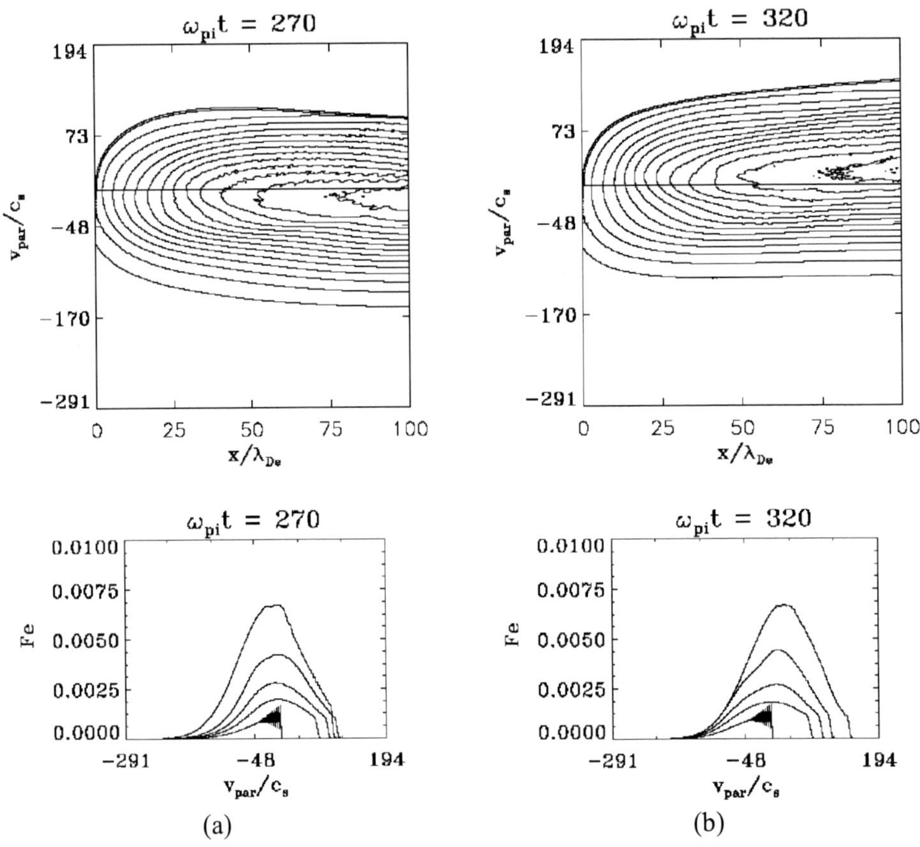

Figure 44. Electron distribution function.

Figs.(44a,b) present the distribution function of the electrons for a half period interval at t=270 and t=320 showing the strong oscillation of the running electrons throughout the domain, especially at the sheath entrance at the right boundary and at the sheath wall at the left boundary, to maintain zero total current normal to the plate. The same oscillation is observed in the gyrating ion population in the x direction in Fig.(45a,b), where the 1D distribution function $F_i(x, v_x)$ is plotted:

$$F_i(x, v_x) = \int dv_y dv_z f_i(x, \vec{v}) \tag{5.40}$$

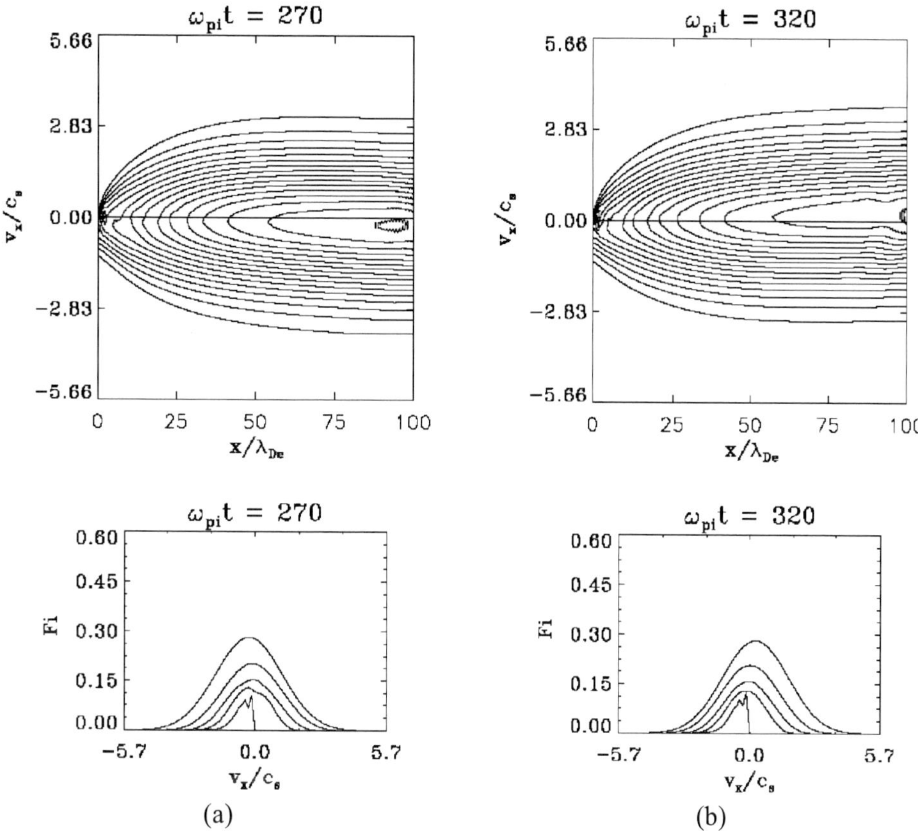

Figure 45. Ion distribution function $F_i(x, \upsilon_x)$.

Fig.(46) and Fig.(47) represents $F_i(x, v_z)$ and $F_i(x, v_y)$ given by:

$$F_i(x, v_{y,z}) = \int dv_x dv_{z,y} f_i(x, \vec{v}) \tag{5.41}$$

Note in Fig.(47) the ions entering the sheath in the direction y (almost parallel to the magnetic field since $\alpha = 0.75^0$) at the sound speed. The peaks of the lower curves in Fig.(47) are very close to the initial ion sound velocity $v_0 = \sqrt{3} = 1.73$ used to inject the ions parallel to the magnetic field. There is however a negligible oscillation in the y direction in front of the plate, compare to what we see in the x and z directions due to the perpendicular ions gyration. The bottom curves in Figs.(44-47) are the distribution functions obtained by cuts in the above contours at (from bottom to top) x=0, x=L/16, x=L/8, x=L/4, and x=L. These distribution functions in front of the wall are not Maxwellian, so that temperature measurements for instance, should be interpreted with care accordingly. Fig.(48) presents the ion temperature $T_{ix}(x)$ (full curve),$T_{iy}(x)$ (broken curve) and $T_{iz}(x)$ (dash-dot curve) (see Eq.(5.11)). Note the cooling observed when approaching the wall for $T_{ix}(x)$ and $T_{iz}(x)$ .We also see in Fig.(46) for instance (for the z-direction // $\vec{E}x\vec{B}$ drift) that when approaching the

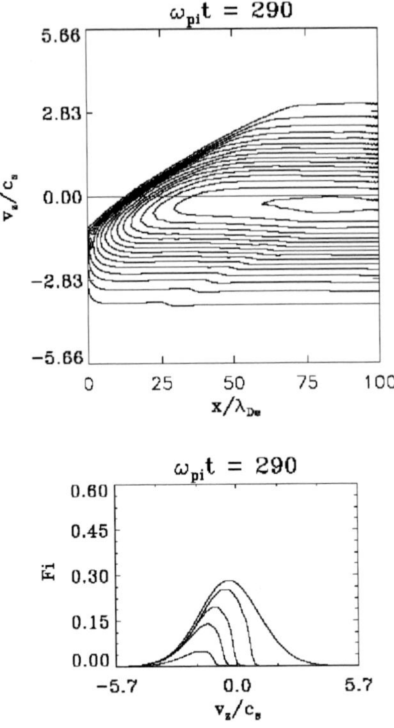

Figure 46. Ion distribution function $F_i(x, \upsilon_z)$.

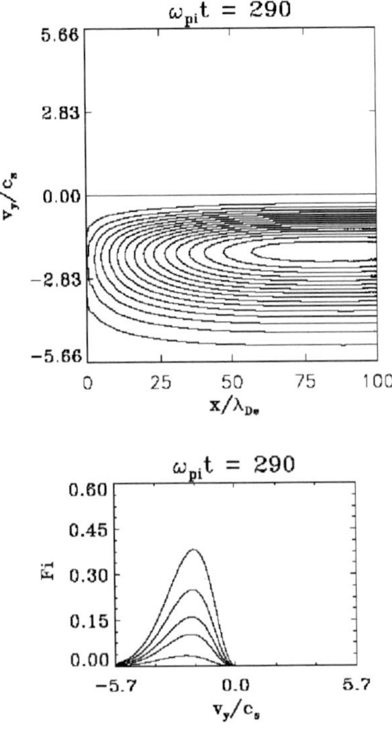

Figure 47. Ion distribution function $F_i(x, \upsilon_y)$.

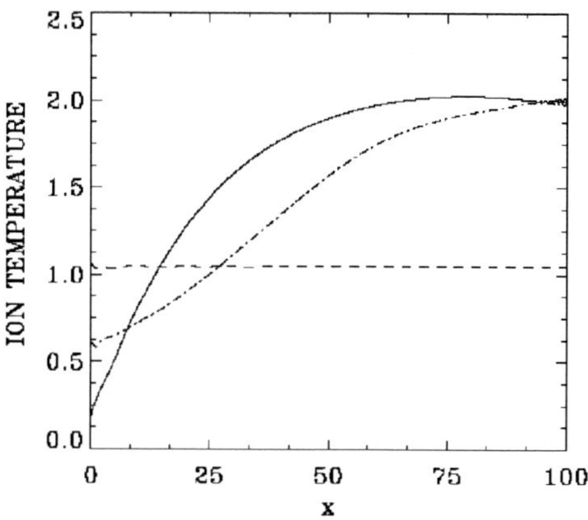

Figure 48. $T_{ix}(x)$ (full curve), $T_{iy}(x)$ (broken curve), $T_{iz}(x)$ (dash-dot curve) at t=275.

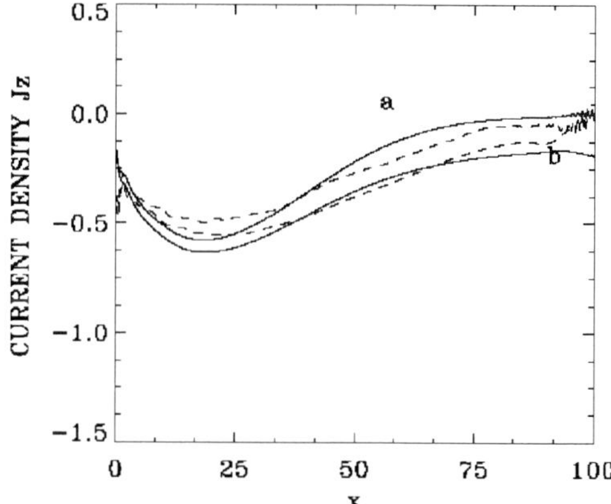

Figure 49. Ion urrent density J_{zi} (full curve), and the drift current $(n_i E_x - \partial n_i T_i / \partial x)\cos\alpha / 0.1$ (broken curve at a) t=225, b) t=275.

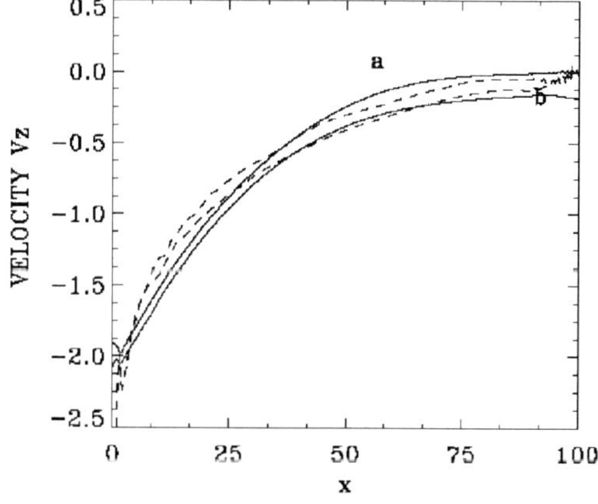

Figure 50. Velocity $< v_z >= J_{zi} / n_i$ (full curves) and drift $(E_x - \frac{1}{n_i}\partial n_i T_i / \partial x)\cos\alpha / 0.1$ at a) t=225, b) t=275.

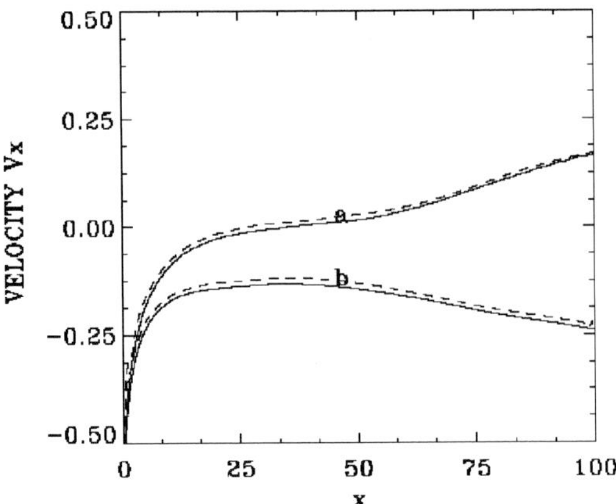

Figure 51. Velocity $<v_x>=J_{xi}/n_i$ (full curves), $<v_{xe}>=J_{xe}/n_e$ (broken curve) at a) *t*=225, b) *t*=275.

wall the width of the 1D distribution functions shrinks in the gyration plane perpendicular to $\vec{B}$. This gyro-cooling effect was discussed in [48], and is due to the fact that the thermal energy of gyro-motion is gradually converted into kinetic energy of the wall-tangential electric drift motion. We show in Fig.(49) the current density J_{zi} (full curve), at a) *t*=225and b) *t*=275:

$$J_{zi} = \int d\vec{v} v_z f_i(x,\vec{v}) \tag{5.42}$$

The broken curves in Fig.(49) is the sum of the macroscopic currents in the *z* direction $n_i\left(\vec{E}\times\vec{B}/B^2+\vec{v}_D\right)$, where the diamagnetic drift $\vec{v}_D=\vec{B}\times\nabla n_iT_i/\left(n_i\,eB^2\right)$, calculated at the same time. In our normalized units this drift current is $(n_iE_x-\partial n_iT_i/\partial x)\cos\alpha/0.1$, where we define $T_i=(T_{xi}+T_{zi})/2$. In Fig.(50), the full curves are the velocity $<v_z>=J_{zi}/n_i$, and the broken curves are the sum of the drifts $(E_x-\frac{1}{n_i}\partial n_iT_i/\partial x)\cos\alpha/0.1$. In Fig.(51) we present the velocity $<v_x>=J_{xi}/n_i$ (full curve), and the velocity $<v_{xe}>=J_{xe}/n_e$ (broken curves) at the same time. Note how the electrons are closely following the ions. Note also in Figs.(50) and (51) that the division by the densities (which are very small close to *x*=0), can be artificially enhancing the velocity curves at *x*=0, where the density is very small, and possibly slightly inaccurate.

Finally Fig.(52) presents the electron temperature $T_e(x)$ defined by:

$$T_e(x)=\frac{1}{n_e}\int dv_{\|}(v_{\|}-<v_{\|}>)^2 f_e(x,v_{\|})\,; \qquad <v_{\|}>=J_{\|}/n_e \tag{5.43}$$

(at *t*=290 and at *t*=340). It shows a regular temperature pulsation taking place during the important oscillation of the electrons, and it also shows a cooling in front of the plate.

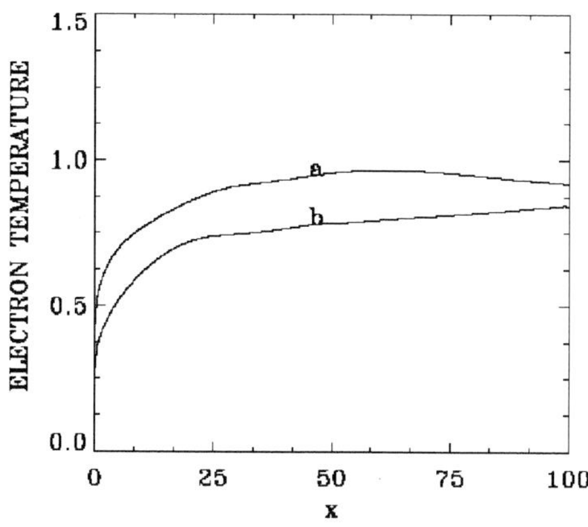

Figure 52. Electron temperature at a) t=275, b) t=325.

We have presented in this section a study of the physics associated with a collisionless plasma sheath with grazing incidence of the magnetic field, when the ions gyroradius is large compared to the Debye length. The numerical scheme presented in section 5.2 applies a method of fractional step coupled to an integration along the characteristics. Below a critical angle of incidence, the physics of the sheath differs from the classical sheath behaviour. In the case we have presented, the ratio of the gyroradius to the Debye length is 20. Low frequency oscillations are observed in the sheath, due to combined effect of electrons running along the magnetic field lines, and ions gyrating across the magnetic field lines. These low frequency oscillations will have tendency to be more complex in the presence of impurity ions (which have larger gyroradii and larger gyro-period), and are not related to edge turbulence. They certainly play an important role in controlling the energy flux normal to the wall (see Fig.(43)). Indeed, below the critical angle, the energy flux delivered to the wall has a tendency to increase, in agreement with the experimental observation in [60]. We have also seen that the distribution functions in front of the wall are non-Maxwellian, and this has to be taken into account when probes are used in temperature measurements.

6. Study of the Formation of a Charge Separation and an Electric Field at a Plasma Edge

We consider the problem of the generation of radial electric fields and poloidal flows to achieve radial force balance at a steep density gradient, in the presence of an external magnetic field. This problem is of particular interest and importance in the steep density gradient of an H-mode pedestal in a tokamak or near the edge of a plasma. In full toroidal geometry, there are "neoclassical" effects which can play a role in this problem [61,62]. We focus for simplicity in the present section on a cylindrical geometry for the problem of the generation of an electric field and poloidal flow at a plasma edge, when the external magnetic field is applied along the axis of the cylinder. Along the steep density gradient at the plasma edge, the electrons, frozen by the magnetic field lines, have a constant density profile which varies rapidly along the gradient over an ion orbit size. The frozen electrons cannot then move across the magnetic field to exactly compensate the ion charge which results from the finite ions' gyroradius along the gradient. Therefore a charge separation appears at the edge of the plasma. This effect is especially important for large values of ρ_i / λ_{De} (where ρ_i is the ions' gyroradius and λ_{De} the Debye length). Accurate calculation of the exact ion orbits is important in this case for the accurate calculation of the charge separation and the self-consistent electric field. We study this problem in the present section using an Eulerian Vlasov code, which has the advantage of very low noise levels, and makes it possible to measure accurately a very small charge separation [5, 32]. The results we present in this section are obtained by applying a method of fractional step to solve a full 2D Vlasov equation in cylindrical geometry, coupled to a numerical solution using the method of characteristics. In this case the presence of centrifugal and Coriolis forces require a 2D interpolation in velocity space, which is done by using a tensor product of cubic *B*-splines [5,32]. The interpolation in space is also done in 2D using a tensor product of cubic *B*-splines [32]. We compare the electric field calculated along the gradient with the macroscopic values calculated from the kinetic code for the gradient of the ion pressure, and we find that this quantity balance the electric field fairly well, a result similar to what has been presented in cartesian geometry in Ref. [63]. The contribution of the Lorentz force term along the gradient is negligible. And in the 2D case presented in this section, we find that the $\vec{E} \times \vec{B}$ drift is balanced fairly well by the diamagnetic drift.

6.1. The Relevant Equations and the Numerical Method for the 2D Problem in Cylindrical Geometry

In the cylindrical coordinate system we are considering, the inhomogeneous direction in the cylindrical plasma is the radial direction r, normal to a vessel surface located at $r = R$. The constant magnetic field is in the z direction which represents the toroidal direction, and θ is the poloidal direction. The ions are described by the normalized 2D Vlasov equation for the ions distribution function $f_i(r, \theta, v_r, v_\theta, t)$:

$$\frac{\partial f_i}{\partial t} + v_r \frac{\partial f_i}{\partial r} + \frac{v_\theta}{r} \frac{\partial f_i}{\partial \theta} + \left[E_r + v_\theta \, \omega_{ci} + \frac{v_\theta^2}{r} \right] \frac{\partial f_i}{\partial v_r} - \left(- E_\theta + \omega_{ci} \, v_r + \frac{v_r v_\theta}{r} \right) \frac{\partial f_i}{\partial v_\theta} = 0 \tag{6.1}$$

Which is coupled to Poisson's equation:

$$\frac{1}{r}\frac{\partial}{\partial r} r \frac{\partial \phi}{\partial r} + \frac{1}{r^2}\frac{\partial^2 \phi}{\partial \theta^2} = -(n_i - n_e) \; ; \; E_r = -\frac{\partial \phi}{\partial r} \; ; \; E_\theta = -\frac{1}{r}\frac{\partial \phi}{\partial \theta} \tag{6.2}$$

In Eq.(6.1), time is normalized to the inverse plasma frequency ω_{pi}^{-1}, velocity is normalized to the acoustic velocity $c_s = \sqrt{T_e / m_i}$, and length to the Debye length

$\lambda_{De} = c_s / \omega_{pi}$. T_e is the electron temperature and m_i is the ion mass. The potential is normalized to T_e / e, and the density is normalized to the peak initial central density. ω_{ci} is the ion cyclotron frequency (normalized to ω_{pi}). The system is solved over a length $L = 150$ Debye lengths in front of the vessel surface, with an initial ion distribution function for the deuterons over the domain $r = \{R - L, R\}$ given in our normalized units, by:

$$f_i(r, \theta, v_r, v_\theta) = n_i(r) \frac{e^{-(v_r^2 + v_\theta^2)/2T_i}}{2\pi T_i} ; \tag{6.3}$$

n_i and n_e are the initial ion and electron densities respectively, with $n_e(r) = n_i(r)$ in the initially neutral system. R is the radius of the cylinder. The electrons are assumed frozen by the external magnetic field, so that the $n_e(r)$ profile remains constant during the simulation. We also use the following parameters:

$$\frac{T_i}{T_e} = 1 ; \qquad \frac{\rho_i}{\lambda_{De}} = \sqrt{\frac{2T_i}{T_e}} \frac{1}{\omega_{ci0} / \omega_{pi}} = 10\sqrt{2} \tag{6.4}$$

We assume the magnetic field B_z along the axis of the cylinder varies as $B_z = B_o/(1+\epsilon \cos \theta)$, where $\epsilon = 0.2$ in the present calculations, and B_o is the field at $\theta = \pi/2$. In this case $\omega_{ci} = \omega_{cio} / (1+ \epsilon \cos \theta)$, and $\omega_{cio} / \omega_{pi} = 0.1$ in the present calculations. With the initial distribution function for the ions in Eq. (6.3), we have $T_{i\theta} = T_{ir} = T_i$ spatially constant, the factor $2T_i$ in Eq. (6.4) in the calculation of the gyroradius takes into account that the perpendicular temperature is $\langle v_\perp^2 \rangle = \langle v_r^2 \rangle + \langle v_\theta^2 \rangle = \frac{2T_i}{m_i}$.

Equation (6.1) is solved by a method of fractional step coupled to a method of characteristics. To advance Eq. (6.1) for one time-step Δt, the splitting of the equation is effected as follows.

Step1 - We solve for $\Delta t/2$ the equation:

$$\frac{\partial f_i}{\partial t} + v_r \frac{\partial f_i}{\partial r} + \frac{v_\theta}{r} \frac{\partial f_i}{\partial \theta} = 0 \tag{6.5}$$

We then solve Poisson's equation in Eq.(6.2) to calculate the electric field.

Step2 We then solve for Δt the equation:

$$\frac{\partial f_i}{\partial t} + \left(E_r + v_\theta \omega_{ci} + \frac{v_\theta^2}{r} \right) \frac{\partial f_i}{\partial v_r} - \left(- E_\theta + \omega_{ci} v_r + \frac{v_r v_\theta}{r} \right) \frac{\partial f_i}{\partial v_\theta} = 0 \tag{6.6}$$

Step3 We repeat **Step1** for $\Delta t/2$

This completes a one time-step cycle Δt. For the solution of Eq. (6.5), we first solve the characteristic equations:

$$\frac{dr}{dt} = v_r \;; \qquad \frac{d\theta}{dt} = \frac{v_\theta}{r} \tag{6.7}$$

at a given (v_r, v_θ) in velocity space. The solution of Eqs (6.7) originating at (r_o, θ_o) at time t, and reaching the grid point (r, θ) at $t = t + \Delta t/2$ can be written as follows, for a half time-step $\Delta t/2$:

$$r_o = r - v_r\,\Delta t/2 \;; \qquad \theta_o = \theta - \frac{v_\theta}{v_r}\ln\frac{r}{r - v_r\,\Delta t/2} \tag{6.8}$$

For $v_r\,\Delta t/2 \ll 1$, the second equation reduces to $\theta_o = \theta - \frac{v_\theta}{r}\Delta t/2$. Therefore the solution of Eq. (6.5) can be written, for half a time-step $\Delta t/2$, as follows:

$$f_i^*\left(r, \theta, v_r, v_\theta, t + \Delta t/2\right) = f_i\left(r - v_r\;\Delta t/2, \theta - \frac{v_\theta}{v_r}\ln\frac{r}{r - v_r\Delta t/2}, v_r, v_\theta, t\right) \tag{6.9}$$

The right hand side of Eq. (6.9) is calculated by interpolation using a tensor product of cubic *B*-splines, in which θ is periodic [5,32]. For each v_r and v_θ, we write:

$$s(r,\theta) = \sum_{j=0}^{N_r}\sum_{k=0}^{N_\theta}\eta_{jk}B_j(r)B_k(\theta) \tag{6.10}$$

taking into account that θ is periodic. The cubic *B*-spline is defined as:

$$B_j(x) = \frac{1}{6h^3}\begin{cases} (x - x_j)^3 & x_j \le x < x_{j+1} \\ h^3 + 3h^2(x - x_{j+1}) + 3h(x - x_{j+1})^2 & \\ \qquad\qquad - 3(x - x_{j+1})^3 & x_{j+1} \le x < x_{j+2} \\ h^3 + 3h^2(x_{j+3} - x) + 3h(x_{j+3} - x)^2 & \\ \qquad\qquad - 3(x_{j+3} - x)^3 & x_{j+2} \le x < x_{j+3} \\ (x_{j+4} - x)^3 & x_{j+3} \le x < x_{j+4} \end{cases} \tag{6.11}$$

and $B_j(x)=0$ otherwise. *h* is the grid size. For the calculation of the coefficients η_{jk} of the cubic *B*-spline interpolation function $s(r,\theta)$ in Eq.(6.10) see details in [5,32].

We then go to **Step2** and solve for Δt the equation:

$$\frac{\partial f_i}{\partial t}+\left(E_r+v_\theta\,\omega_{ci}+\frac{v_\theta^2}{r}\right)\frac{\partial f_i}{\partial v_r}-\left(-E_\theta+\omega_{ci}v_r+\frac{v_r v_\theta}{r}\right)\frac{\partial f_i}{\partial v_\theta}=0 \qquad (6.12)$$

The electric field E_θ remained very small and negligible, inspite of the small azimuthal variation of the charge, since $E_\theta=-\frac{1}{r}\frac{\partial\phi}{\partial\theta}$, and $r\sim R$ =10000 towards the edge in our calculation. For the solution of Eq.(6.12), we first solve the characteristics equations by neglecting E_θ :

$$\frac{dv_r}{dt}=E_r+v_\theta\omega_{ci}+\frac{v_\theta^2}{r};\qquad \frac{dv_\theta}{dt}=-\omega_{ci}v_r-\frac{v_r v_\theta}{r} \qquad (6.13)$$

To an order $O(\Delta t)^2$, the solution of Eq. (6.13) yields the following solution to Eq.(6.12):

$$f_i^{**}(r,\theta,v_r,v_\theta,t+\Delta t)=f_i^{*}(r,\theta,v_r-2a,v_\theta-2b,t) \qquad (6.14)$$

The 2D interpolation in Eq. (6.14) is done using again a tensor product of cubic *B*-splines [5], as explained in Eq.(6.10), and *a* and *b* are given, to an order $O(\Delta t)^2$, by the expressions [5]:

$$a=\frac{\Delta t}{2}\left\{E_r+\omega_{ci}\left(v_\theta+\frac{\Delta t}{2}\left(\omega_{ci}\,v_r+\frac{v_r\,v_\theta}{r}\right)\right)+\frac{v_\theta^2}{r}+\Delta t\frac{v_\theta}{r}\left(\omega_{ci}v_r+\frac{v_r\,v_\theta}{r}\right)\right\} \qquad (6.15)$$

$$b=-\frac{\Delta t}{2}\left\{\begin{array}{l}\omega_{ci}\left(v_r-\frac{\Delta t}{2}\left(E_r+v_\theta\omega_{ci}+\frac{v_\theta^2}{r}\right)\right)+\frac{v_r v_\theta}{r}+\frac{\Delta t}{2}\frac{v_r}{r}\left(\omega_{ci}v_r+\frac{v_r v_\theta}{r}\right)\\ -\frac{\Delta t}{2}\frac{v_\theta}{r}\left(E_r+v_\theta\omega_{ci}+\frac{v_\theta^2}{r}\right)\end{array}\right\}$$

We then repeat the solution of Eq. (6.9) for $\Delta t/2$ using f_i^{**} from Eq.(6.14) to calculate f_i^{n+1}.

Two cases will be considered. In the first case the density profile is given by:

$$n_i(r)=n_e(r)=0.5(1+\tanh((R-r-L/4)/7)) \qquad (6.16)$$

The profile in Eq. (6.16) situates the steep gradient at a distance of approximately $L/4$ from the wall of the vessel, which put the plasma relatively close to the floating wall of the vessel. The wall of the vessel will collect the charge coming from the plasma. In the second case, the density profile is given by:

$$n_i = n_e = 0.5(1 + \tanh\,((R - r - L/2)/4) \tag{6.17}$$

which situates the steep gradient at approximately a distance $L/2$ from the vessel (in the center of the plasma edge domain we are studying). In this case the plasma is further away from the wall of the vessel with respect to the case in Eq.(6.16) and there is no charge collected on the wall of the vessel, and the electric field at the vessel is set equal to zero.

It is clear that the final equilibrium depends on the initial profiles and on the initial values in Eqs.(6.3-6.4). It was pointed out in the analysis of H-mode power threshold in [64] that the changes in n_e and ∇n_e in the transition to H-mode are small, and changes in T_e are barely perceptible in the data. So it will be sufficient for the purpose of our study to assume that the magnetized electrons are frozen along the magnetic field lines with a constant profile given by Eq. (6.16) or (6.17). In this case the electrons cannot move across the magnetic field in the gradient region to compensate the charge separation which results due to the finite ion orbits. To determine this charge separation along the gradient, it is important to calculate the ion orbits accurately by using an Eulerian Vlasov code, which is the subject of the present section. The larger the ions' gyroradius, the bigger the charge separation and the self-consistent electric field at the edge. (Hence the important role played by even small fractions of impurity ions, because of their large gyroradii).

We use N=200 grid points in space in the radial direction, and 32 grid points in the azimuthal direction. 80 grid points are used in each velocity direction. The velocity extrema in our normalized units are $\pm 4\sqrt{T_e / T_i}$, with $T_e/T_i = 1$ in the present simulations.

6.2. Results

Case1.

In this case the density profile is given in Eq.(6.16), where the steep gradient is located at a distance about $L/4$ from the floating wall of the vessel. We assume in the present calculation that the deuterons hitting the wall surface at $r = R$ are collected by the floating cylindrical vessel. Since the magnetized electrons do not move in the r direction across the magnetic field, there is no electron current collected at the floating vessel. Therefore we have only an ion current, and at $r = R$ we have the relation:

$$\left.\frac{\partial E_r(r,\theta)}{\partial t}\right|_{r=R} = -J_{ri}\big|_{r=R} \quad \text{or} \quad E_r(r,\theta)\big|_{r=R} = -\int_0^t J_{ri}\big|_{r=R}\,dt \tag{6.18}$$

Where
$$J_{ri}(r,\theta) = \int v_r f_i\left(r,\theta,v_r,v_\theta,t\right) dv_\theta dv_r \tag{6.19}$$

The electric field E_r is directed towards the center to the interior of the plasma. The electric field E_θ remained very small and negligible, inspite of the small azimuthal variation of the charge, since $E_\theta = -\frac{1}{r}\frac{\partial\phi}{\partial\theta}$, and $r \sim R$ =10000 towards the edge. Integrating Eq. (6.2) over the domain (*R-L*, *R*), and neglectring the azimuthal variation $\frac{\partial\phi}{\partial\theta}$, we get the total charge σ:

$$R\, E_r\big|_{r=R} - (R-L)E_r\big|_{r=R-L} = \int_{R-L}^{R} (n_i - n_e) r\, dr = \sigma \tag{6.20}$$

We assume that the plasma ions are allowed to enter or leave at the left boundary. So the difference between the electric fields at the left boundary $r = R - L$ and at the wall $r = R$ in Eq. (6.20) must be equal to the total charge appearing in the system, and Eq.(6.20) must be satisfied at every time-step in every direction θ. In the present simulation, $E_r\big|_{r=R}$ is calculated from Eq. (6.18), which defines the derivative of the potential at the right boundary in Eq. (6.2). We fix the potential to be zero at the left boundary and solve Poisson's equation (6.2) for the potential. The electric field E_θ remained negligibly small and is used in the solution of Eq.(6.2) only as a very small correction. Then the resulting electric field at $E_r\big|_{r=R-L}$, calculated at $r = R - L$ from Eq. (6.2), must satisfy Eq. (6.20). As we shall see in the following results, the value of $E_r\big|_{r=R-L}$ remained very small, so that $E_r\big|_{r=R} \approx \sigma / R$, *i.e.* the charge appearing in the system is equal to the electric field appearing on the floating wall of the vessel. We use a large value of the cylindrical plasma radius *R* (*R* = 10000 Debye lengths in the present calculation), so that the system behaves essentially as a Cartesian system. Indeed we recover results similar to those which have been found in Cartesian geometry in [63].

Fig.(53) shows the electric field at $\theta = 0$ (full curve), $\theta = \pi/2$ (broken line) and $\theta = \pi$ (dot-dash line), at time $t = 520$. The results remained symmetric from $\theta = \pi$ to $\theta = 2\pi$. The electric field at the vessel wall $r = R$ was calculated using Eq. (6.18) at 32 points over the 2π circle. We see from Fig. (53) that the electric field at $\theta = 0$ is lower than the one at $\theta = \pi$. We will comment later on this point. The value of the electric field E_r at the wall at $\theta = 0$, $\pi/2$ and π are respectively -0.268, -0.308 and -0.387 respectively. Fig.(54) gives at time $t = 520$ and for $\theta = 0$ the electric field E_r (full curve). Also at $\theta = 0$ the dash-dotted curve in Fig.(54) gives the Lorentz force, which in our normalized units is given by $-< v_\theta > \omega_{ci} / \omega_{pi} = -0.1 < v_\theta > /(1.+0.2)$, and the broken curve gives the pressure force $\nabla P_i / n_i$, $P_i = 0.5\, n_i (T_{ir} + T_{i\theta})$, with the following definition:

$$T_{ir,\theta}(r,\theta) = \frac{1}{n_i} \int dv_r\, dv_\theta\, (v_{r,\theta} - < v_{r,\theta} >)^2\, f_i(r,\theta, v_r, v_\theta) \tag{6.21}$$

$$< v_{r,\theta} > = \frac{1}{n_i} \int dv_r\, dv_\theta\, v_{r,\theta}\, f_i(r,\theta, v_r, v_\theta) \; ; \; n_i(r,\theta) = \int dv_r\, dv_\theta\, f_i(r,\theta, v_r, v_\theta) \tag{6.22}$$

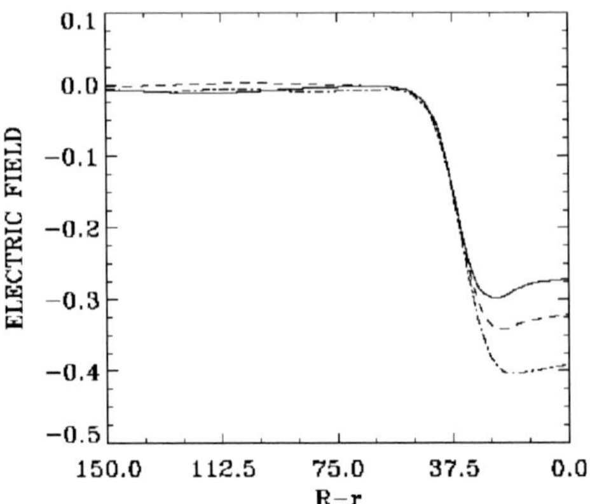

Figure 53. Electric field E_r at $\theta = 0$ (full curve), $\theta = \pi / 2$ (broken curve), and $\theta = \pi$ (dash-dot curve), at t=520.

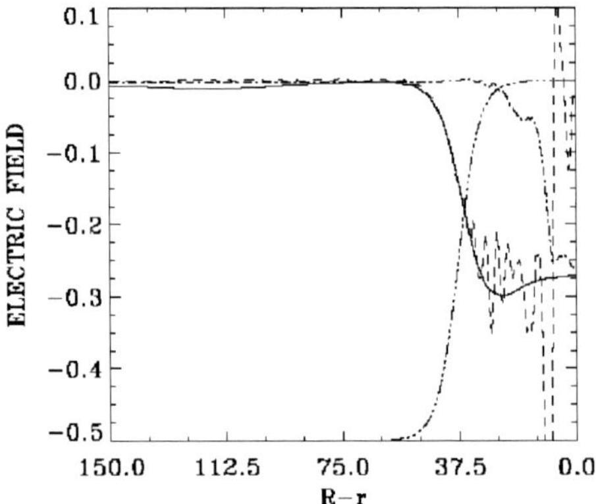

Figure 54. Plot at $\theta = 0$ of the electric field E_r (full curve), the Lorentz force $-0.1\langle v_\theta\rangle/(1+0.2\cos\theta)$ (dash-dot. curve), and the pressure force $\nabla P_i / n_i$ (broken curve). The curve $-n_i/2$ is plotted for reference (dash- 3 dots curve). At time t =520.

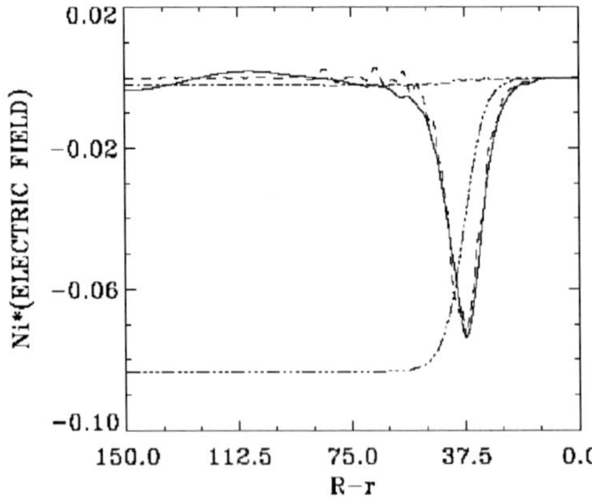

Figure 55. Plot at $\theta = 0$ of $n_i E_r$ (solid curve), $-0.1 n_i\langle v_\theta\rangle/(1+0.2\cos\theta)$ (dash-dot curve), and ∇P_i (broken-curve). The curve $-n_i/12$ is plotted for reference (dash-3 dots curve). At time t=520.

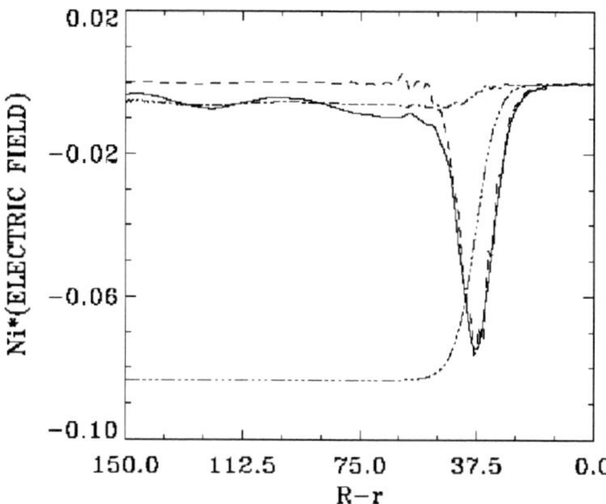

Figure 56. Plot at $\theta = \pi$ of $n_i E_r$ (solid curve), -0.1n_i<v_θ>/(1+0.2 cosθ) (dash-dot curve), and ∇P_i (broken-curve). The curve $-n_i/12$ is plotted for reference (dash-3 dots curve). At time t=520.

In steady state the transport $< v_r >$ vanishes. The $\nabla P_i / n_i$ term (broken curve) shows a very good agreement along the gradient with the solid curve for E_r, and the Lorentz force appears negligible along the gradient. In a region of about one to two gyroradii from the wall (around 20 Debye lengths from the wall), we have small irregular oscillations in space (and time), the accuracy of $\nabla P_i / n_i$ being degraded by the division with a very low value of the density n_i and a large ∇T_i appearing close to the vessel surface. To avoid this problem, we plot in Fig. (55) the quantities -0.1n_i<v_θ>/(1+0.2 cosθ), $n_i E_r$, ∇P_i, at $\theta = 0$ (note that $J_{\theta i} = n_i < v_\theta >$). We see that there is a very nice agreement for the relation $n_i\ E_r \approx \nabla P_i$ along the gradient (the density $-n_i/12$ is also plotted in Fig. (55) to locate the different profiles with respect to the gradient). The Lorentz force term remains negligible in the gradient region, and gives a very small contribution in the bulk at $\theta = 0$, close to the left boundary. The $\nabla P_i / n_i$ term is zero in the bulk at the left boundary since n_i and T_i are flat in that region. We note that $R = 10000$, and the centrifugal force term gives negligible contribution at the boundary. The total charge σ / R appearing in the system and calculated by the code by integrating the charge as in Eq. (6.20) and is essentially equal to the electric field $E_r\big|_{r=R}$. The difference $E_r\big|_{r=R} - \sigma / R$ as calculated from Eq. (6.20) gives a value of $E_r\big|_{r=R-L}$ which is negligible, as can be verified from Figs.(54) and (56). We note at the left boundary inside the plasma at $\theta = \pi$ in Fig.(56), in the flat part of the density where $\nabla P_i = 0$, the electric field $n_i E_r$ is compensated by the Lorentz force due to the poloidal drift $-0.1 < J_{\theta i} > /(1.+0.2\cos\theta) = -0.1 < J_{\theta i} > /(1-0.2)$, which is small, while along the gradient the electric field $n_i E_r$ is essentially balanced by ∇P_i . Fig.(57) shows the

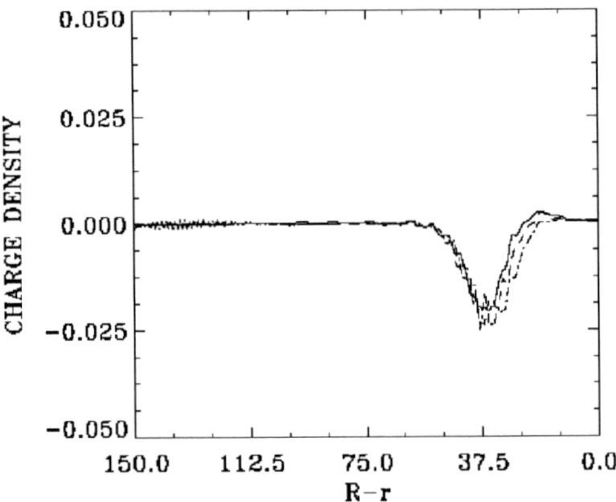

Figure 57. Charge density at $\theta = 0$ (full curve), $\theta = \pi / 2$ (broken curve) and $\theta = \pi$ (dash-dot curve), at *t*=520.

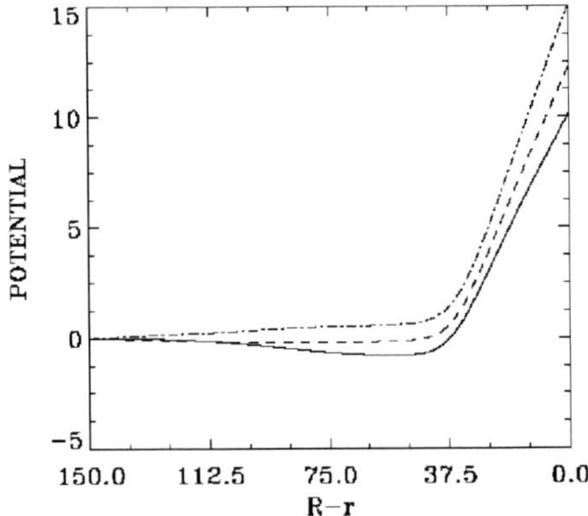

Figure 58. Potential at $\theta = 0$ (full curve), $\theta = \pi / 2$ (broken curve) and $\theta = \pi$ (dash-dot curve) at *t*=550.

charge density $n_i - n_e$ at the time $t = 520$ and $\theta = 0$ (full curve), $\theta = \pi / 2$ (broken curve) and $\theta = \pi$ (dash-dot curve). The charge is important along the gradient at the plasma edge. The electrons, which are frozen to the magnetic field lines, cannot compensate along the gradient the charge separation caused by the finite ion gyroradius. Figure (58) presents at $t = 550$ the potential at $\theta = 0$ (full curve), $\theta = \pi/2$ (broken curve) and $\theta = \pi$ (dash-dot curve), and Fig. (59) the corresponding curves at the time $t = 520$. There is a small oscillation appearing, which is more pronounced on the full curve corresponding to the solution at $\theta = 0$, while at $\theta = \pi/2$ and $\theta = \pi$ this oscillation has a smaller amplitude. The period of this oscillation is about 60, which is close to the ion gyroperiod in the center, given in our normalized units as 62.8. Fig. (60) gives at $t = 520$ the temperature T_{ir} as defined in Eq. (6.21), for $\theta = 0$ (full curve), $\theta = \pi/2$ (broken curve) and $\theta = \pi$ (dash-dot curve). The dash-3 dots curve is for $n_i T_{ir}$, which is essentially the same for all angles θ. Fig. (61) presents similar results for $T_{i\theta}$. Note in Fig.(60) and (61) the division by the small density at the edge

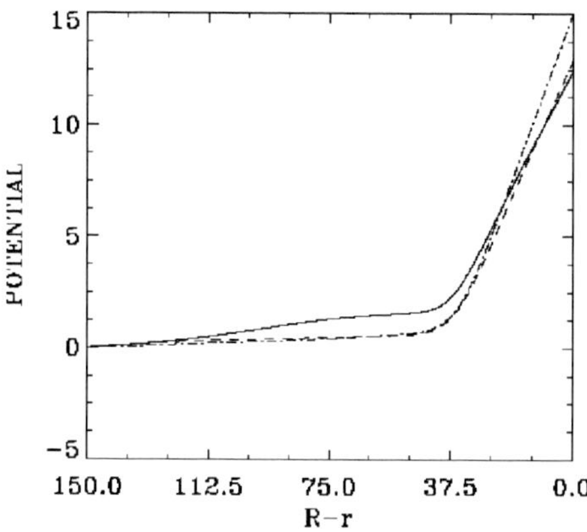

Figure 59. Potential at $\theta = 0$ (full curve), $\theta = \pi / 2$ (broken curve) and $\theta = \pi$ (dash-dot curve), at t= 520.

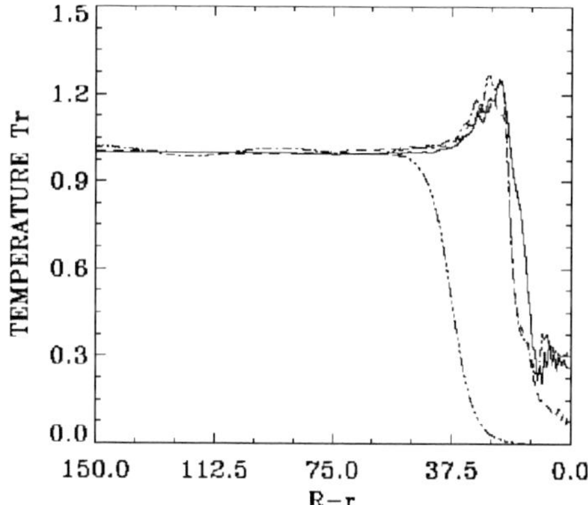

Figure 60. Temperature T_{ir} at $\theta = 0$ (full curve), $\theta = \pi / 2$ (broken curve) and $\theta = \pi$ (dash-dot curve) at t=520. The dash -3 dots curve is for $n_i T_{ir}$, which is essentially the same for all θ. Time t=520.

gives irregular oscillations. However the dash-3 dots curve in Figs(60) and (61) is very smooth, and is for $n_i T_{ir}$ and $n_i T_{i\theta}$ which removes the problem of the division by the very small density at the edge. Fig.(62) presents the plot of the $\vec{E} \times \vec{B} / B^2$ drift, which in our normalized units is written - $E_r(1+0.2 \cos\theta)/0.1$, at $\theta = 0$ (full curve), $\theta = \pi/2$ (broken curve) and $\theta = \pi$ (dash-dot curve). We have noted in Fig. (64) that E_r is smaller at the edge at $\theta = 0$. We note however in Fig. (62) that $\vec{E} \times \vec{B} / B^2$ depends very weakly on θ, so the azimuthal variation of this drift is negligible at the edge, and at the interior of the plasma, and a weak dependence on θ appears along the gradient (from r=40 to r=30). We plot in Fig. (63) the total poloidal current $n_i \left(\vec{E} \times \vec{B} / B^2 + \vec{v}_D\right)$, where the diamagnetic drift is $\vec{v}_D = \vec{B} \times \nabla n_i T_i / \left(n_i eB^2\right)$. We see that the total current is essentially zero, *i.e.* the profile is

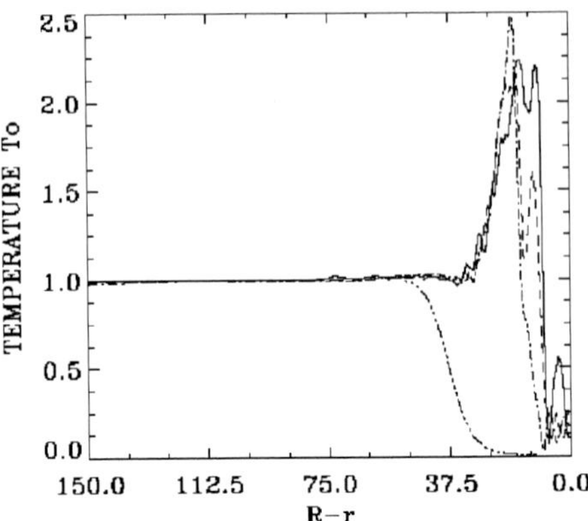

Figure 61. Temperature $T_{i\theta}$ at $\theta = 0$ (full curve), $\theta = \pi/2$ (broken curve) and $\theta = \pi$ (dash-dot curve) at *t*=520. The dash-3 dots curve is for $n_i T_{i\theta}$, which is essentially the same for all θ. Time *t*=520.

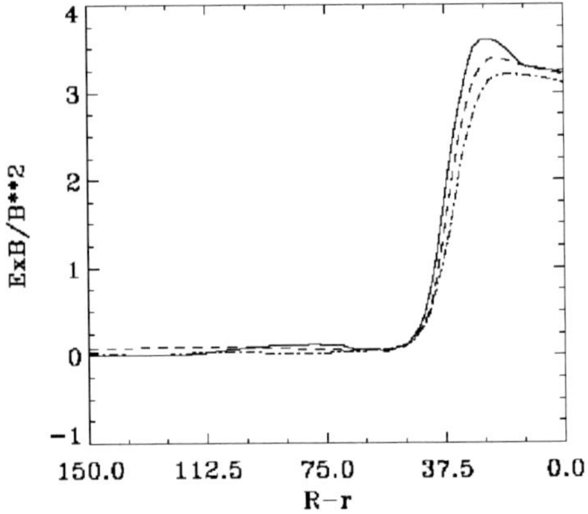

Figure 62. Plot of $-E_r \ (1+0.2\cos\theta)/0.1$ for $\theta = 0$ (full curve), $\theta = \pi/2$ (broken curve) and $\theta = \pi$ (dash-dot curve). At time *t*=520.

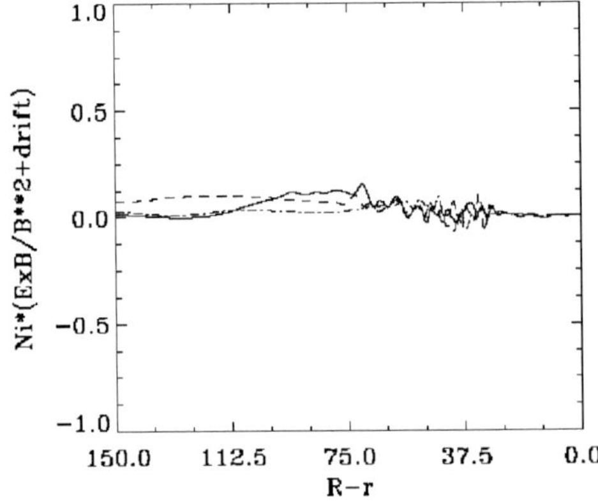

Figure 63. Plot of the total poloidal current $(-n_i E_r + \nabla P_i)\ (1+0.2\cos\theta)/0.1$ for $\theta = 0$ (full curve), $\theta = \pi/2$ (broken curve) and $\theta = \pi$ (dash-dot curve). At time *t*=520.

adjusting itself so that the $\vec{E} \times \vec{B} / B^2$ drift and the diamagnetic drift are essentially equal and opposite (in our units, the total poloidal current is $(-n_i E_r + \nabla p_i)\ (1 + 0.2 \cos\theta)/0.1$, essentially equal to zero in Fig.(63), or at most very small at the left boundary). This is the result we get if we calculate the poloidal current

$$J_{\theta i}(r,\theta) = \int v_\theta f_i(r,\theta,v_r,v_\theta,t)\, dv_\theta dv_r \tag{6.23}$$

We get indeed a negligible value for $J_{\theta i}$. This absence or very weak poloidal dependence of the $\vec{E} \times \vec{B} / B^2$ drift at the edge, can perhaps explain the absence of turbulence at the edge of the H-mode in a tokamak.

Case2.

In this case the density profile is given in Eq.(6.17), where the steep gradient is located around the center of the plasma at a distance about $L/2$ from the floating wall of the vessel. There is no deuterons hitting the wall surface at $r = R$, there is no charge collected at the wall of the cylindrical vessel, which is maintained at zero potential. The electric field $E_\theta$remained very small and negligible and azimuthal variation of the charges very small. We assume that the gyrating plasma ions are allowed to enter or leave at the left boundary as in the previous section. The electric field at $r = R$ in Eq. (6.20) is equal to zero. So the electric fields at the left boundary $r = R - L$ in Eq. (6.20) must be such that for the total charge appearing in the system, Eq. (6.20) satisfies $-(R-L)E_r\big|_{r=R-L} = \sigma$. The distribution of the charge in the domain is such that the total charge $\sigma/(R-L)$ is essentially zero (in the simulation, the total charge $\sigma/(R-L)$ calculated numerically by simple summation rule was of the order of 10^{-4}), so the electric field at the left boundary remained negligibly small. Fig.(64) shows the electric field at $\theta = 0$ (full curve), $\theta = \pi/2$ (broken line) and $\theta = \pi$ (dot-dash line), at $t = 520$. The results remained symmetric from $\theta = \pi$ to $\theta = 2\pi$. The electric field at the vessel wall $r = R$ is zero, as well as at $r = R\text{-}L$, calculated at 32 points over the 2π circle. So the electric field exists only in a layer around the gradient. Fig.(65) gives at $\theta = 0$ the electric field E_r (full curve) at time t=520. The dash-dotted curve gives the Lorentz force, which in our normalized units is given by $-0.1\langle v_\theta\rangle/(1+0.2\cos\theta)$, and the broken curve gives the pressure force $\nabla P_i / n_i$, $P_i = 0.5\, n_i (T_{ir} + T_{i\theta})$, as defined in Eqs.(6.21-6.22).. The $\nabla P_i / n_i$ term (broken curve) shows a very good agreement along the gradient with the solid curve for E_r, and the Lorentz force appears negligible. In the region close to the wall we have irregular oscillations in space (and time), the accuracy of $\nabla P_i / n_i$ being degraded by the division with a very low value of the density n_i. We plot in Fig. (66) the quantities $n_i E_r$, ∇P_i, and $-0.1 n_i\langle v_\theta\rangle/(1+0.2\cos\theta)$ at $\theta = 0$ (note that $J_{\theta i} = n_i < v_\theta >$).

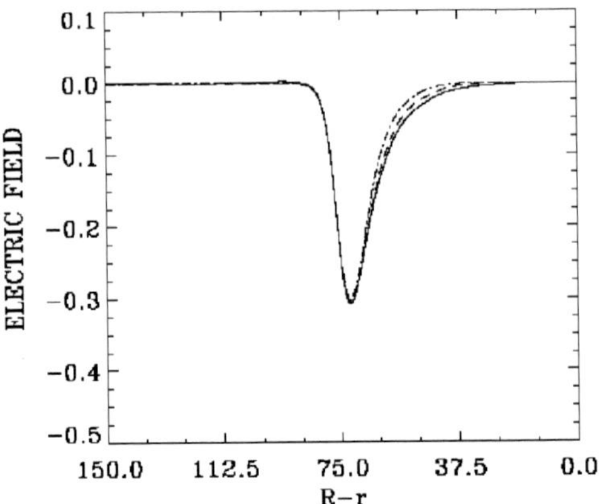

Figure 64. Electric field E_r at $\theta = 0$ (full curve), $\theta = \pi / 2$ (broken curve), and $\theta = \pi$ (dash-dot curve), at *t*=520.

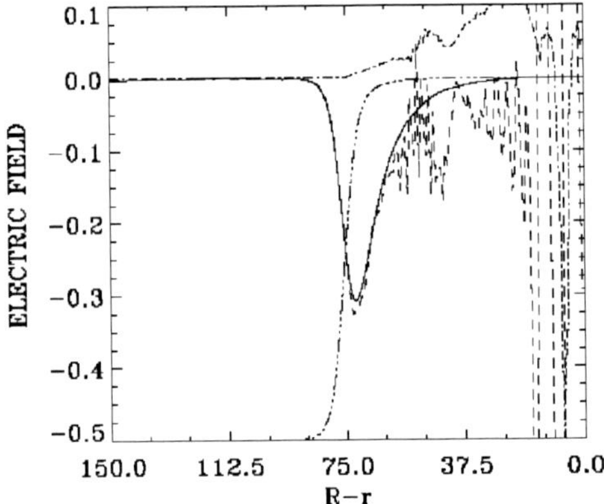

Figure 65. Plot at $\theta = 0$ of the electric field E_r (full curve), the Lorentz force $-0.1\langle v_\theta\rangle/(1+0.2\cos\theta)$ (dash-dot curve), and the pressure force $\nabla P_i / n_i$ (broken curve). The curve $-n_i/2$ is plotted for reference (dash- 3 dots curve). At time t =520.

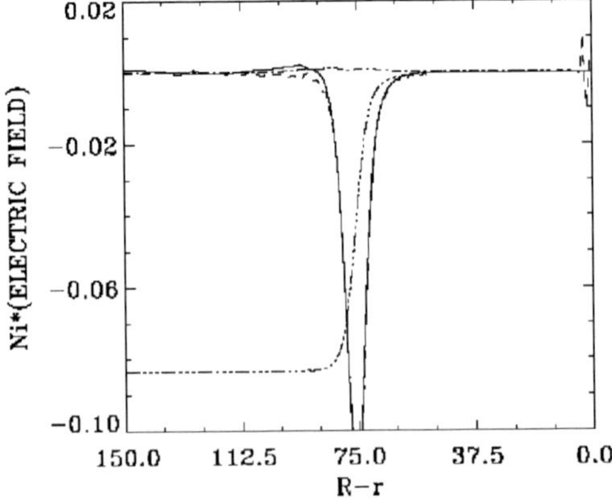

Figure 66. Plot at $\theta = 0$ of $n_i E_r$ (solid curve), $-0.1 n_i\langle v_\theta\rangle/(1+0.2\cos\theta)$ (dash-dot curve), and ∇P_i (broken-curve). The curve $-n_i/12$ is plotted for reference (dash-3 dots curve). At time *t*=520.

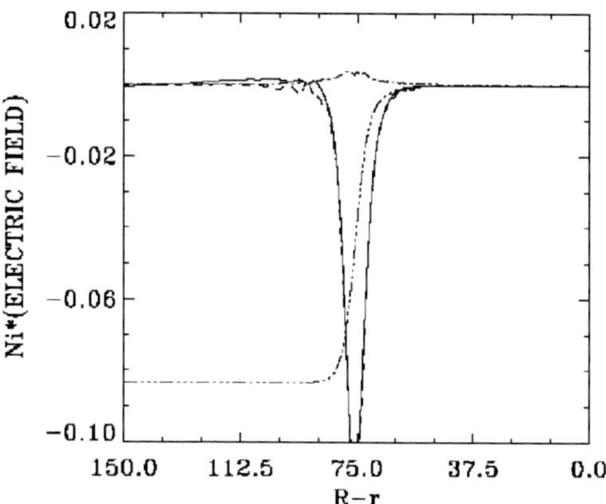

Figure 67. Plot at $\theta = \pi$ of $n_i E_r$ (solid curve), $-0.1 n_i \langle v_\theta \rangle / (1+0.2\cos\theta)$ (dash-dot curve), and ∇P_i (broken-curve). The curve $-n_i/12$ is plotted for reference (dash-3 dots curve). At time t=520.

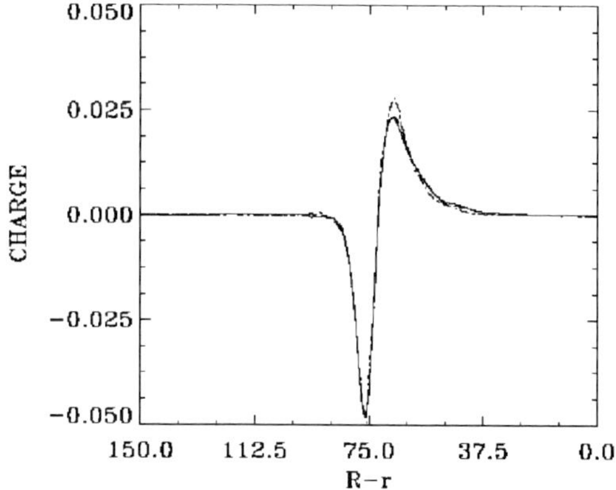

Figure 68. Charge density at $\theta = 0$ (full curve), $\theta = \pi/2$ (broken curve) and $\theta = \pi$ (dash-dot curve), at t=520.

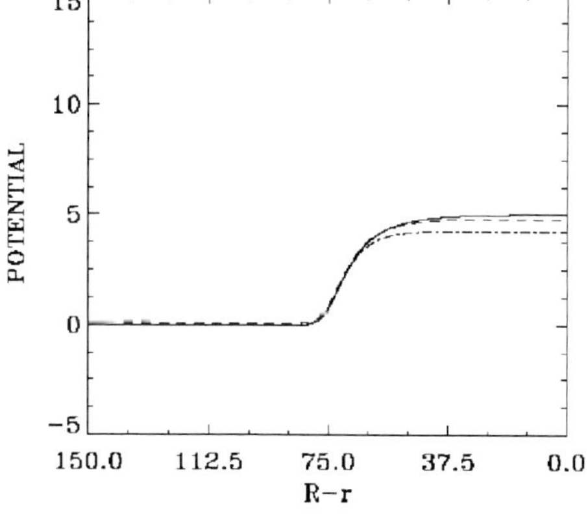

Figure 69. Potential at $\theta = 0$ (full curve), $\theta = \pi/2$ (broken curve) and $\theta = \pi$ (dash-dot curve) at t= 520.

We see that there is a very nice agreement for the relation $n_i\ E_r \approx \nabla P_i$ along the gradient (the density $-n_i/12$ is also plotted in Fig.(66) to locate the gradient with respect to the different profiles). A similar plot is presented in Fig.(67) at $\theta = \pi$. The Lorentz force term remains negligible. The ∇P_i term is zero in the bulk at the left boundary since n_i and T_i are flat in that region. Fig.(68) shows the charge density $n_i - n_e$ at time $t = 520$ at $\theta = 0$ (full curve), $\theta = \pi/2$ (broken curve) and $\theta = \pi$ (dash-dot curve). The charge is important along the gradient at the plasma edge, and the dependence of the charge on θ is weak. The electrons, which are frozen to the magnetic field lines, cannot compensate along the gradient the charge separation caused by the finite ion gyroradius. Note that the distribution of the charge density is such that the total charge is essentially zero (numerically, it was of the order of 10^{-4}). This agrees which Eq.(6.20) where, as we have seen, the electric fields at both boundaries are zero. Fig. (69) presents at time $t = 520$ the potential at $\theta = 0$ (full curve), $\theta = \pi/2$ (broken curve) and $\theta = \pi$ (dash-dot curve). A very weak oscillation is present at the right boundary, much weaker than what is observed in **Case1** in this section, presented in Figs(58) and (59). Note how the slope of the potential is zero at both boundaries, since the electric field is essentially zero at both boundaries. Fig. (70) gives at time $t = 520$ the temperature T_{ir} as defined in Eq. (6.21) for $\theta = 0$ (full curve), $\theta = \pi/2$ (broken curve) and $\theta = \pi$ (dash-dot curve). We are truncating the curves at the edge, since the division by the very small densities gives meaningless results. The dash-3 dots curve is for n_iT_{ir}, which is essentially the same for all angles θ. Fig.(71) presents similar results for $T_{i\theta}$ and $n_iT_{i\theta}$. Fig.(72) presents the plot of the $\vec{E} \times \vec{B}/B^2$ drift, which in our units is written - $E_r(1+0.2\cos\theta)/0.1$, at $\theta = 0$ (full curve), $\theta = \pi/2$ (broken curve) and $\theta = \pi$ (dash-dot curve). We note in the present case this $\vec{E} \times \vec{B}/B^2$ drift is essentially zero on either side of the gradient, and depends weakly on θ in a layer around the region of the gradient. We plot in Fig.(73) the total poloidal current $n_i\left(\vec{E} \times \vec{B}/B^2 + \vec{v}_D\right)$, where the diamagnetic drift $\vec{v}_D = \vec{B} \times \nabla n_i T_i / \left(n_i eB^2\right)$.

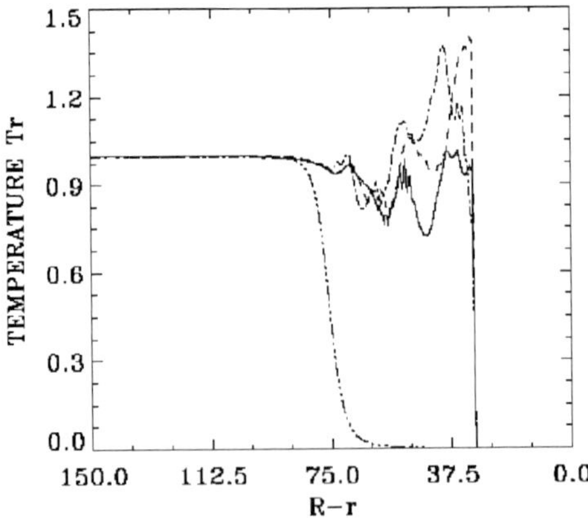

Figure 70. Temperature T_{ir} at $\theta = 0$ (full curve), $\theta = \pi/2$ (broken curve) and $\theta = \pi$ (dash-dot curve) at t=520. The dash 3 dots curve is for $n_i\ T_{ir}$, which is essentially the same for all θ. Time t=520.

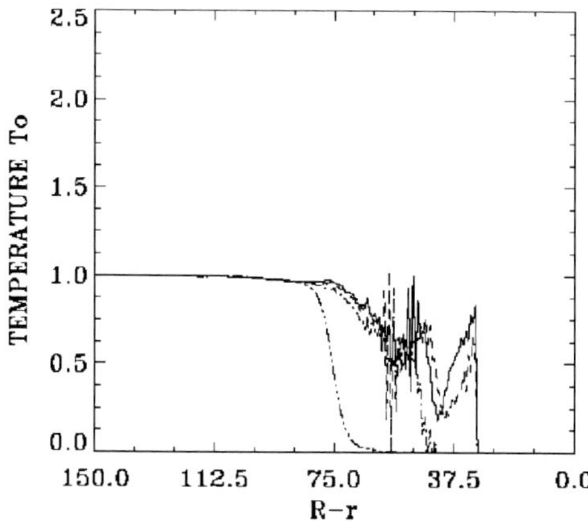

Figure 71. Temperature $T_{i\theta}$ at $\theta = 0$ (full curve), $\theta = \pi / 2$ (broken curve) and $\theta = \pi$ (dash-dot curve) at *t*=520. The dash 3 dots curve is for $n_i T_{i\theta}$, which is essentially the same for all θ. Time *t*=520.

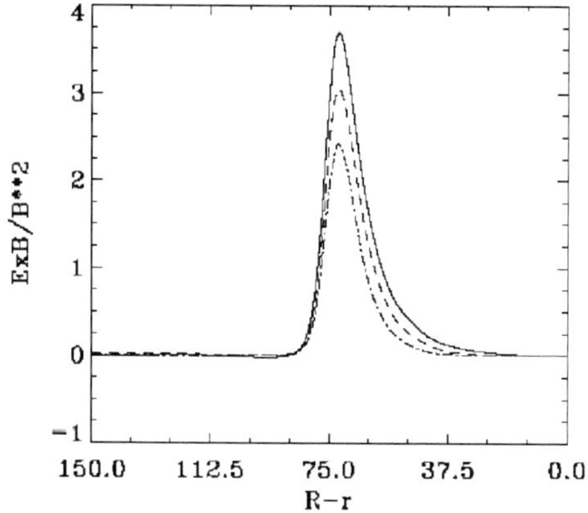

Figure 72. Plot of $-E_r \ (1 + 0.2\ \cos\theta)/0.1$ $\theta = 0$ (full curve), $\theta = \pi / 2$ (broken curve) and $\theta = \pi$ (dash-dot curve). At time *t*=520.

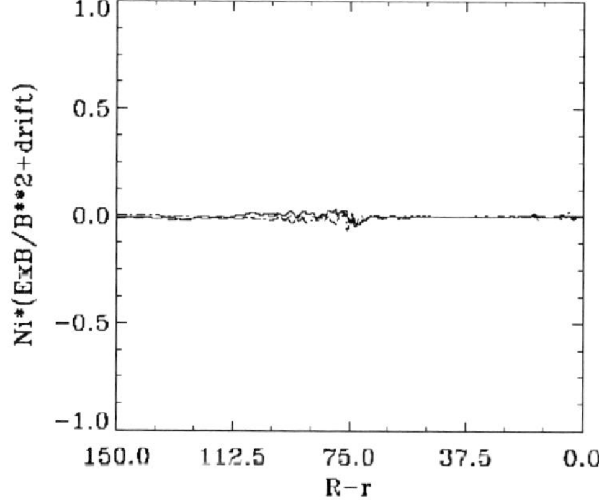

Figure 73. Plot of the total poloidal current $(-n_i E_r + \nabla P_i)$ $(1 + 0.2\ \cos\theta)/0.1$ for $\theta = 0$ (full curve), $\theta = \pi / 2$ (broken curve) and $\theta = \pi$ (dash-dot curve). At time *t*=520.

We see that the total current is essentially zero, *i.e.* the profile is adjusting itself so that the $\vec{E} \times \vec{B} / B^2$ drift, and the diamagnetic drift are essentially equal and opposite in the region of the gradient (in our units, the total poloidal current is $(-n_i E_r + \nabla P_i)$ $(1 + 0.2 \cos\theta)/0.1$). This is the result we get if we calculate the poloidal current as in Eq.(6.23). We get indeed a value of zero for $J_{\theta i}$.

We have presented in this section the self-consistent kinetic solution for the problem of the generation of a charge separation and an electric field at a plasma edge, under the combined effect of a large ratio ρ_i / λ_{De} and a steep density gradient, when the electrons which are bound to the magnetic field cannot compensate along the gradient the charge separation due to the finite ions' gyroradius. This problem is of great importance in the study of the H-mode physics in tokamaks. Two cases have been presented. In the first case, the gradient in the density profiles is located in front of a floating vessel. So the charge appearing in the system is essentially equal to the charge collected on the walls of the floating vessel. In the second case presented in this section, the gradient in the density profiles is located away from the wall of the vessel. No charge is collected on the wall of the vessel, and the charge appearing along the gradient is distributed in a layer along the gradient in such a way that the total charge is zero. Note the difference in the electric field profiles in Figs.(53) and (64), and the difference in the charge distribution along the gradient in Figs(57) and (68). In both cases, the electric field along the gradient is balanced by the gradient of the pressure, and the total poloidal current is zero. The results are close to what has been reported in Cartesian geometry in [63,65]. The absence or negligible shear in the poloidal drift, and the negligible poloidal current can explain the absence of turbulence at the edge. The numerical method used for the solution of the Vlasov equation in cylindrical geometry, based on an integration of the equation along the characteristics coupled to a two-dimensional interpolation applied successively in configuration space and in velocity space, is producing accurate results.

7. Numerical Simulation of Wake-Field Acceleration

Large amplitude wake fields can be produced by propagating ultrahigh power, short laser pulses in plasmas. When the laser power is high enough, the electron oscillation (quiver) velocity becomes relativistic, and large amplitude wake fields are generated which support acceleration gradients much greater than those obtained in conventional linear accelerators. In the laser wake-field accelerator concept, a correctly placed trailing electron bunch can be accelerated by the longitudinal electric field and focused by the transverse electric field of the wake plasma waves. Some important aspects of this problem and other nonlinear problems related to large amplitude laser-plasma interactions have been discussed using fluid quantities assumed to satisfy the cold relativistic fluid equations (see for instance [66-68]). Numerical simulations however remain the only alternative to study the kinetic effects in this highly relativistic and highly nonlinear problem. Kinetic effects (e.g. particle trapping and acceleration) in short-pulse laser plasma interactions are often simulated numerically using particle-in-cell (PIC) codes. However several numerical effects in PIC codes can lead to phase-space errors and unphysical numerical heating in the simulation, and hence the detailed phase-space structure and kinetic effects will be poorly approximated in the simulation. It was indeed reported in [69] that the results obtained by the PIC codes show a momentum spread inside the laser pulse which is excessively and unphysically large. At high laser intensities, this can lead to spurious trapping of erroneously large levels. In the present work, we study the problem of the laser wake-field accelerator by using an Eulerian Vlasov code for the numerical solution of the 1D relativistic Vlasov-Maxwell equations. A fully nonlinear 1D relativistic Vlasov-Maxwell model to study the self-consistent interaction of intense laser pulses with plasmas can be found, for instance, in [70,71]. A characteristic parameter of a high power laser beam is the normalized vectror potential $\left|\vec{a}_{\perp}\right| = \left|e\vec{A}_{\perp} / m_e c^2\right| = a_0$, where $\vec{A}_{\perp}$ is the vector potential, e and m_e the electronic charge and mass respectively, and c the speed of light. We are interested in the regime $a_0 \geq 1$. The code we use applies a method of characteristics where the numerical scheme is based on a two-dimensional advection technique, of second order accuracy in time-step, and where the value of the distribution function is advanced in time by interpolating in two dimensions along the characteristics using a tensor product of cubic B-splines [1,5,32].

7.1. The Relevant Equations

The 1D relativistic Vlasov-Maxwell model

Time t is normalized to the inverse electron plasma frequency ω_{pe}^{-1}, length is normalized to $l_0 = c\omega_{pe}^{-1}$, velocity and momentum are normalized respectively to the velocity of light c and to $m_e c$. The general form of the Vlasov equation is written for the present problem in a 4D phase-space for the electron distribution function $F_e(x, p_{xe}, p_{ye}, p_{ze}, t)$ and the ion distribution function $F_i(x, p_{xi}, p_{yi}, p_{zi}, t)$ (one spatial dimension) as follows :

$$\frac{\partial F_{e,i}}{\partial t} + M_{e,i} \frac{p_{xe,i}}{\gamma_{e,i}} \frac{\partial F_{e,i}}{\partial x} \mp (\vec{E} + \frac{\vec{p} \times \vec{B}}{\gamma_{e,i}}) \cdot \frac{\partial F_{e,i}}{\partial \vec{p}_{e,i}} = 0 \tag{7.1}$$

with

$$\gamma_{e,i} = \left(1 + M_{e,i}^2 (p_{xe,i}^2 + p_{ye,i}^2 + p_{ze,i}^2)\right)^{1/2} \tag{7.2}$$

(the upper sign in Eq.(7.1) is for electrons and the lower sign for ions, and subscripts e or i denote electrons or ions respectively). In our normalized units $M_e = 1$ and $M_i = \frac{m_e}{m_i}$.

We write the hamiltonian of a particle in the electromagnetic field of the wave:

$$H_{e,i} = \frac{1}{M_{e,i}} (\gamma_{e,i} - 1) \mp \varphi . \tag{7.3}$$

where φ is the scalar potential. Eq.(7.1) can be reduced to a two-dimensional phase-space Vlasov equation as follows. The canonical momentum $\vec{P}_{ce,i}$ is connected to the particle momentum $\vec{p}_{e,i}$ by the relation $\vec{P}_{ce,i} = \vec{p}_{e,i} \mp \vec{a}$. $\vec{a} = e\vec{A}/m_e c$ is the normalized vector potential. From Eq.(7.3), we can write:

$$H_{e,i} = \frac{1}{M_{e,i}}\left(\left(1 + M_{e,i}^2 (\vec{P}_{ce,i} \pm \vec{a})^2\right)^{1/2} - 1 \right) \mp \varphi . \tag{7.4}$$

Choosing the Coulomb gauge ($div\vec{a} = 0$), we have for our one dimensional problem $\frac{\partial a_x}{\partial x} = 0$, hence $a_x = 0$. The vector potential $\vec{a} = \vec{a}_\perp(x,t)$, and we also have the following relation along the longitudinal direction:

$$\frac{dP_{cxe,i}}{dt} = -\frac{\partial H_{e,i}}{\partial x} \tag{7.5}$$

And since there is no transverse dependence :

$$\frac{d\vec{P}_{c\perp e,i}}{dt} = -\nabla_\perp H_{e,i} = 0 . \tag{7.6}$$

This last equation means $\vec{P}_{c\perp e,i} =$const. We can choose this constant to be zero without loss of generality, which means that initially all particles at a given *(x,t)* have the same perpendicular momentum $\vec{p}_{\perp e,i} = \pm\vec{a}_\perp(x,t)$. The Hamiltonian now is written:

$$H_{e,i} = \frac{1}{M_{e,i}}\left(\left(1 + M_{e,i}^2 p_{xe,i}^2 + M_{e,i}^2 a_\perp^2(x,t)\right)^{1/2} - 1 \right) \mp \varphi(x,t) . \tag{7.7}$$

The 4D distribution function $F_{e,i}(x, p_{xe,i}, \vec{p}_{\perp e,i}, t)$ can now be reduced to a 2D distribution function $f_{e,i}(x, p_{xe,i}, t)$:

$$F_{e,i}(x, p_{xe,i}, \vec{p}_{\perp e,i}, t) = f_{e,i}(x, p_{xe,i}, t)\delta(\vec{p}_{\perp e,i} \mp \vec{a}_\perp) . \tag{7.8}$$

$f_{e,i}(x, p_{xe,i}, t)$ verify the relation:

$$\frac{df_{e,i}}{dt} = \frac{\partial f_{e,i}}{\partial t} + \frac{\partial H_{e,i}}{\partial p_{xe,i}}\frac{\partial f_{e,i}}{\partial x} - \frac{\partial H_{e,i}}{\partial x}\frac{\partial f_{e,i}}{\partial p_{xe,i}} = 0 . \tag{7.9}$$

Which gives the following Vlasov equations for the electrons and the ions::

$$\frac{\partial f_{e,i}}{\partial t}+M_{e,i}\frac{p_{xe,i}}{\gamma_{e,i}}\frac{\partial f_{e,i}}{\partial x}+(\mp E_x-\frac{M_{e,i}}{2\gamma_{e,i}}\frac{\partial a_{\perp}^2}{\partial x})\frac{\partial f_{e,i}}{\partial p_{xe,i}}=0. \qquad (7.10)$$

where $\gamma_{e,i}=\left(1+\left(M_{e,i}p_{xe,i}\right)^2+\left(M_{e,i}a_{\perp}\right)^2\right)^{1/2}$.

$$E_x=-\frac{\partial\varphi}{\partial x}\text{ and }\vec{E}_{\perp}=-\frac{\partial\vec{a}_{\perp}}{\partial t} \qquad (7.11)$$

and Poisson's equation is given by:

$$\frac{\partial^2\varphi}{\partial x^2}=\int f_e(x,p_{xe})dp_{xe}-\int f_i(x,p_{xi})dp_{xi} \qquad (7.12)$$

The transverse electromagnetic fields E_y, B_z and E_z, B_y for the circularly polarized wave obey Maxwell's equations. With $E^{\pm}=E_y\pm B_z$ and $F^{\pm}=E_z\pm B_y$, we have:

$$(\frac{\partial}{\partial t}\pm\frac{\partial}{\partial x})E^{\pm}=-J_y.;\ (\frac{\partial}{\partial t}\mp\frac{\partial}{\partial x})F^{\pm}=-J_z \qquad (7.13)$$

Which are integrated along their vacuum characteristic $x=t$. In our normalized units we have the following expressions for the normal current densities:

$$\vec{J}_{\perp}=\vec{J}_{\perp e}+\vec{J}_{\perp i}\ ;\qquad \vec{J}_{\perp e,i}=-\vec{a}_{\perp}M_{e,i}\int_{-\infty}^{+\infty}\frac{f_{e,i}}{\gamma_{e,i}}dp_{xe,i}. \qquad (7.14)$$

The numerical scheme

The numerical scheme to advance Eq.(7.10) from time t_n to t_{n+1} necessitates the knowledge of the electromagnetic field $E^{\pm}$ and $F^{\pm}$ at time $t_{n+1/2}$. This is done using a centered scheme where we integrate Eq.(7.13) exactly along the vacuum characteristics with $\Delta x=\Delta t$, to calculate $E^{\pm n+1/2}$ and $F^{\pm n+1/2}$ as follows:

$$E^{\pm}(x\pm\Delta t,t_{n+1/2})=E^{\pm}(x,t_{n-1/2})-\Delta t J_y(x\pm\Delta t/2,t_n) \qquad (7.15)$$

with $$J_y(x\pm\Delta t/2,t_n)=\frac{J_y(x\pm\Delta x,t_n)+J_y(x,t_n)}{2}$$

A similar equation can be written for $F^{\pm n+1/2}$ with J_z. From Eq.(7.11) we also have $\vec{a}_\perp^{n+1} = \vec{a}_\perp^{n} - \Delta t \vec{E}_\perp^{n+1/2}$, from which we calculate $\vec{a}_\perp^{n+1/2} = (\vec{a}_\perp^{n+1} + \vec{a}_\perp^{n})/2$. To calculate $E_x^{n+1/2}$, two methods have been used. A first method calculates E_x^n from $f_{e,i}^n$ using Poisson's equation, then we use a Taylor expansion::

$$E_x^{n+1/2} = E_x^n + \frac{\Delta t}{2}\left(\frac{\partial E_x}{\partial t}\right)^n + 0.5\left(\frac{\Delta t}{2}\right)^2\left(\frac{\partial^2 E_x}{\partial t^2}\right)^n ;$$
$$\text{with } \left(\frac{\partial E_x}{\partial t}\right)^n = -J_x^n ; \qquad \left(\frac{\partial^2 E_x}{\partial t^2}\right)^n = -\left(\frac{\partial J_x}{\partial t}\right)^n . \tag{7.16}$$

and $J_x^n = M_i \int\limits_{-\infty}^{+\infty} \frac{p_{xi}}{\gamma_i} f_i^n dp_{xi} - M_e \int\limits_{-\infty}^{+\infty} \frac{p_{xe}}{\gamma_e} f_e^n dp_{xe}$

A second method to calculate $E_x^{n+1/2}$ is to use Ampère's equation: $\frac{\partial E_x}{\partial t} = -J_x$, from which $E_x^{n+1/2} = E_x^{n-1/2} - \Delta t J_x^n$. Both methods gave the same results. (We have used this second method in the results presented in this section). Now given $f_{e,i}^n$ at mesh points (we stress here that the subscript i denotes the ion distribution function), we calculate the new value $f_{e,i}^{n+1}$ at mesh points from the relation (see [1,5,32] for details):

$$f_{e,i}^{n+1}(x, p_{xe,i}) = f_{e,i}^n(x - 2\Delta_{xe,i}, p_{xe,i} - 2\Delta_{p_{xe,i}}) \ ; \quad . \tag{7.17}$$

$\Delta_{xe,i}$ and $\Delta_{p_{xe,i}}$ are calculated from the solution of the characteristics equations for Eq.(7.10), which are given by:

$$\frac{dx}{dt} = M_{e,i}\frac{p_{xe,i}}{\gamma_{e,i}}$$
$$\frac{dp_{xe,i}}{dt} = \mp E_x - \frac{M_{e,i}}{2\gamma_{e,i}}\frac{\partial a_\perp^2}{\partial x} . \tag{7.18}$$

This calculation is effected as follows. We rewrite Eq.(7.17) in the vectorial form:

$$f_{e,i}^{n+1}(\mathbf{X}_{e,i}) = f_{e,i}^n(\mathbf{X}_{e,i} - 2\Delta_{\mathbf{X}e,i}) \ ; \quad . \tag{7.19}$$

$\mathbf{X}_{e,i}$ is the two dimensional vector $\mathbf{X}_{e,i} = (x, p_{xe,i})$, and $\Delta_{\mathbf{X}e,i} = (\Delta_{xe,i}, \Delta_{p_{xe,i}})$ is the two dimensional vector calculated from the implicit relation:

$$\Delta_{\mathbf{X}e,i} = \frac{\Delta t}{2}\mathbf{V}_{e,i}(\mathbf{X}_{e,i} - \Delta_{\mathbf{X}e,i}, t_{n+1/2}) \ . \tag{7.20}$$

$\mathbf{V}_{e,i} = \left(M_{e,i}\dfrac{p_{xe,i}}{\gamma_{e,i}}, \mp E_x^{n+1/2} - \dfrac{M_{e,i}}{2\gamma_{e,i}}\dfrac{\partial (a_{\perp}^{(n+1/2)})^2}{\partial x} \right)$. Eq.(7.20) for $\Delta_{\mathbf{X}e,i}$ is implicit and is solved iteratively: $\Delta^{k+1}_{\mathbf{X}e,i} = \dfrac{\Delta t}{2}\mathbf{V}_{e,i}(\mathbf{X}_{e,i} - \Delta^{k}_{\mathbf{X}e,i}, t_{n+1/2})$, where we start the iteration with $\Delta^{0}_{\mathbf{X}e,i} = 0$ for k=0. Usually two or three iterations are sufficient to get a good convergence. Then $f_{e,i}^{n+1}$ is calculated from $f_{e,i}^{n}$ in Eq.(7.17) by calculating the shifted value using two-dimensional interpolation in the two dimensional phase-space $(x, p_{xe,i})$. Similarly in Eq.(7.20) the shifted value was calculated at every iteration using a two-dimensional interpolation. These two-dimensional interpolations are effected using a tensor product of cubic B-splines (see for details [1,5,32]) .

7.2. Results

The case of a circular polarization

A generous number of grid points has been used in the simulation, to reproduce accurately the fine details which develop in phase-space, especially the thin beams which are accelerated. N_x =13000 is the number of grid points in space, for a length L = 50.444. N_{pxe} =1600 is the number of grid points in momentum for the electrons (p_{xmaxe} =20, p_{xmine} = -20). And for the ions N_{pxi} =256 (p_{xmaxi} = 11, p_{xmini} = -11.). We assume the frequency of the laser pulse $\omega_0/\omega_p >> 1$ ($\omega_0/\omega_p = 10$ in the present calculation), and the radiation envelope of the laser pulse changes on a time-scale which is long compared to the wave period. The spatial length of the envelope of the laser pulse is $L = \lambda_p = 2\pi c/\omega_p$, much longer than the laser field wavelength λ. The model is similar to what has been presented in [66,67], with the addition that in the present simulation we include a kinetic 1D relativistic Vlasov equation, and this is done for both electrons and ions. The evolution of the circularly polarized laser pulse is calculated self-consistently with Maxwell's equations. The validity of the 1D model requires that the laser beam transverse dimension $r >> \lambda_p$.

The system is initially neutral ($n_e = n_i$). The density in our normalized units is equal to 1 in the flat central part, with steep gradients and vacuum at both ends. The length of the vacuum region is 0.6 on each side, and the length of the transition region for the density from 0 to the flat

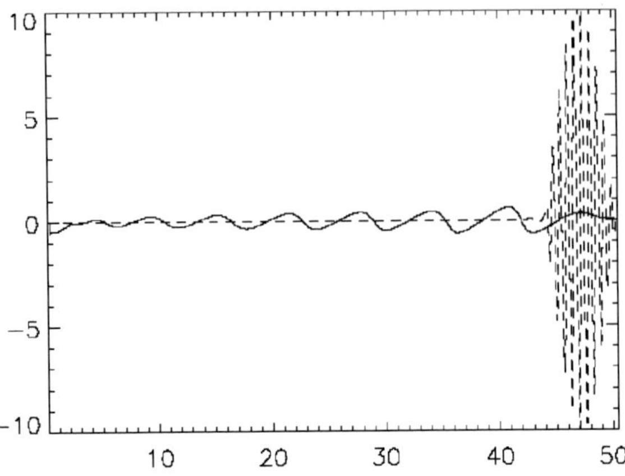

Figure 74. Laser pulse (dash curve) and wake field E_x (full curve) at t=50.444.

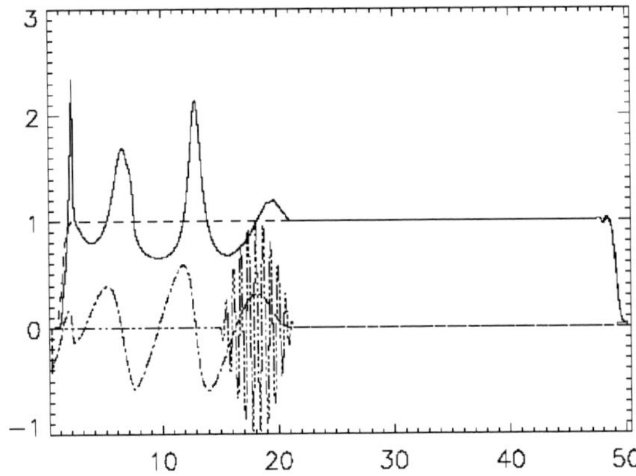

Figure 75. Plot at t = 21.34 of the electron density (full curve), the ion density (dash curve), the axial wake field E_x (dash-dot curve). The laser pulse (amplitude divided by 10) has been also added for reference.

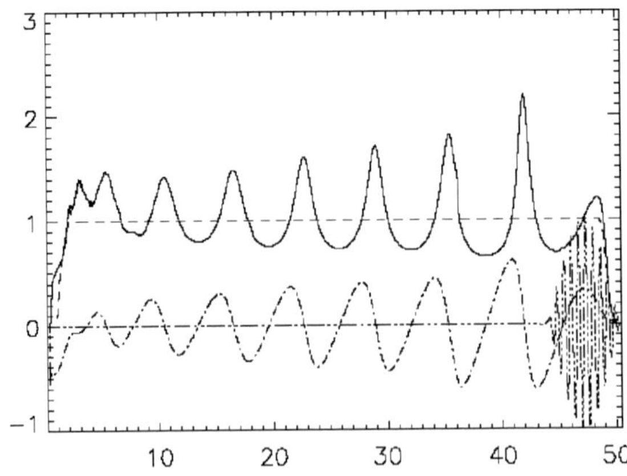

Figure 76. Plot at t = 50.444 of the electron density (full curve), the ion density (dash curve), the axial wake field E_x (dash-dot curve). The laser pulse (amplitude divided by 10) has been also added for reference.

value of 1 is 1.4. The electrons and ions have initially a Maxwellian distribution, with a temperature T_e = 3keV for the electrons and T_i = 1keV for the ions. The forward propagating circularly polarized laser pulse is penetrating from the vacuum at the left boundary, and propagate towards the right, and is written in our normalized units as $E^+ = 2E_0 \sin(\pi\xi / L)\sin(k_0\xi)$ and $F^- = 2E_0 \sin(\pi\xi / L)\cos(k_0\xi)$ for $-L \le \xi \le 0$, $\xi = x - t$, and E_0 =0 otherwise. In vacuum we have for the electromagnetic (EM) wave $k_0 = \omega_0 = 10$ (so in our normalized units the wavelength $\lambda = 2\pi / k_0 = 0.628$). We have ten oscillations of the EM wave in the length $L = 2\pi$ of the pulse envelope. We choose for the amplitude of the potential vector $a_0 = 1$, so that $E_0 = \omega_0 a_0 = 10$. Since the envelope is slowly varying, we can write for the corresponding vector potential for $t \le 2\pi$: $a_y = -a_0 \sin(\pi\xi / L)\cos(k_0\xi)$, $a_z = a_0 \sin(\pi\xi / L)\sin(k_0\xi)$. At $t = 2\pi$, the entire envelope of length L of the forward propagating pulse has penetrated the domain, and is left to evolve self-consistently using Eqs.(7.13), where $\vec{a}_\perp$ is calculated as indicated in the previous section. Fig.(74) shows the results for the laser pulse at t= 50.444 (dash curve), after crossing the whole domain and reaching the right boundary, which is followed by the wake field E_x (full curve). For the present set of parameters, the pulse has propagated through the plasma with little deformation. Fig.(75) shows at t= 21.34 the plot of the electron density (full curve), the ion density (dash curve) and again the axial wake field E_x (dash-dot curve). The laser pulse has been also added, with its amplitude divided by 10, for reference. The amplitude of E_x reaches since the beginning of the simulation a maximum peak of 0.6 just behind the pulse. This is close to the projected theoretical value for saturation for cold plasma [66,67] given by $E_{x\max} = (\gamma_0^2 - 1)/\gamma_0 = 0.717$, where $\gamma_0 = \sqrt{1 + a_0^2} = \sqrt{2}$. The electron density (initially equal to 1 in the central region) is forming spikes surrounded by depleted regions, and the electric field E_x is rapidly changing sign at these spikes. Fig.(76) shows the equivalent results at t = 50.444, when the laser pulse has reached the right boundary. We see again that the peak electric field behind the pulse is still 0.6, and decays slowly as we move away from the pulse. This is in contrast with the results reported for a cold plasma [66,67], which showed the electric field reaching a constant amplitude throughout the domain. This decay of the amplitude of the electric field agrees however with the results reported in [72] for the plasma wake-field accelerators. Note the electric field E_x shows a steeper variation at the right of Fig.(76) compared to the profile at the left, and the period of E_x is slightly longer for the oscillation at the right of Fig.(76), compared to the period of oscillation of E_x at the left of Fig.(76). The density peak at the right in Fig.(76) (full curve) is reaching a value of 2.2, the same value as the equivalent peak in Fig.(75). So the front peaks seem to be following the laser pulse with little deformation. Note that the first thin peak density at the left in Fig.(75) results from the interaction of the pulse with the region of the sheath along the edge gradient. This peak varies rapidly and remained localized in the sheath region at the edge, and did not interfere with the stable pattern which appears on the flat top of the density

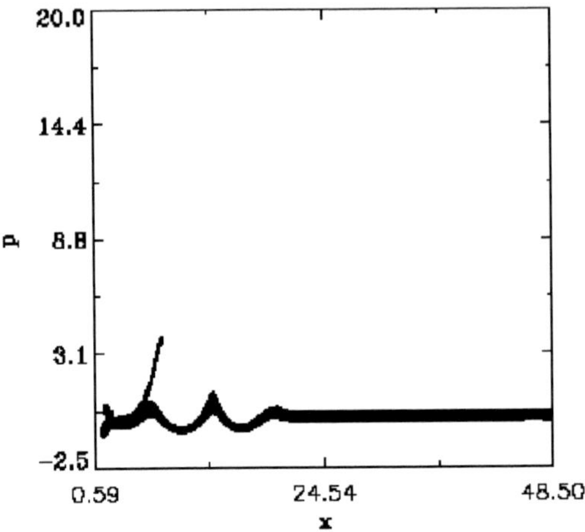

Figure 77. Phase-space contour plot of the electron distribution function at $t = 21.34$.

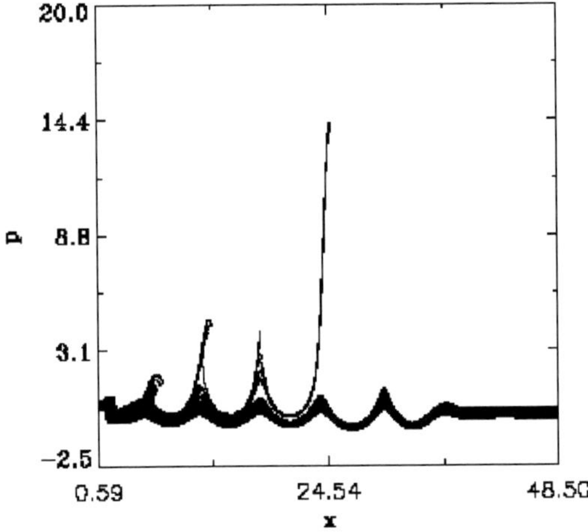

Figure 78. Phase-space contour plot of the the electron distribution function at $t = 38.88$.

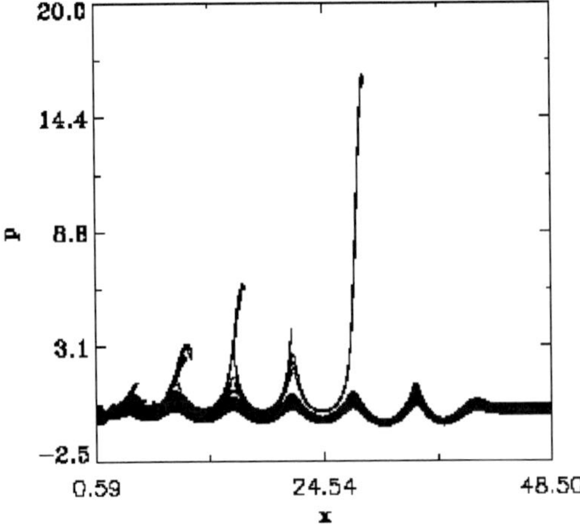

Figure 79. Phase-space contour plot of the electron distribution function at $t = 42.68$.

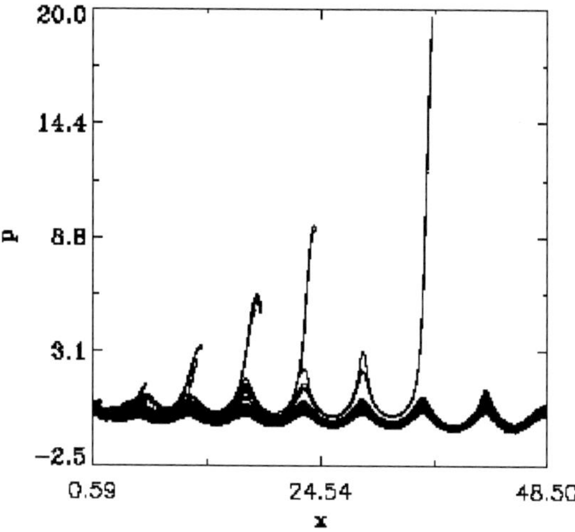

Figure 80. Phase-space contour plot of the electron distribution function at $t = 50.44$.

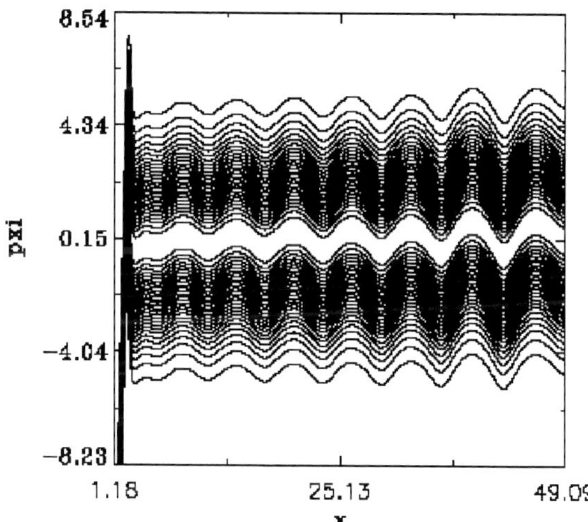

Figure 81. Phase-space contour plot of the ion distribution function at $t = 50.44$.

profile. Figs(77-80) show the phase-space for the electron distribution function at t =21.34, 38.88, 42.68 and 50.44. There is a population which seems to detach itself from the bulk and follows the modulation of the bulk, and then accelerates when it reaches the position of a peak of the electric field. Fig.(81) shows the ion distribution function at t=50.44. Although the ion density profile (broken curve in Figs.(75-76)) appears constant, Fig.(81) shows a modulation in the ion distribution function contour plot. For the relatively short run we have presented, it shows a modulation which has a tendency to increase as we move to the right.

Behind the peak showing the first accelerated beam in Figs(77-80), there is a peak which did not show an accelerated beam, followed by others peaks where accelerated beams are present. We show in Figs.(82-83) the results obtained in another simulation done with the same parameters [74], but with L = 40.355 and different grid sizes N_x = 10000, N_{pxe} =1100 (p_{xmaxe} = 12, p_{xmine} = -12, so the grid size in momentum space was slightly smaller for the results in Fig.(82)). We see in Fig.(82) particles accelerating at every peak behind the first accelerated beam (note the difference in the vertical scale p_{xe}). The reason for this difference is for the moment not very clear and needs further investigation. The peak of the wake electric field behind the laser pulse in Fig.(83) is also also 0.6, and the front density peak is also 2.2. We used the same plotter in both cases (emphasizing slightly the low density regions to make the accelerated beams more visible). We show in Fig.(84) a contour plot emphasizing the region of the tip of the front beam in Fig.(79). The lower part in Fig.(84) is a 3D view of the tip of the beam and the tail preceding it. It clearly shows at the front edge at the top a well localized thin beam .

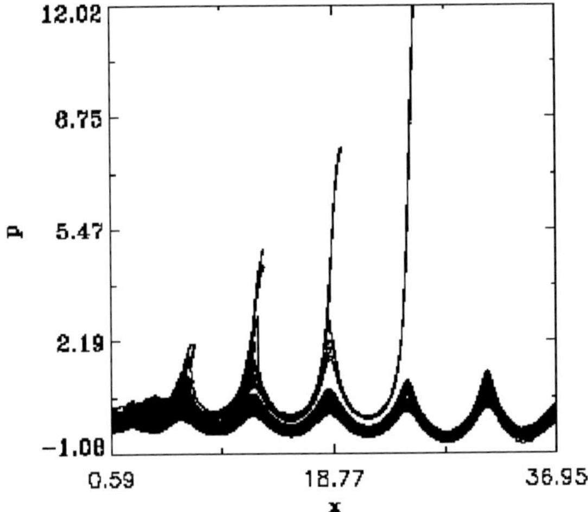

Figure 82. Phase-space contour plot of the electron distribution function at t = 40.35 (the simulation presented in [74]).

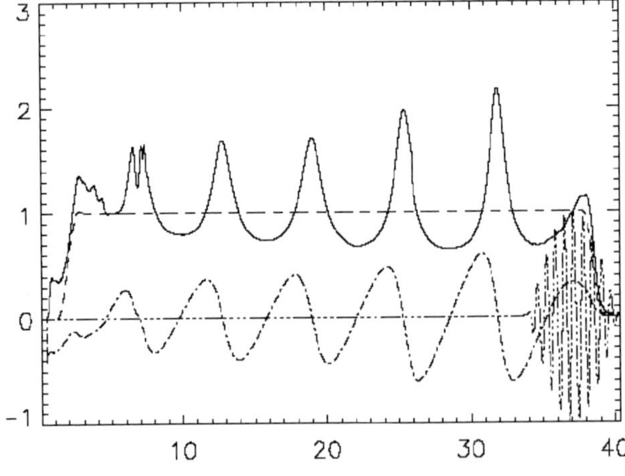

Figure 83. Plot at t = 40.35 of the electron density (full curve), the ion density (dash. curve), the axial wake field E_x (dash-dot curve). The laser pulse (amplitude divided by 10) has been also added for reference (from the simulation presented in [74]).

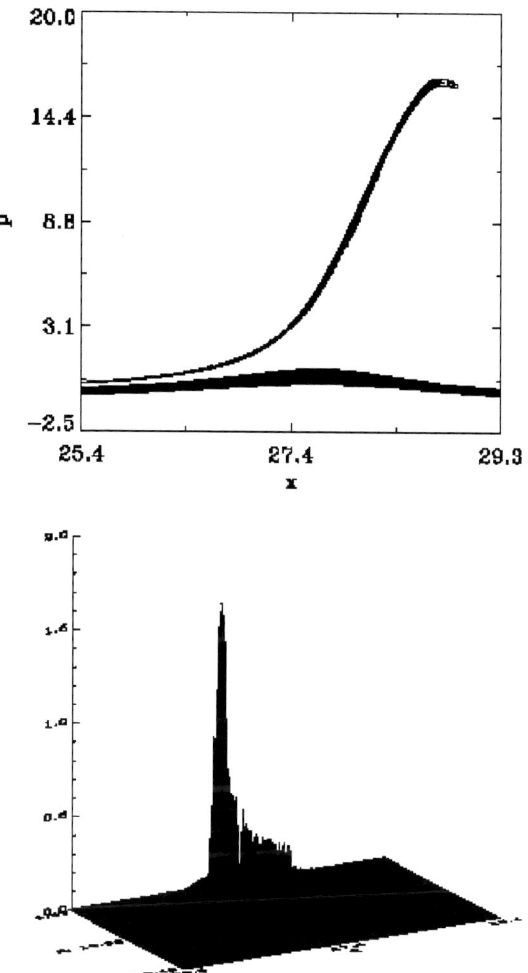

Figure 84. Phase-space contour-plot of the electron distribution function presented in Figure (79) at t=42.68, concentrating on the region of the front beam. Note the sharp beam structure appearing in the lower 3D plot of the front end of the beam.

The case of a linear polarization

We use L = 40.355 and for the grid points N_x = 10000, N_{pxe} =1600 (p_{xmaxe} = 8, p_{xmine} = -8), and for the ions we have the same number of grid point in velocity space N_{pxi} =256. We also use T_e =0.05 kev and T_i =0.01 kev. In this case only E^+ is excited as before, and F^- =0. Only a_y is excited initially, then $a_\perp^2 - a_y^2$, and a_0 is still the amplitude of the vector potential, except that $a_\perp$ is now modulated in time. Fig.(85) shows at t=38.38 the plot of the electron density (full curve) and of the ion demsity (broken curve), plotted against distance. Note at the right the modulation caused in the electron density caused by the by the linearly polarized pulse (which also indicates the position of the pulse at t=38.38). Fig.(86) shows the plasma wake field plotted against distance at t=38.38. Fig.(87) show the phase-space plot for the electron distribution function at t=38.38, in the region of the original flat top (eliminating the edge regions), showing how the beams detach from the bulk of the plasma, and then accelerate.

Note in the right of the figure the oscillation of the particles in the field of the linearly polarized wave.

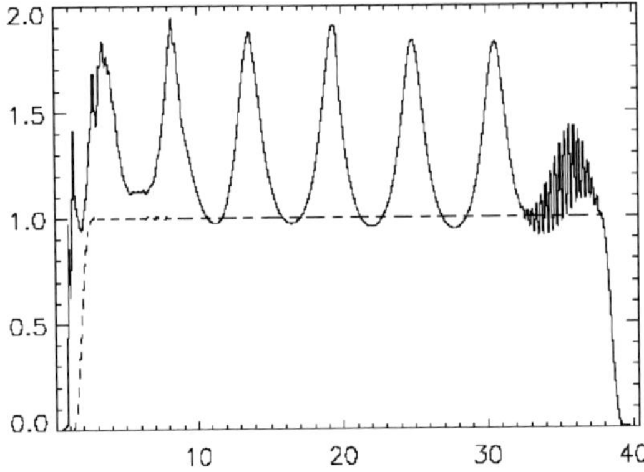

Figure 85. Plot of the electron density (full curve) and the ion density (broken curve) at t=38.38.

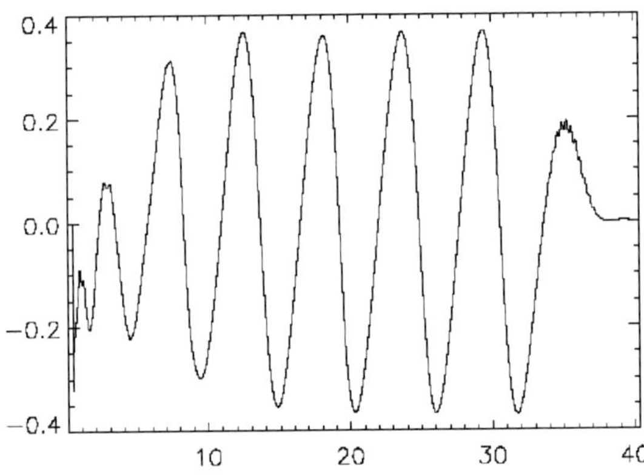

Figure 86. Plot of the longitudinal electric field in the plasma E_x at t=38.38.

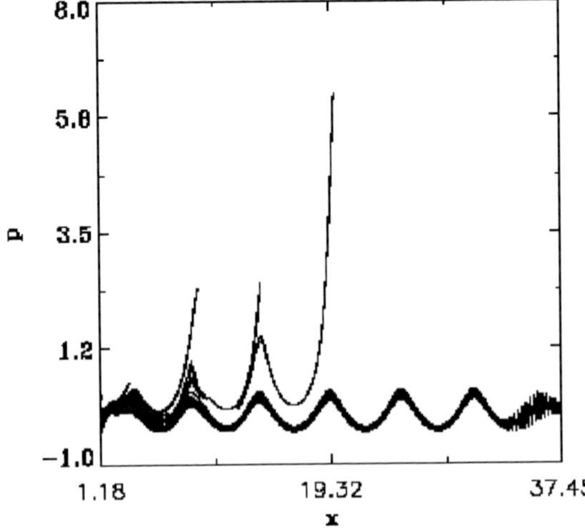

Figure 87. Phase-space contour plot of the electron distribution function at t = 38.38.

Results obtained with PIC codes for this problem can lead to unphysical numerical heating in the simulation, and show a momentum spread inside the laser pulse which is unphysically large [69], hence the detailed kinetic effects, especially for the accelerated beam, will be poorly approximated in the PIC simulation. However, Eulerian Vlasov codes applying the method of characteristics have been successfully used in recent years to study several problems in plasma physics associated with wave-particle interactions. The present simulations show in Fig.(84) a beam well localized in phase-space and which has little numerical diffusion, and offer one more example of a problem of wave-particle interaction. See also the recent results presented in [74].

8. Interaction of a High Intensity Laser Field Incident on an Overdense Plasma

We study in this section the interaction of an intense laser wave incident on an overdense plasma, when the frequency of the wave is below the plasma frequency. If the intensity of the wave is sufficiently high to make the oscillations of the electrons relativistic, the plasma frequency is modified by relativistic mass variation, which allows some interesting interaction of the wave with the overdense plasma edge. An electric field is generated at the plasma edge, in which a strong acceleration of the ions takes place. The acceleration process of the ions is the consequence of an important charge separation induced under the action of the intense laser field on the surface of the plasma. The laser field acts on the high density target, and appears to be literally pushing the electrons and the plasma edge in front of the laser wave. Electrons accumulate at the plasma edge, forming a quasi-stationary peak which induces an electric field. This electric field formed at the plasma edge due to the laser produced electron peak accelerates the ions. Laser-induced ions acceleration seems one of the most promising applications of laser pulses interacting with solid matter [75,83].

The equations and the numerical method are those presented in section 7. We keep in mind that in the present simulation we have a kinetic equation for both the electrons and the ions. A right circularly polarized wave is penetrating the domain at x=0, with $E^+(x=0,t)=2E_0P(t)\cos\omega t$, $E^-(x=0,t)=0$, $F^+(x=0,t)=0$, $F^-(x=0,t)=-2E_0P(t)\sin\omega t$ (see Eqs.(7.13-7.15)). $P(t)$ is a rise time profile given by $P(t)=\sin^2\pi t/2\tau$ for $t<\tau$ ($\tau=50$ in the present calculation), and $P(t)=1$ for $t>\tau$. We use an overdense plasma with $n_0/n_{crit}=1.5$. This would correspond to a frequency $\omega=0.814$ (normalized to ω_{pe}). We use for the wave a normalized quiver momentum $a_0=\sqrt{3}$. This corresponds to an irradiation $I\lambda^2=8.2\text{x}10^{18}$ W cm^{-2} μ m^2. The relativistic factor is then $\gamma_0=\sqrt{1+a_0^2}=2$, leading to $n_0/\gamma_0 n_{crit}=0.75$. The number of grid points in phase-space is $N_x\,N_{pxe}$=6000x1200 for the electrons, and $N_x\,N_{pxi}$ = 6000x1000 for the ions. The time-step is $\Delta t=0.01796$. Initially there is a vacuum region of length 4.52 on both sides of the plasma, and a steep parabolic gradient of length 5.80 where the electron and ion densities vary rapidly from zero to the flat density of 1. This flat density of 1 is located starting from x=10.32. So the plasma is initially neutral. The wave is incident from the left boundary at x=0, where only E^+ and F^- are excited in vacuum as we mentioned above, and propagates to the right. Fig.(88) shows the forward field E^+ after 2000 time-steps, at t=35.92 (full curve). The horizontal axis is length. The field did not reach yet its peak amplitude (since the rise time $P(t)$=1 at t=50). However, the front of the field, strongly damped, has reached the flat top at x=35.92. The broken curve in Fig.(88) is the reflected wave E^- returning from the plasma to the vacuum. Similarly, Fig.(89) represents at t=35.92 the forward wave F^- (full curve), and the backward wave F^+ (broken curve) reflected from the pasma to the vacuum. At that time the wave has already reached the flat top of density 1. Figs.(88,89) show an observable penetration up to x=18. The penetration on the flat top of density 1, which starts at x=10.32, is about 7, which is about one free-space wavelength of the wave. But is it really a penetration ? Fig.(99) show the density profiles at t=35.92. The profiles are originally symmetric. At t=35.92 the density profiles at the right boundary remained unchanged. At the left boundary the electron and ion density profiles seem to be literally pushed to the right by the wave, resulting in an appreciable electron accumulation forming a peak at the edge and top of the flat density profile. The motion of the ions, however, results more from the acceleration in the electric field established by the electrons, as it will become more clear in

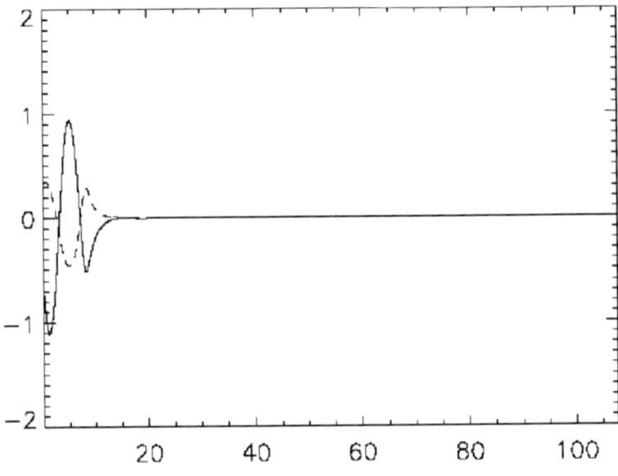

Figure 88. Forward wave E^+ (full curve), and backward wave E^- (broken curve) at t=35.92.

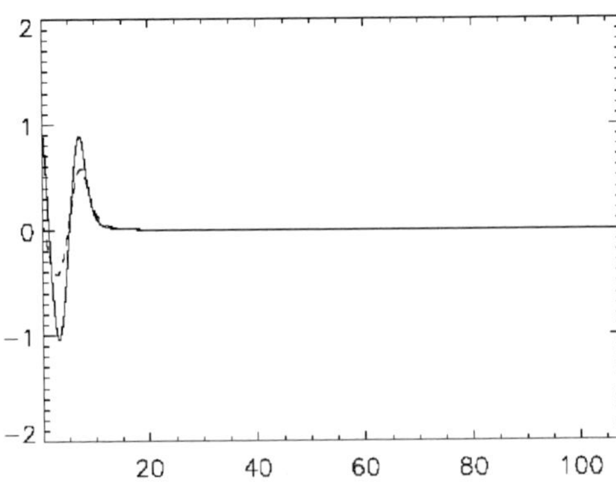

Figure 89. Forward wave F^- (full curve) and backward wave F^+ (broken curve) at t=35.92.

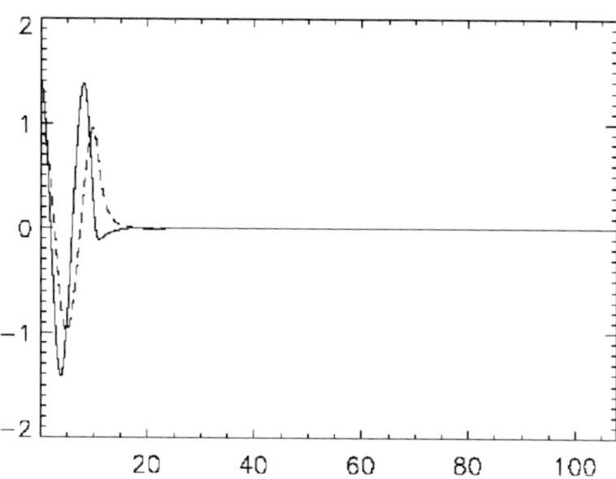

Figure 90. Forward wave E^{+} (full curve), and backward wave E^{-} (broken curve) at *t*=53.88.

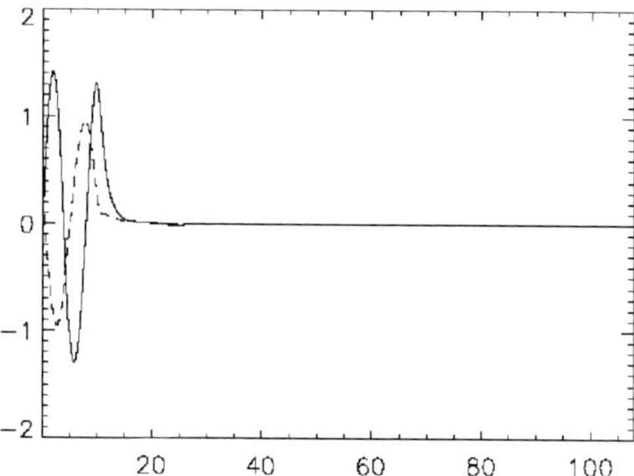

Figure 91. Forward wave F^{-} (full curve) and backward wave F^{+} (broken curve) at *t*=53.88.

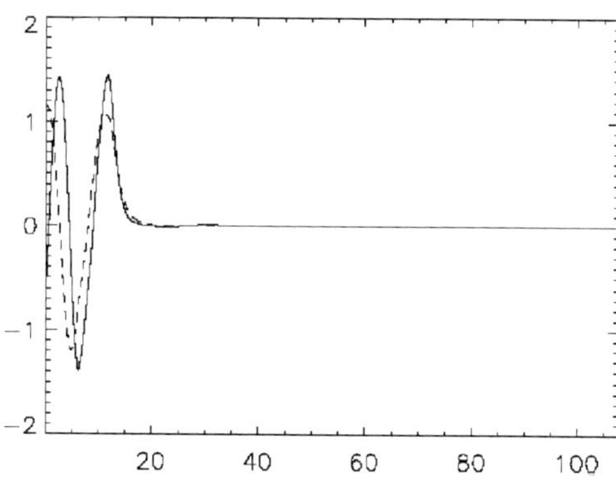

Figure 92. Forward wave E^{+} (full curve), and backward wave E^{-} (broken curve) at *t*=71.84.

the following results. Figs.(90) and (91) show the corresponding results for respectively E^{+} and E^{-}, and for F^{-} and F^{+}, at *t*=53.88 (after 3000 time steps). The field has now reached its peak value with *P(t)*=1. Note the peak of the reflected waves (broken curves) reaching

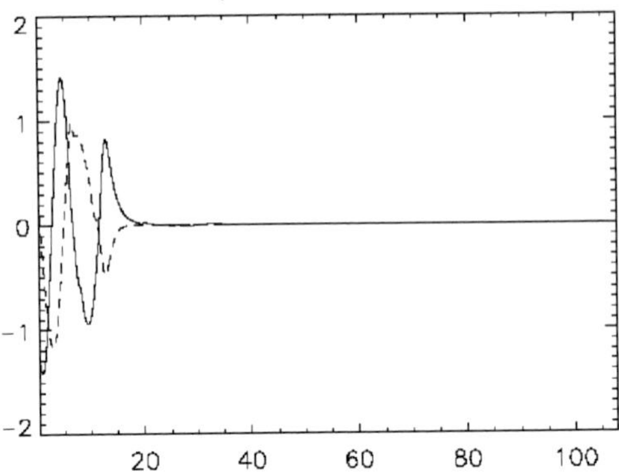

Figure 93. Forward wave F^- (full curve) and backward wave F^+ (broken curve) at t=71.84.

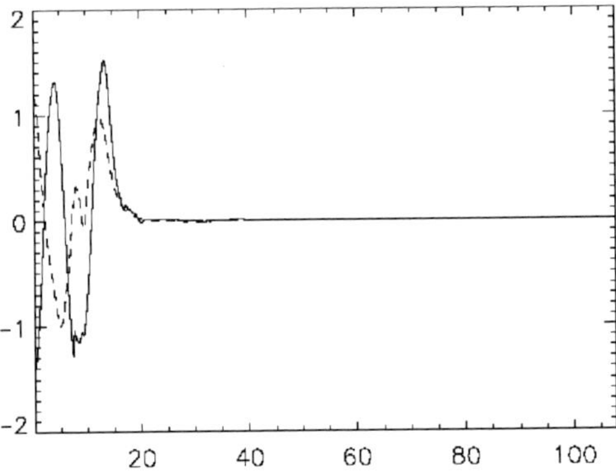

Figure 94. Forward wave E^+ (full curve), and backward wave E^- (broken curve) at t=80.82.

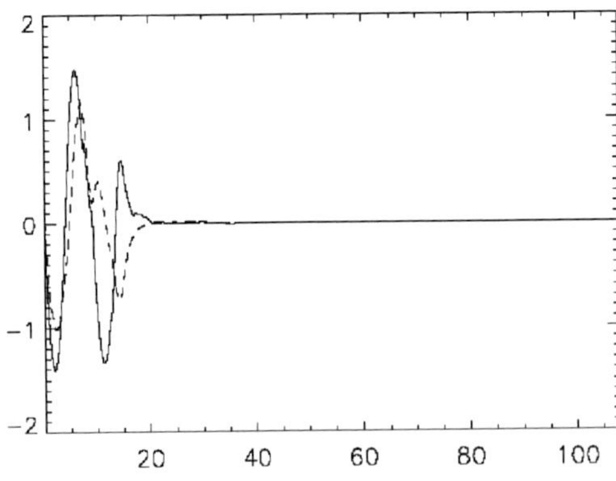

Figure 95. Forward wave F^- (full curve) and backward wave F^+ (broken curve) at t=80.82.

higher values. Figs.(92) and (93) show the corresponding results for respectively E^+ and E^-, and for F^- and F^+, at t=71.84 (after 4000 time steps). Figs.(94) and (95) show the corresponding results for respectively E^+ and E^-, and for F^- and F^+, at t=80.82 (after 4500 time steps). Figs.(96) and (97) show the corresponding results for respectively E^+ and E^-, and for F^- and F^+, at t=89.8 (after 5000 time steps). There is very little penetration of

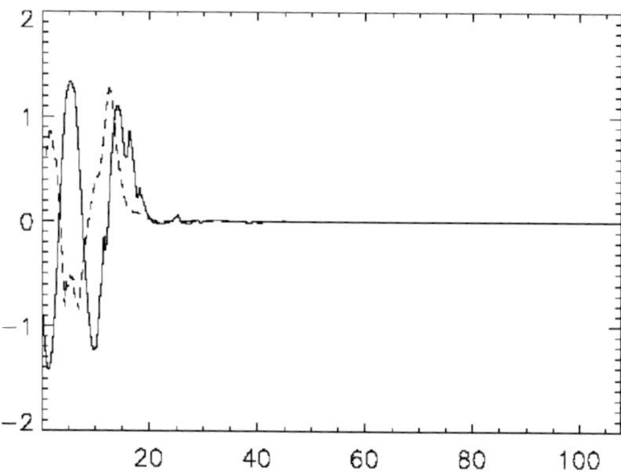

Figure 96. Forward wave E^+ (full curve), and backward wave E^- (broken curve) at t=89.8.

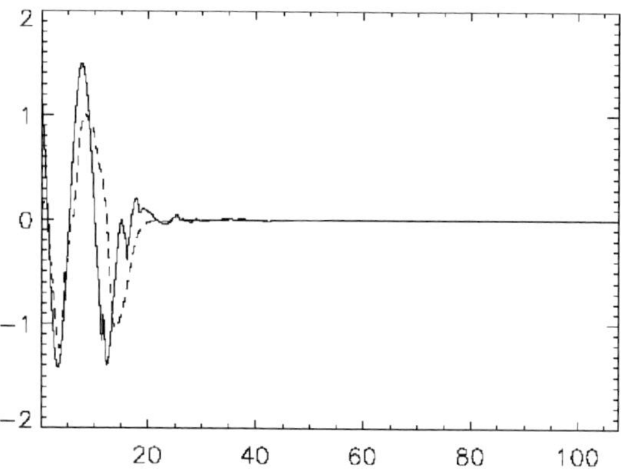

Figure 97. Forward wave F^- (full curve) and backward wave F^+ (broken curve) at t=89.8.

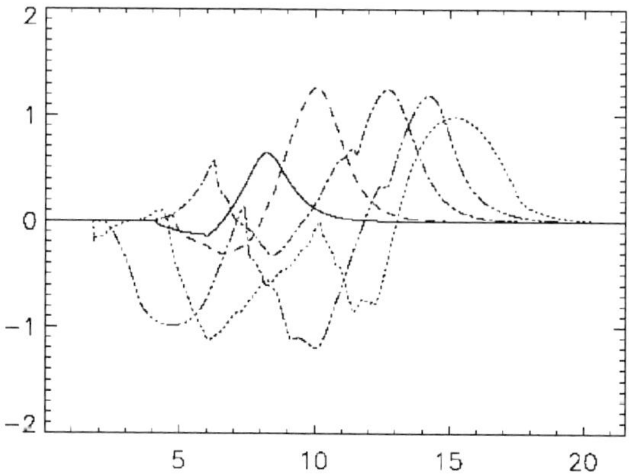

Figure 98. Longitudinal electric field E excited in the plasma, (as function of distance) at t=35.92 (full curve), t=53.88 (broken curve), t=71.88 (dash-dot curve), t=80.82 (dash-three dots curve), and t=89.8 (dot curve).

the laser wave between t=35.92 in Figs.(88,89), and t=89.8 in Figs.(96,97). One can hardly speak of a wave propagation in the plasma, however this limited penetration which goes hardly beyond x=20 (the initial flat top of 1 started at x=10.32) is very reach in physics. Fig.(98) show the longitudinal field E_x excited at the plasma edge (we are presenting the portion close to the left boundary, since this field is zero otherwise). In Fig.(98) the full curve is at t=35.92, the broken curve is at t=53.88, the dash-dot curve is at t=71.88, the dash-3 dots curve is at t= 80.82, the dot curve is at t=89.8. There is indeed a stable quasi-stationary profile of the longitudinal electric field which is moving slowly following the electron density peak being pushed by the penetrating wave. Figs.(99-102) show the density profiles (full curve electrons, broken curves ions), at respectively t=35.92, t=53.88, t=62.86, and t=71.88. The density profiles were initially symmetric as we mentioned before. The right boundary remained unchanged during the simulation, and has no effect on the results since we stopped the simulation long before the laser wave reached the right boundary. We see in Fig.(99) the deformation caused by the wave on the left boundary: in the electron density (full curve), electrons are accumulating forming a huge peak at the flat top at the position where the observation shows the wave has penetrated. In the underdense region at the bottom of the steep gradient, we have two local sharp peaks surrounding a region where trapped electrons

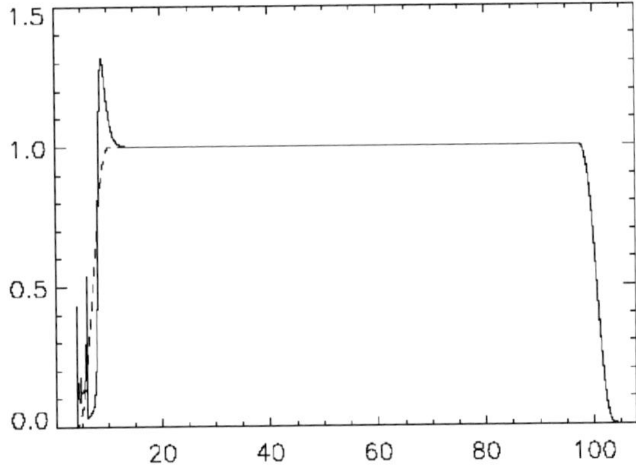

Figure 99. Density profiles for the electrons (full curve) and the ions (broken curve) at t=35.92.

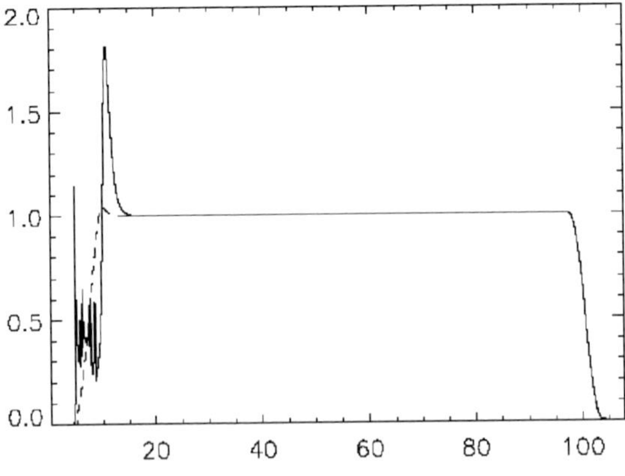

Figure 100. Density profiles for the electrons (full curve) and the ions (broken curve) at t=53.88.

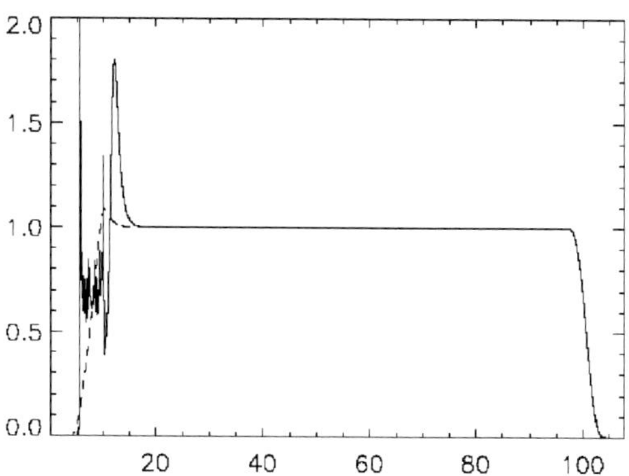

Figure 101. Density profiles for the electrons (full curve) and the ions (broken curve) at t=62.86.

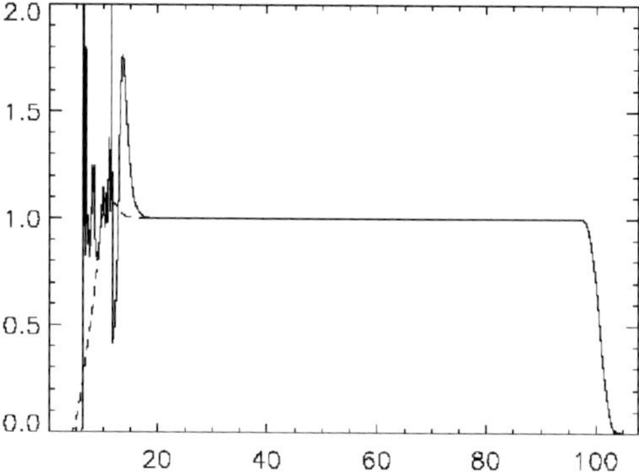

Figure 102. Density profiles for the electrons (full curve) and the ions (broken curve) at t= 71.84.

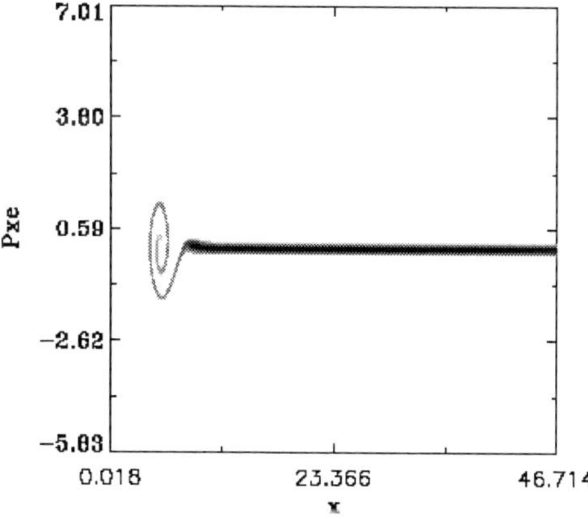

Figure 103. Phase-space contour plot of the electron distribution function t=35.92.

are forming a vortex (which can be seen in Fig.(103) at t=35.92, where we are increasing artificially the plot of the inner spiral of the electron vortex in order to make the contour plot of the low density spiral visible). Fig.(100) shows the density profile at t=53.88. Again a huge electron bump at the top of the flat profile, and then in the underdense region the electron

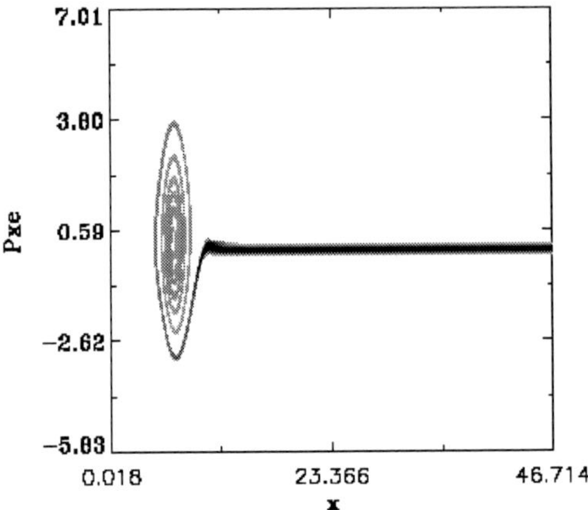

Figure 104. Phase-space contour plot of the electron distribution function t=53.88.

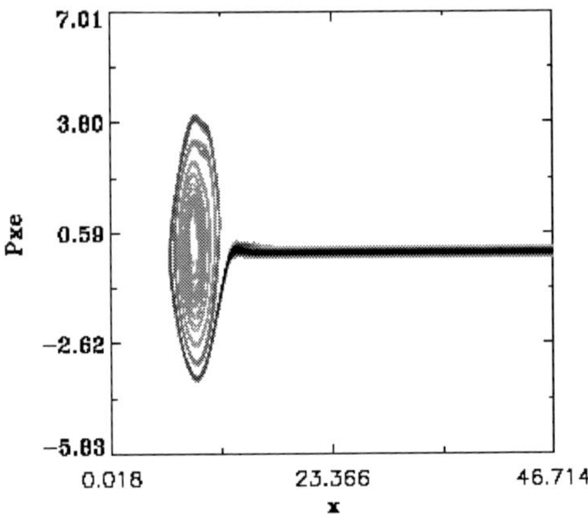

Figure 105. Phase-space contour plot of the electron distribution function *t*=71.84.

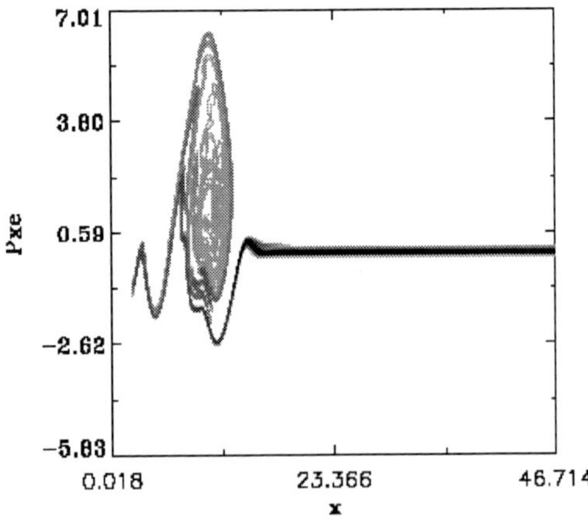

Figure 106. Phase-space contour plot of the electron distribution function t=80.82.

vortex (see Fig.(104)) is surrounded by two steep spikes, especially the one adjacent to the vacuum., and the entire structure (including the density curve of the ions, the broken line) seems to be pushed to the right. The broken density curve of the ions shows the ions accelerating in both direction, as can be verified from the contour plots in Figs(109-114). The trapped electron are also pushed to the right, leaving nothing behind. The same picture appears in Fig.(101) at t=62.86 and Fig.(102) at t=71.84. At t=71.84 however, the vortex structure (see Figs.(102,105)) is lifted up the gradient and has reached the level of the overdense plasma. The vortex structure becomes unstable (see Fig.(106) at t=80.82), and there appears to be two vorticities now produced in the system, in Fig.(107) at t=89.8 and Fig.(108) at t=96.78. The physical results are difficult to follow at this point, and require probably a finer grid and smaller time-step to follow. However, the stable electron peak in Figs(99-102) and the stable electron vortex in Figs.(103-105) are establishing a stable electric field (see Fig.(98)), which is accelerating the ions. The story of the evolution of the ions is given in Figs.(109-114), showing how the self-consistent quasi-stationary longitudinal electric field given in Fig.(98) is accelerating the ions. There is an acceleration of ions in both positive and negative directions, more important in the positive direction. The acceleration in the positive direction seems to saturate. As we can see, the distribution function showing the accelerated ions remains stable for sufficiently long time.

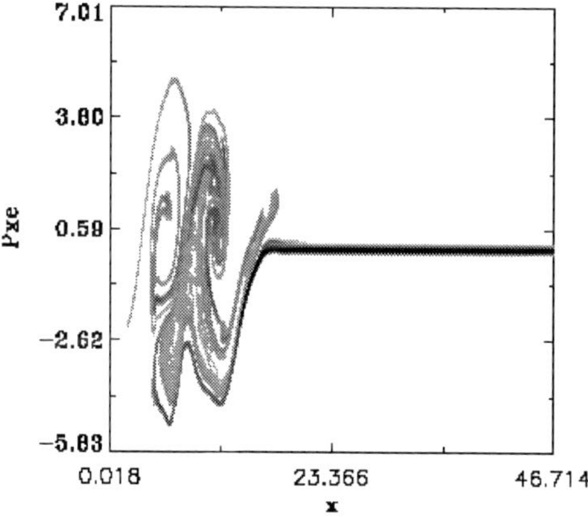

Figure 107. Phase-space contour plot of the electron distribution function t=89.8.

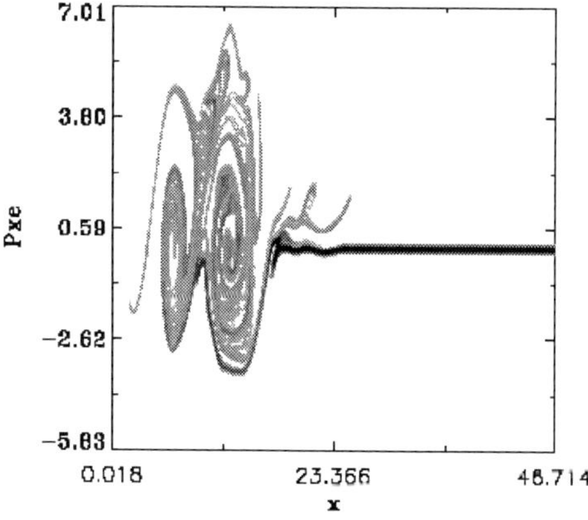

Figure 108. Phase-space contour plot of the electron distribution function t=96.78.

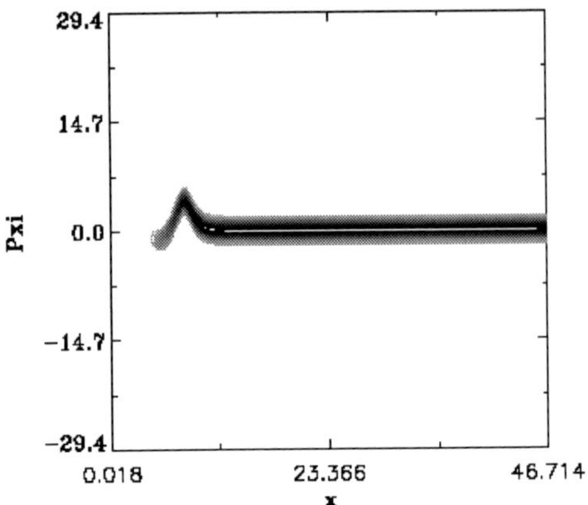

Figure 109. Phase-space contour plot of the ion distribution function t=35.92.

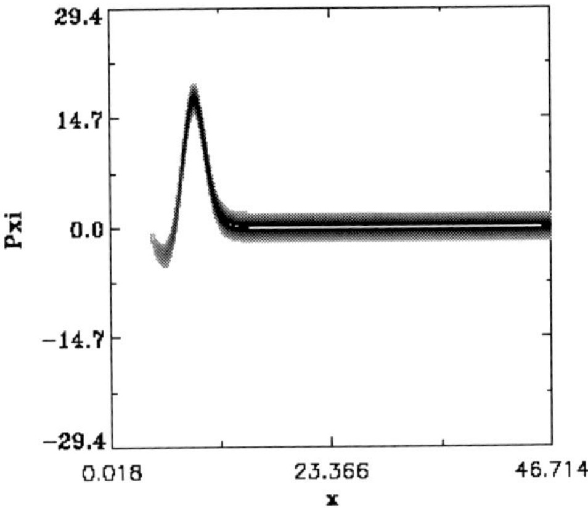

Figure 110. Phase-space contour plot of the ion distribution function t=53.88.

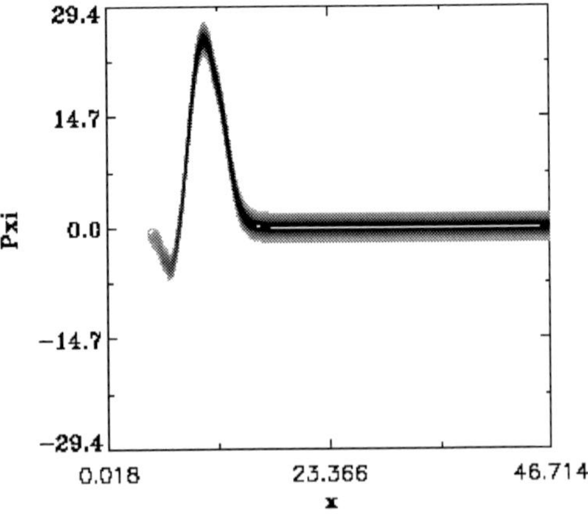

Figure 111. Phase-space contour plot of the ion distribution function t=71.84.

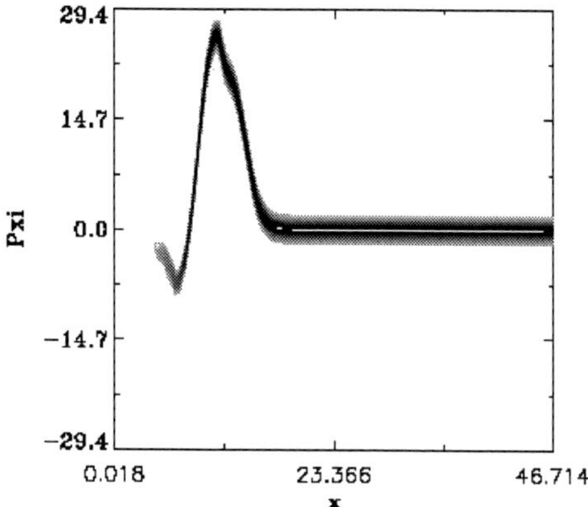

Figure 112. Phase-space contour plot of the ion distribution function t=80.82.

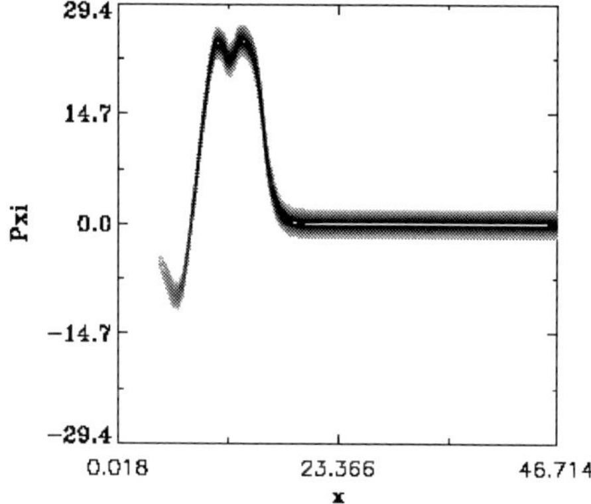

Figure 113. Phase-space contour plot of the ion distribution function t=89.8.

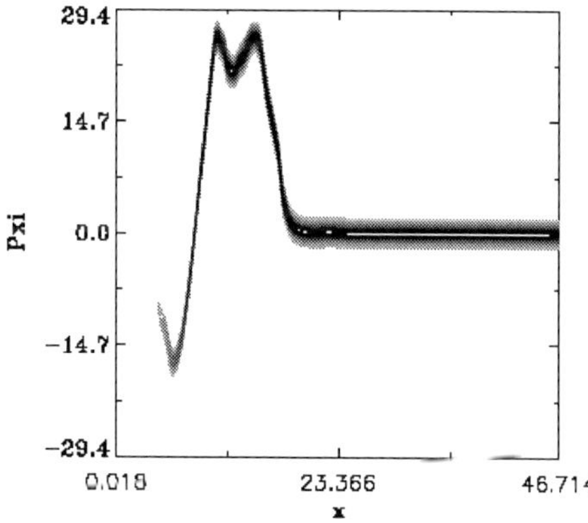

Figure 114. Phase-space contour plot of the ion distribution function t=96.78.

The parameters we have used in the present simulation are those used in [84] to study transparency. We note however that the results presented in [84], where ions were treated as an immobile background, are flawed since as we have seen the strong acceleration of the ions is an important part in the self-consistent solution of this problem of the interaction of a laser wave incident on an overdense plasma. In the present results, the penetration of the wave remained confined close to the plasma edge, so one cannot speak of transparency. What appears rather is that the plasma is being pushed to the right, and the build-up of the electron and ion densities at the edge in Figs.(99-102) has the tendency to make the plasma more opaque.

9. Fuid Equations

The application of the method of characteristics for fluid equations has been discussed in [1], where applications have been presented for the numerical solution of the shallow water equations [14-16], and to magnetohydrodynamic equations. We give here two rapid examples to illustrate the applications of this method when applied to fluid-type equations. The first example will discuss the application of the method to a one-dimensional model of blood flow in the aorta. The second example will discuss the application of the method to the equations of acoustic wave propagation.

9.1. A One-Dimensional Model for the Blood Flow in the Aorta

The blood flow in human arteries is regarded to be convective and mainly driven by the pressure pulse generated by the beating heart. The equations of flow in the elastic arteries are hyperbolic and admit nonlinear, wavelike solutions for the mean velocity U and the pressure P. The present discussion will be primarily concerned with the numerical solution of the mathematical one-dimensional flow model for this problem. The method of characteristics provides the necessary numerical tool to investigate the nonlinear equations for nonsteady blood flow, as recently presented in [17].

The aortic valve, assumed located at the origin of the x-axis, is closed during diastole so that initial flow velocity is equal to zero. Let P_0 be the pressure in the aorta during diastole. As long as the pressure inside the left ventricle is below P_0, the aortic valve remains closed. As the myocardium contracts, the ventricular pressure rises, and when it exceeds P_0, the valve opens. The pressure pulse therefore propagates in the aorta. The equations of flow can be written in terms of P and U as follows:

$$\frac{\partial P}{\partial t} + U\frac{\partial P}{\partial x} + \rho c^2 \frac{\partial U}{\partial x} = 0 \, . \tag{9.1}$$

$$\frac{\partial U}{\partial t} + \frac{1}{\rho}\frac{\partial P}{\partial x} + U\frac{\partial U}{\partial x} = 0 \, . \tag{9.2}$$

c is the wave speed. We assume the density of the blood ρ is constant. Changes in the pressure P and velocity U due to forward wave fronts are in the same sense, so that a forward wave front with positive pressure change (termed a compression wave front) accelerates the flow, whereas a forward wave front with negative pressure change (termed an expansion wave front) decelerates the flow. Conversely, backward wave fronts induce opposite changes in pressure and velocity. The characteristics equations for Eqs.(9.1-9.2) are defined by $\frac{dx}{dt} = U \pm c$, where the positive sign refers to the forward characteristic direction and the negative sign refers to the backward characteristic direction. Along these characteristics directions, the differential equations in Eqs.(9.1-9.2) can be rewritten using the Riemann invariants $R_{\pm}$:

$$\left(\frac{\partial}{\partial t} + (U + c)\frac{\partial}{\partial x}\right)R_{+} = 0 . \tag{9.3}$$

$$\left(\frac{\partial}{\partial t} + (U - c)\frac{\partial}{\partial x}\right)R_{-} = 0 . \tag{9.4}$$

The Riemann invariants $R_{\pm}$ are given by:

$$R_{\pm} = U \pm \int^{P} \frac{dP}{\rho c} . \tag{9.5}$$

Equations (9.3) and (9.4) state that $R_{\pm}$ are constant along their characteristics. Any changes in pressure or velocity which are imposed on the flow will cause a change in $R_{\pm}$ which will propagate along their characteristic directions. For the case where ρ and c are constant, U and P are reconstructed at every time-step from the relations $U = (R_{+} + R_{-})/2$ and $P = (R_{+} - R_{-})\rho c/2$. The numerical solution of Eqs.(9.4) and (9.5) is done following the discussion in section 1. Because of the hyperbolic nature of the equations, shock waves are to be expected. The solution of this problem has been recently presented in [17].

9.2. Acoustic Waves

The equation of continuity and the equation of motion in an acoustic medium are expressed using the pressure P and velocity $\vec{\upsilon}$ as follows (see for instance the recent work [85]):

$$\frac{\partial P}{\partial t} + \kappa \nabla . \vec{\upsilon} = 0 . \tag{9.6}$$

$$\frac{\partial \vec{\upsilon}}{\partial t} + \frac{1}{\rho} \nabla P = 0 . \tag{9.7}$$

κ is the bulk modulus and ρ is the density. Using a method of fractional step for the numerical solution, we can rewrite Eqs.(9.6-9.7) in each Cartesian direction as follows:

<u>*x* direction</u>

$$\frac{\partial P}{\partial t}+\kappa\frac{\partial \upsilon_x}{\partial x}=0\ . \tag{9.8}$$

$$\frac{\partial \upsilon_x}{\partial t}+\frac{1}{\rho}\frac{\partial P}{\partial x}=0\ . \tag{9.9}$$

<u>*y* direction</u>

$$\frac{\partial P}{\partial t}+\kappa\frac{\partial \upsilon_y}{\partial y}=0\ . \tag{9.10}$$

$$\frac{\partial \upsilon_y}{\partial t}+\frac{1}{\rho}\frac{\partial P}{\partial y}=0\ . \tag{9.11}$$

<u>*z* direction</u>

$$\frac{\partial P}{\partial t}+\kappa\frac{\partial \upsilon_z}{\partial z}=0\ . \tag{9.12}$$

$$\frac{\partial \upsilon_z}{\partial t}+\frac{1}{\rho}\frac{\partial P}{\partial z}=0\ . \tag{9.13}$$

We can rewrite Eqs.(9.8-9.9) using Riemann invariants [86], as follows:

$$\frac{\partial(P+\sqrt{\rho\kappa}\upsilon_x)}{\partial t}+\sqrt{\kappa/\rho}\frac{\partial(P+\sqrt{\rho\kappa}\upsilon_x)}{\partial x}=0\ . \tag{9.14}$$

$$\frac{\partial(-P+\sqrt{\rho\kappa}\upsilon_x)}{\partial t}-\sqrt{\kappa/\rho}\frac{\partial(-P+\sqrt{\rho\kappa}\upsilon_x)}{\partial x}=0\ . \tag{9.15}$$

$P+\sqrt{\rho\kappa}\upsilon_x$ is the forward propagating wave and $-P+\sqrt{\rho\kappa}\upsilon_x$ is the backward propagating wave in the x direction. We have similar equations for Eqs.(9.10-9.11) and Eqs.(9.12-9.13), by substituting respectively x by y and x by z in Eqs.(9.14-9.15). These equations represent the characteristic equations of the acoustic wave propagation in three dimensions. Equations (9.14-9.15) can be solved directly by cubic spline interpolation as discussed in the introduction of this chapter, and as in [86]. The method used in [85], called the CIP method (cubic interpolated profile) necessitates, in addition to Eqs.(9.14-9.15), the equations for the derivatives, so that one has to solve for the three dimensional problem twelve equations. In addition, it has been pointed out [19] that the CIP method is diffusive. We refer the reader to [85] for more information on the application of the CIP method to this problem of acoustic wave propagation. However, the direct solution of Eqs.(9.14-9.15) along the characteristics, as explained in the introduction, and applied in the many references cited in this chapter (as for instance in [86]), requires only six equations similar to Eqs.(9.14-9.15) to be solved for the three dimensional problem and shows very little numerical diffusion.

10. Conclusion

Eulerian Vlasov codes applying the method of characteristics have been successfully applied in recent years to study several problems in plasma physics, especially problems associated with wave-particles interaction. Interest in Eulerian grid-based Vlasov solvers associated with the method of characteristics for the numerical solution of hyperbolic type differential equations arise from the very low noise level associated with these codes, which allows accurate representation of the low density regions of the phase-space. This is obviously important if the physics of interest is in the low density region of the phase-space or in the high energy tail of the distribution function. (see references cited in the introduction, see also [87-95]). In addition, these methods have been successfully applied to fluid-type equations [1,14-16,34,35,43].

In the present chapter, the method of characteristics for the numerical solution of hyperbolic type differential equations has been applied in sections 2-5 to problems pertinent to industrial plasmas (sheaths, capacitive coupling, ion extraction). These problems were studied using models based on the Vlasov-Poisson system, with appropriate boundary conditions. While previous studies of these problems have been generally done using an adiabatic law to simplify the equations of the electrons, in the study presented in this chapter both electrons and ions have been treated with a kinetic equation. Detailed representations of the distribution functions in phase-space have been presented, and the associated macroscopic quantities have been calculated accurately. In section 5, the problem of a sheath at grazing incidence of an external magnetic field has been presented. We have shown that if the grazing angle of incidence is below a critical value, low frequency steady state oscillations exist, associated with enhanced energy fluxes to the wall. This problem is important to understand the heat load on limiters and divertors in tokamaks, and for the interpretation of probes measurements. In section 6, we have applied the method of characteristics to the more complicated cylindrical geometry to study the solution of the Vlasov-Poisson system for the problem of the formation of an electric field at a plasma edge. We have presented a self-consistent kinetic solution for the problem of the generation of a charge separation at a plasma edge, under the combined effect of a large ratio ρ_i / λ_{De} and a steep gradient. This problem is of fundamental importance for the study of the H-mode in tokamaks. The results show that for a large ratio of the ion gyro-radius to the Debye length ρ_i / λ_{De}, the electrons frozen by the magnetic field applied along the cylinder axis cannot move along the gradient to compensate the charge separation resulting from the large gyro-radius of the gyrating ions. The electric field which appears along the steep gradient is balanced by the gradient of the pressure term. The steeper the gradient, the higher the electric field. In addition, the Lorentz force term remains negligible along the gradient. The $\vec{E} \times \vec{B} / B^2$ drift compensates exactly the diamagnetic drift along the gradient, so that the total poloidal current is essentially zero. The present results point out the important role of the finite ions gyro-radius effect in many problems of plasma-wall transition with steep parameters [96]. They also suggest an important role played by large gyro-orbit impurity ions, which give an important contribution to the charge at the edge. Even a small fraction of large gyro-orbit impurity ions can significantly affect the electric field at the plasma edge, as previously suggested in [97]. Accurate calculations of the ion orbits is important for the accurate calculation of the charge separation at a plasma edge. Eulerian Vlasov codes applying the method of characteristics, having a low noise level, offer a powerful tool for the study of these problems. In sections 7 and 8, we have applied the method of characteristics for the solution of a one-dimensional relativistic Vlasov-Maxwell system, to study two important problems. In section 7 we have studied the problem of wake-field accelerators. The simulations show the production and acceleration of thin beams, well localized in phase-space. This method gives sensible results even in the very relativistic regimes. Usually, numerical results obtained with PIC codes for this problem can lead to unphysical numerical heating in the simulation, and show a momentum spread inside the laser pulse which is unphysically large [69]. In section 8 we have used the relativistic Vlasov-Maxwell equations to study the interaction in a overdense plasma of an intense laser wave, whose frequency is less than the plasma frequency. In this interaction an electric field is formed at the boundary due to the spatial distribution of the laser produced electrons accumulating at the edge, which accelerates the ions. Laser-induced ion acceleration seems one of the most promising applications of laser pulses interacting with solid matter and high density targets [75,83]. All these problems have fundamentally required nothing more than cubic spline interpolations, and we have chosen to do this work on fixed Eulerian grids.

The applications we have presented are but few examples for the method of characteristics applied to the kinetic equations of plasmas and to fluid equations. There is a rich and abundant literature recently published on these problems. The kinetic effects in an inductively coupled plasmas have been recently studied in [98]. Sheath problems have been also studied in [99]. Solutions of the kinetic Boltzman and Fokker-Planck equations applied to plasmas were presented in [45,100]. See also the review article in [101]. Gyrokinetic codes in tokamak physics are of special importance to study many aspects of the physics of fusion confined plasmas, and have been recently developed using Eulerian codes applying the method of characteristics [6-12,41,102]. Of special importance is also the recent applications of these methods to Boltzman-Poisson and Fokker-Planck equations in semi-conductor physics [103,104].

The method of characteristics has also been successfully applied for the numerical solution of fluid-type equations. This method is widely used in wheather forecast and climate modeling (called the semi-Lagrangian method [34,35]). More references on this subject can be found in [105]. Results have been already presented in [1] for the numerical solution of the shallow water equations [14-16] and the magnetohydrodynamic equations. Further applications can be found in [43]. We have outlined in section 9 some recent results for a model to study the flow of the blood in the aorta [17], and for the propagation of acoustic waves [86].

Acknowledgments

The fruitful discussions with Dr Vladimir Kolobov concerning the results in section 2.3 are gratefully acknowledged. The author is also grateful to the Centre de calcul scientifique de l'IREQ (CASIR) for computer time used to do part of this work. The constant support and interest of Dr. André Besner is gratefully acknowledged.

References

[1] Shoucri, M. *The Method of Characteristics for the Numerical solution of Hyperbolic Differential Equations*; in Computer Physics Research Trends, S.J. Bianco Ed., p.1 Nova Science Publishers: New York, 2007

[2] Ghizzo, A., Bertrand, P., Shoucri, M., Johnston, T., Feix, M., Fijalkow, E. *J. Comp. Phys.* 1990, 90, 431-457

[3] Ghizzo, A., Bertrand, P., Shoucri, M., Johnston, T., Fijalkow, E., Feix, M.R., Demchenko, V.V., *Nucl. Fusion* 1992, 32, 45-65

[4] Strozzi, D., Shoucri, M., Bers, A., Williams, E., Langdon, A.B. *J. Plasma Physics.* 2006, 72, 1299-1302

[5] Shoucri, M., Gerhauser, H., Finken, K-H. *Comp. Phys. Comm.* 2003, 154, 65-75

[6] Manfredi, G., Shoucri, Shkarofsky, I., Ghizzo, A., Bertrand, P., Fijalkow, E., Feix, M., Karttunen, S., Pattikangas, T., Salomaa, R. *Fusion Tech.* 1996, 29, 244-260

[7] Manfredi, G., Shoucri, M., Dendy, R.O., Ghizzo, A., Bertrand, P. *Phys. Plasmas* 1996, 3, 202-217

[8] Shoucri, M. *Czech. J. Phys.* 2001, 51, 1139-1151

[9] Pohn, E., Shoucri, M. *Proc. Vlasovia workshop*, in *Commun. Nonlinear Sci. Numer. Simul.* 2008, 13, 183-188

[10] Grandgirard, V., Brunetti, M., Bertrand, P., Besse, N., Garbet, X., Ghendrih, P., Manfredi, G., Sarazin, Y., Sauter, O., Sonnendrücker, E., Vaclavick, J., Villard, L. *J. Comp. Phys.* 2006, 217, 395-423

[11] Jenko, J. *Comp. Phys. Comm.* 2000, 125, 196-209

[12] Candy, J., Waltz, R.E. *J. Comp. Phys.* 2003, 186, 545-581

[13] Ghizzo, A., Shoucri, M., Bertrand, P., Johnston, T., Lebas, J. *J. Comp. Phys.* 1993, 108, 373-376

[14] Shoucri, M. *Comp. Phys. Comm.* 2007, 176;. 23-32

[15] Imai, Y., Aoki, T., Shoucri, M. *J. Appl. Meteo. Climat.* 2007, 46, 388-395

[16] Shoucri, M. *Comp. Phys. Comm.* 2004, 164;. 396-401

[17] Shoucri, R., Shoucri, M. *Cardiovasc. Eng.* 2007, 7, 1-6

[18] Shoucri, M., Pohn, E., Knorr, G., Bertrand, P., Kamelander, G., Manfredi, G., Ghizzo, A. Phys. Plasmas 2000, 7, 2517-2525

[19] Pohn, E.; Shoucri, M.; Kamelander, G. *Comp. Phys. Comm.* 2001, 137, 380-395;*ibid* 2001, 137, 396-404

[20] Pohn, E.; Shoucri, M.; Kamelander, G. *Comp. Phys. Comm.* 2005, 166, 81-93

[21] Yanenko, N.N. *The Method of Fractional Steps*, Springer-Verlag, New York, 1971

[22] Cheng, C.Z., Knorr, G. *J. Comp. Phys.* 1976, 22, 330-351

[23] Shoucri, M., Gagné, R. *J. Comp. Phys.* 1997, 24, 445-449; *Phys. Fluids* 1978, 21, 1168-1175, *IEEE Plasma Science* 1978, PS-6, 245-248; *J. Comp. Phys.* 1978, 27, 315-322

[24] Shoucri, M. *Phys. Fluids* 1978, 21, 1359-1365; ibid 1979, 22, 2038; ibid 1980, 23, 2030; *J. de Physique* 1979, 40, 38-39

[25] Shoucri, M., Storey, O. 1986, 29, 262-265; Simon, A., Short, R.W. *Phys Fluids* 1988, 31, 217

[26] Cheng, C.Z. *J. Comp. Phys.* 1977, 24, 348-360

[27] Shoucri, M., Gagné, R. *J. Comp. Phys.* 1978, 27, 315-322; Shoucri, M. In Modeling and Simulation; *Proc. 10th Annual Pittsburgh Conference*; Publisher: Instrument Society of America, Pittsburgh, 1979; Vol. 10; 1187-1192

[28] Shoucri, M. *IEEE Plasma Science*, 1979, PS-7, 69-72

[29] Johnson, L.E. *J. Plasma Phys.* 1980, 23, 433-452

[30] Rickman, J.D. *IEEE Plasma Science* 1982, PS-10, 45-56

[31] Shoucri, M. *Proc. Vlasovia Workshop, Comm. Nonlinear Sci. Numer. Simul.* 2008, 13, 174-182

[32] Shoucri, M., Gerhauser, H., Finken, K-H. *Comp. Phys. Comm.* 2004, 164, 138-149

[33] Sonnendrücker, E., Roche, J., Bertrand, P., Ghizzo, A. *J. Comp. Phys.* 1998, 149, 201

[34] Makar, P.A., Karpik, S.R. *Mon. Wea. Rev.* 1996, 124, 182-199

[35] Staniforth, A., Côté, J. *Mon. Wea. Rev.* 1991, 119, 2206-2223

[36] de Boor, C. *A Practical Guide to Splines*; Applied Mathematics 27; Springer-Verlag: New-York, N.Y., 1978

[37] Ahlberg, J.H., Nilson, E.N., Walsh, J.L. *The Theory of Splines and their Applications*; Academic Press: New York, N.Y., 1967

[38] Abbott, B.A. *An Introduction to the Method of Characteristics*; Thames and Hudson: London, 1966

[39] Mangeney, A., Califano, F., Cavazzoni, C., Travnicek, P. *J. Comp. Phys.* 2002, 179, 475-490

[40] Brunner, S., Valeo, E. *Phys. Rev. Lett.* 2004, 93, 145003-1 -145003-4

[41] Watanabe, T.-H., Sugama, H. *Trans. Theory Stat. Phys.* 2005, 34, 287-309

[42] Watanabe, T.-H., Sugama, H., Sato, T. *J. Phys. Soc. Japan*, 2001, 70, 3565-3576

[43] Shoucri, M., Lebas, J., Knorr, G., Bertrand, P., Ghizzo, A., Manfredi, G., Christopher, I., *Phys. Scripta* 1997, 55, 617-627; ibid 1998, 57, 283-285

[44] Manfredi, G., Shoucri, M., Bertrand, P., Ghizzo, A., Lebas, J.,Knorr, G., Sonnendrücker, E., Bürbaumer, H., Entler, W., Kamelander, G. *Phys. Scripta* 1998, 58, 159-175

[45] Batishchev, O., Shoucri, M., Batishcheva, A., Shkarofsky, I. *J. Plasma Physics* 1999, 61, 347- 364

[46] Arber, T.D., Vann, R.G.L *J. Comp. Phys.* 2002, 180, 339-350

[47] Stangegy, P.C. *Plasma Boundary of Magnetic Fusion Devices*, Institute of Physics: Bristol, 2000

[48] Gerhauser, H., Claassen, H.A. *Contrib. Plasma Phys.* 1998, 38, 331-334

[49] Devaux, S., Manfredi, G. *Phys. Plasmas* 2006, 13, 083504-1-12

[50] Bohm, D. *The characteristics of Electrical Discharges in Magnetic Fields*, Ed. A. Guthry and R.K. Wakerling, McGraw-Hill: New-York, 1949, Chap.3, 77

[51] Riemann, K.-U. *J. Phys. D* 1991, 24, 493-518

[52] Shoucri, M. *Jap. J. Appl. Phys.* 2007, 46, 3045-3051

[53] Chabert, P., Levif, P., Raimbault, J.-L., Rax, J.-M., Turner, M.M., Lieberman, M.A. *Plasma Phys. Contr. Fusion* 2006, 48, B231-B237

[54] Turner, M.M., Chabert, P. *Phys. Rev. Lett.* 2006, 96, 205001-1-4

[55] Gozadinos, G., Vender, D., Turner, M.M. *J. Comp. Phys.* 2001, 172, 348-355

[56] Ben Abdallah, N., Mas-Gallic, S., Raviart, P.-A. *Asymptotic Anal.* 1995, 10, 1-28

[57] Degond, P., Raviart, P.-A. *Asymptotic Anal.* 1991, 4, 187-214

[58] Shoucri, M., Gerhauser, H., Finken, K.-H. *Czech. J. Phys.* 2002, 52, 1121- 1131

[59] Shoucri, M., Gerhauser, H., Finken, K.-H. *Phys. Scr.* 2007, 75, 712-721

[60] Koch, B., Bohmeyer, W., Fussman, G. *J. Nucl. Mater.* 2005, 337-339, 211-215

[61] Stix, T. *Phys. Fluids* 1973, 16, 1260-1269

[62] Hirschman, S.P. *Nucl. Fus.* 1978, 18, 917-927

[63] Shoucri, M., Gerhauser, H., Finken K.H. 2003, *Proc. 30th European Physical Society Conf. Controlled Fusion and Plasma Physics,* St Petersburg, in: ECA, vol.27A, P-1.170

[64] Groebner, R.J., Thomas, D.M., Deranian, R.D. *Phys.Plasmas* 2001, 8, 2722-2732

[65] Shoucri, M. *Phys.Plasmas* 2002, 9, 731-734

[66] Sprangle, P., Esarey, E., Ting, A. *Phys.Rev.* 1990, A41, 4463-4469

[67] Ting, A., Esarey, E., Sprangle, P. *Phys.Fluids* 1990, B2, 1390-1394

[68] Bingham, R., Mendonca, J.T., Shukla, P.K. *Plasma Phys.Control.Fusion* 2004, 46, R1-R23

[69] Shadwick, B.A., Tarkenton, G.M., Esarey, E., Schroeder, B.C. *Phys.Plasmas* 2005, 12, 056710-1-056710-8

[70] Akhiezer, A.I., Polovin, R.V. *Sov.Phys.JETP* 1956, 3, 696-703

[71] Guérin, S., Laval, G., Mora, P., Adam, J.-C., Héron, A., Bendib, A. *Phys.Plasmas* 1995, 2, 2807-2818

[72] Krall, J., Joyce, G., Esarey, E. 1991, Naval Research Lab. Memorandum Report 6772

[73] Shoucri, M. 2007, *Proc. 34th European Physical Society Conf. Controlled Fusion and Plasma Physics,* Warsaw, (to be published).

[74] Shoucri,M. *Comm.Comp.Phys.* 2008, 4, 703-718

[75] Snavely, R.A., Key, M.H., Hatchett, S.P. *et al. Phys.Rev.Lett.* 2000, 85, 2945-2948

[76] Lontano, M., Passoni, M. *Phys.Plasmas* 2006, 13, 042102-1-042102-7

[77] Yogo, A., Daido, H., Bulanov, S.V. *et al. Phys.Rev. E* 2008, 77, 016401-1-016401-6

[78] Bulanov, S.S., Brantov, A., Bychenkov, V.Yu. *et al. Med.Phys.* 2008, 35, 1770-1776

[79] Robson, L., Simpson, P.T., Clarke, R.J.*et al. Nature Physics* 2007, 3, 58-62

[80] Cerchez, M., Jung, R., Osterholz, J. *et al. Phys.Rev.Lett.* 2008, 100, 245001-1-245001-4

[81] Akli, K.U., Hansen, S.B., Kemp, A.J. *et al Phys.Rev.Lett.* 2008, 100, 165002-165002-4

[82] Salamin, Y.I., Harman, Z., Keitel, C.H. *Phys.Rev.Lett.* 2008, 100, 155004-1-155004-4

[83] Klimo, O., Psikal, J., Limpouch, J. *et al. Phys.Rev. Special Topics-Accel. Beams* 2008, 11, 031301-1-031301-14

[84] Huot, F., Ghizzo A., Bertrand, P., Sonnendrücker, E., et al *J. Comp. Phys.* 2003, 185, 512-531

[85] Shiraishi, K., Matsuoka, T. *Comm.Comp.Phys.* 2008, 3, 121-135

[86] Shoucri, M. *J.Comp.Phys.* 1986, 63, 240-245

[87] Bertrand, P., Ghizzo, A., Feix, M., Fijalkow, E., Mineau, P., Suh, N.D., Shoucri, M. In *Nonlinear Phenomena in Vlasov Plasmas;* Doveil, F.,Ed.; Proc. Cargèse Workshop; *Les Editions de Physique*: Les Ulis, France, 1988; 109-125

[88] Bertrand, P., Ghizzo, A., Johnston, T., Shoucri, M., Fijalkow, E., Feix, M. *Phys. Fluids* 1990, B2, 1028-1037

[89] Johnston, T., Bertrand, P., Ghizzo, A., Shoucri, M., Fijalkow, E., Feix, M. *Phys. Fluids* 1992, B4, 2523-2537

[90] Shoucri, M., Bertrand, P., Ghizzo, A., Lebas, J., Johnston, T., Feix, M., Fijalkow, E. *Phys. Letters A*, 1991, 156, 76-80

[91] Bertrand, P., Ghizzo, A., Karttunen, S., Pättikangas, T., Salomaa, R., Shoucri, M. *Phys. Plasmas* 1995, 2, 3115-3129; *Physical Rev. E* 1994, 49, 5656-5659

[92] Ghizzo, A., Bertrand, P., Bégué, M.L., Johnston, T., Shoucri, M. *IEEE Plasma Science* 1996, 24, 370-378

[93] Shoucri, M., Manfredi, G., Bertrand, P., Ghizzo, A., Knorr, G. *J. Plasma Phys* 1999, 61, 191

[94] Sircombe, N.J., Arber, T.D., Dendy, R.O. *Phys.Plasmas* 2005, 12, 0122303-1-012303-8

[95] Pohn, E., Shoucri, M., Kamelander, G. *Comp. Phys. Comm.* 2005, 166, 81-93; *J. Plasma Physics*. 2006, 72, 1139-1143

[96] Manfredi, G., Shoucri, M., Shkarofsky, I., Bertrand, P., Ghizzo, A., Krasheninnikov, S., Sigmar, D., Batishchev, O., Batishcheva, A. *J. Nucl. Mat.* 1999, 266-269, 873-876

[97] Shoucri, M., Pohn, E., Bertrand, P., Knorr, G., Kamelander, G., Manfredi, G., Ghizzo, A. *Phys. Plasmas* 1999, 6, 1401-1404

[98] Shoucri, M., Matte, J.P., Côté A. *J.Phys.D :Appl.Phys.* 2003, 36, 2083-2089

[99] Shoucri, M., Cardinali, A., Matte, J.P., Spigler, R. *Eur.Phys.J. D* 2004, 30, 81-92

[100] Shoucri, M., Shkarofsky, I., Stansfield, B., Boucher, C., Pacher, G., Décoste, R., O. Batishchev, A. Batishcheva, Krashenennikov, S., Sigmar, D. *Contrib.Plasma Phys.* 1998, 38, 225-230

[101] Kolobov, V., Arslanbekov, R. *IEEE Trans.Plasma Science* 2006, 34, 895-909

[102] Idomura, Y., Watanabe, T.-H., Sugama, H. *C.R. Physique* 2006, 7, 650-669

[103] Kolobov, V. *Comp.Mat.Science* 2003, 28, 302-320

[104] Carrillo, J., Majorana, A., Vecil, F. *Comm.Comp.Phys.* 2007, 2, 1027-1054

[105] Durran, D.R. *Numerical Methods for Wave Equations in Geophysical Fluid Dynamics*; Text in Applied Mathematics 32; Springer:New-York,N.Y., 1998

In: Numerical Simulation Research Progress
Editors: Simone P. Colombo et al, pp. 99-139
ISBN 978-1-60456-783-0

Chapter 2

MIXED FINITE DIFFERENCE-SPECTRAL NUMERICAL APPROACH FOR KINETIC AND FLUID DESCRIPTION OF NONLINEAR PHENOMENA IN PLASMA PHYSICS

Francesco Valentini, Marco Onofri and Leonardo Primavera
Università della Calabria. Italy

Abstract

Several approaches are commonly used to study a system made of charged particles embedded in electromagnetic fields (commonly called a plasma): the kinetic one, where the dynamics of the system is studied through a statistical description in terms of the probability distribution function of finding a particle in a given volume of the phase space; the fluid one, which involves the solution of a system of partial differential equations similar to that describing neutral fluids.

Both in astrophysics and laboratory plasmas, one is often concerned with plasma equilibria, namely the instabilities which can develop in an ionized gas when an initial, inhomogeneous, equilibrium configuration is perturbed by fluctuations. In such problems, the nonlinear phase of the instability can be studied theoretically, in an effective way, only with the use of numerical simulations.

In such problems, the main numerical difficulty is provided by the fact that a very broad range of spatial and temporal scales are excited, requiring rather high resolutions both in space and time to approximate the numerical solution adequately. On the contrary, the geometries of the problem are generally rather simple (Cartesian, cylindrical or spherical), so that no need for special techniques for dealing with complicated or irregular domains, like finite element methods, are required. Furthermore, in the case of plasma instability problems, the solutions are rather regular, therefore no special techniques, like shock-capturing methods, or finite volume methods, are necessarily required.

The task of building equation solvers which implement the solution of a nonlinear set of partial differential equations with both inhomogeneous and periodic directions is often carried out with the use of mixed finite difference and spectral techniques. This permits a simple treatment of the boundary conditions by preserving a good numerical precision or, at least, the possibility to control effectively the numerical truncation errors.

In this work, we review the typical problems which are encountered when studying the effects of plasma instabilities by using either a kinetic or a fluid approach. We illustrate in detail some numerical codes which are currently used in the plasma physics domain. We further give examples of application of such codes to typical plasma problems and show how the numerical results obtained through the use of the mixed finite difference and spectral method give a description of physical phenomena which are in good agreement with the theoretical predictions.

1. Introduction

A gas composed of charged particles is commonly called a *plasma*. Such a system shares several characteristics with ordinary, neutral fluids, though it exhibits a much richer and more complicated behavior due to the fact that electromagnetic forces play a fundamental role in determining the evolution of the system. This is because electric and magnetic fields influence the motion of particles which, in turn, produces electromagnetic forces that modify the overall field structure. For instance, even in the Magnetohydrodynamics (MHD) approximation, in which the plasma is described by an equation set very similar to the continuity, Navier-Stokes and energy equations of ordinary fluids [1], the presence of an external magnetic field involves four normal modes for linear waves (Alfvén, fast and slow magnetoacoustic and entropy waves) of which the usual sound and entropy waves of fluid mechanics are particular cases. Moreover, even in collisionless plasmas, where the interactions among the particles can be neglected, the presence of electromagnetic fields may induce phenomena like collisionless dissipation or plasma acceleration through electromagnetic waves.

Plasmas are found everywhere in nature, whenever the temperature of a system is high enough to ensure the necessary degree of ionization. The plasma state is the usual condition in astrophysical systems, like the interior or the atmosphere of stars, the interstellar medium, and so on. However, plasmas are more and more studied also in laboratories, for the increasing interest raised by the thermonuclear fusion experiments. Although most astrophysical plasmas are highly supersonic and sometimes their study even need a relativistic description, in many problems, both concerning the astrophysical and laboratory plasmas, the main goal is to describe the basic nonlinear properties of the system, such as the development of instabilities and the transition to turbulent regimes in the charged fluid.

Although the underlying physics of a plasma is far richer than the one of a neutral fluid, however, the basic equations describing such systems share many similarities with the ones of ordinary fluids, that is, a nonlinear system of equations which is *hyperbolic* and, possibly, *parabolic* (in the usual case where dissipative effects are kept into account). Moreover, some fields usually are to satisfy a solenoidality condition, which in turn requires the solution of an elliptic equation. Thus, numerical techniques generally adopted for studying fluid dynamics problems can be suitable also for dealing with plasma systems, provided that particular care has to be devoted to peculiar aspects of the problems under study. For instance, in the fluid description of a plasma, a typical numerical problem is to preserve the divergenceless condition for the magnetic field, which is to be solenoidal at each time step [2]. In the kinetic description of a plasma, instead, a typical issue is to ensure the energy conservation of the numerical schemes (*symplectic schemes*), which otherwise would lead to unphysical phenomena.

In general, the solution of the hyperbolic-parabolic system of equations is carried out through a finite difference time stepping. Usually, standard explicit methods, such as high-order Runge-Kutta schemes are used in the fluid approach. In the kinetic description, however, particular techniques have to be applied for the time evolution of the distribution function and to ensure the energy conservation of the numerical scheme.

Many plasma problems deal with the instability and development of turbulence in unstable, inhomogeneous configurations driven by fluctuations of the quantities describing the system. That is, many plasma equilibrium configurations in presence of inhomogeneities may act as a reservoir of free energy for many instabilities. Therefore, even small amplitude fluctuations of the background quantities may perturb the equilibrium, which decades to a status with minimum energy. The energy released from the background, large scale, configuration is typically transferred to the perturbing fluctuations, or may excite characteristic, resonant modes of the system. In both cases, such fluctuations may grow during time and, when their amplitude is large enough that nonlinear interactions take place, a transfer of energy from the large scale to the small scale structures (and, possibly, also in the opposite direction) occurs. This usually leads to a very complex, nonlinear dynamics of the system with possibility of transitions to chaos, bifurcations, space-time intermittency, and so on. Hence, unstable plasma systems often involve the presence of one inhomogeneous direction, which provides the unstable configuration, together with periodic directions, along which the fluctuations propagate.

Both in astrophysical and plasma laboratory simulations, the geometry of the computational domains is kept quite simple (Cartesian, cylindrical or spherical). In the study of plasma instabilities, techniques like finite element methods, largely adopted in both fluid mechanics and aerospatial studies, are seldom used. Such methods allow the simple treatment of complex geometries and boundary conditions, at the price of a rather raw precision in the approximation of the equations. This goes in rather the opposite direction with respect to the plasma studies where, as said, the geometries are simple, and it is often fundamental, instead, to describe the nonlinear dynamics of the system with a precision as high as possible, since too rough approximations can lead to unphysical phenomena and enhanced dissipation.

Another technique often used in fluids is the finite volume approach. In this case, the control volume occupied by the fluid during its motion is divided in volumes of simple form and the equations written in conservative form. In this framework, the variation of the fluid quantities inside the volume is computed by the evaluation of the relative fluxes through the boundaries of the volumes, by means of appropriate average operations of the quantities on the gridpoints closest to the boundaries. This technique allows the constructions of numerically stable codes even in presence of strong gradients in the quantities and are therefore often used for the treatment of highly supersonic flows. However, such schemes turn out to involve very high numerical dissipation which, again, could severely modify the nonlinear dynamics of the system. In most studies of plasma instabilities, both through the fluid and kinetic description, however, the plasma is generally considered subsonic or only moderately supersonic. Therefore, techniques which allow a higher precision in the description of the turbulent dynamics are to be preferred when dealing with subsonic plasma instabilities, although in several astrophysical context, in which supersonic plasmas are involved (solar and stellar outflows and winds, galactic jets, etc.), finite volume techniques are often the only ones through which numerical results can be obtained.

In conclusion, plasma instabilities often involve the use of regular geometrical domain. Under these conditions, standard (pseudo) spectral methods, or high order finite difference methods, still represent a suitable choice in terms of both precision and simplicity. Spectral methods allow to treat periodic boundary conditions easily, since the Fourier basis for the decomposition of the functions automatically satisfies the boundary conditions (Galerkin method). This is, however, not true for the inhomogeneous direction, where more general boundary conditions are to be imposed. A typical method to tackle this problem is through the use of Chebyshev polynomials along the inhomogeneous direction [27]. Such an approach has been largely used both in fluid dynamics and plasma physics [23]. However, this approach has some non trivial drawbacks one has to deal with: 1) the Chebyshev grid points have to be chosen *a priori*, through well defined spatial distributions ; 2) for the same reason, spatial resolution of the Chebyshev grid is strongly enhanced in specific zones of the computational domain: this allows a higher degree of precision in the numerical approximation of the quantities, but prevents the use of explicit methods for the time advancement, due to the (often) prohibitive Courant-Freidricks-Lewy (CFL) condition where the spatial resolution is enhanced; 3) although the Chebyshev polynomials allow the use of fast methods like, for example, the Fast Fourier Transform (FFT) routines, they are not easily parallelizable for the use on multi-processor machines, due to the huge amount of communications required.

On the other hand, the exclusive use of finite difference techniques, even of high order, is also not optimal for all problems. For instance, when dealing with the solution of elliptic equations, the finite difference implementation of the Poisson equation in more than one direction leads to the inversion of large sparse matrices, which is a numerically formidable task to be performed efficiently and (that is even worse) very difficult to parallelize.

For all the above considerations, a method which is largely employed both in fluid dynamics and plasma physics is the use of mixed finite difference-spectral techniques.

Usually, finite difference approximations for the spatial derivatives are preferred along the inhomogeneous directions, due to their simplicity in imposing the boundary conditions, whereas spectral methods, as Fast Fourier Transform (FFT) algorithms, are often employed in the periodic directions due to the fact they identically satisfy the periodic boundary conditions.

In the framework of kinetic theory, Particle in Cell (PIC) [4, 5] methods represent the most adopted approach to numerically describe the plasma dynamics. PIC approach is a Lagrangian approach where the equations of motion of a large number of macro-particles are integrated in time through a standard explicit scheme, under the effects of the self-consistent and external electric and magnetic fields. At each time step, the particles are collected on a uniformly spaced grid to obtain the numerical density to be used in the integration of the Maxwell equations for fields. In the same statistical way, the phase space particle distribution can be evaluated. Besides the Lagrangian PIC method, the so-called Vlasov approach [6] has nowadays become extensively adopted. Vlasov approach is an Eulerian approach, where Vlasov equation is numerically solved by time-advancing the particle distribution function at each time step on a uniform fixed grid used to sample the phase space. Maxwell equations are integrated at each time step using current, density and velocity evaluated by straight numerical integration as moments of the particle distribution function. Such Eulerian approach avoids the statistical noise caused by the fact that macro-particles in PIC codes represent values of the distribution function at randomly selected points of the phase space.

The biggest limitations of the Eulerian Vlasov method are due to the large amount of computing resources and execution time needed to advance the distribution function, that in principle is a six-dimensional variable in phase space. Nowadays, a $5-D$ phase space description can be obtained (see for example [6, 7]), exploiting the computational power of modern parallel supercomputers, but the six-dimensional description is still a hard goal. Nevertheless, in many situations, the understanding of complex plasma dynamics requires a simplified description, and interesting nonlinear phenomena can be analyzed in a phase space of lower dimensionality. This is the typical case where the very low noise Eulerian Vlasov algorithms are extremely efficient and accurate in the analysis of kinetic effects (for example wave-particle interaction) or in the description of the tails of the particle distribution. Furthermore, the Vlasov approach allows for a clean description of the evolution of a single-mode perturbation, at variance with PIC codes where each possible mode in the simulation domain has a "small" amount of energy deriving from the statistical noise. Finally, a low noise algorithm is very useful also in the case of low amplitude initial perturbations problems, such as for instance the linear stage of the instabilities, where the amplitude of each mode in the spectrum could be of the same order of the noise introduced by PIC codes.

In conclusion, the goal of this work is to give a short review about the numerical techniques commonly used in the study of plasma instabilities, pointing out advantages and limitations of the applications of the different methods to the kinetic and fluid descriptions. This chapter is organized as follows. In Sec. 2 we describe the kinetic approach used in plasma physics and the numerical algorithms which represent the state of the art. Analogously, in Sec. 3 we review the fluid approach and the numerical methods commonly used in this framework. In Sections 4 and 5 we discuss some problems of interest in plasma physics, as examples of applications of the algorithms. We conclude in Sec. 6.

2. Kinetic Point of View in Plasma Physics

The kinetic theory includes the dynamics of the individual particle in the evolution of a plasma system. The basic element in this kind of description of a plasma is the distribution function $f(\vec{x},\vec{v},t)$, that carries the information about the position of the particles in both physical space and velocity space. Specifically, the distribution function f represents the number density of particles found near a point in the six-dimensional phase space $(\vec{x},\vec{v})$, at a given instant of time t. In simpler words, the number of particles located in a volume element $d^3\vec{x}d^3\vec{v}$ of the phase space is defined to be $f(\vec{x},\vec{v},t)d^3\vec{x}d^3\vec{v}$.

The principle of particle conservation states that the time rate of change for the number of particles in a particular volume is given by the flux of particles across the surface of the volume, in absence of either sources or sinks of the particles. In collisionless approximation, phase space density obeys a continuity equation in a six-dimensional space. This can be expressed as a differential equation:

$$\frac{\partial f}{\partial t} + \nabla \cdot (\vec{v} f) + \nabla_v \cdot (\vec{a} f) = 0 \tag{1}$$

where ∇_v is the gradient in the velocity space and $\vec{a}$ is the acceleration. Any position in phase space is specified by $\vec{x}$ and $\vec{v}$, then $\vec{v}$ is not dependent on $\vec{x}$. Moreover, in the case of the Lorentz force, $\nabla_v \cdot \vec{a} = 0$, then Eq. (1) becomes:

$$\frac{\partial f}{\partial t} + \vec{v} \cdot \nabla f + \vec{a} \cdot \nabla_v f = 0 \tag{2}$$

$$\vec{a} = \frac{q}{m}\left(\vec{E} + \frac{\vec{v} \times \vec{B}}{c}\right) \tag{3}$$

where $\vec{E}$ and $\vec{B}$ are the electric and magnetic fields and c the speed of light.

Equation (1) is called *Vlasov equation* (or collisionless Boltzmann equation) and describes the time evolution of the distribution function of plasma particles under the effect of external fields and self-consistent microfields, due to the collective motion of the particles themselves.

The interest of the kinetic approach resides principally in the possibility of studying the fundamental Hamiltonian aspects underlying the nonlinear dynamics of collisionless plasma systems, taking into account particle effects (like wave-particle interactions), which are ruled out from fluid description, but play a fundamental role in particle acceleration and wave absorption phenomena, especially in collisionless plasmas.

The Vlasov equation is a nonlinear partial derivative differential equation, whose analytical solution is available only in a few simplified linear cases, but the nonlinear regime, including the most interesting physical phenomena, must be investigated numerically. The numeric integration of the Vlasov equation requires, in the most general case, to perform calculations in a six-dimensional (6-D) phase space $(\vec{x}, \vec{v})$, for both electrons and protons, which is out of the range of the presently available computing resources, unless some simplifications allowed by physical assumptions and symmetries of the model are invoked. In the past years, the integration of the system of coupled Vlasov-Maxwell equations has been almost always limited to 1-D spatial (electrostatic) cases only (usually assuming two-dimensional (2-D) phase space (x, v)). Nowadays, thanks to the development of high performance computing systems, implementation of numeric schemes based on Vlasov equation in phase space of higher dimensions is possible.

In the next section, we will discuss the most efficient numerical techniques used to integrate the Vlasov equation in multi-dimensional phase space and we will illustrate few examples of application of these techniques.

2.1. Hyperbolic Equations of Conservation Law Type

For sake of simplicity, we consider the one-dimensional problem, but all the results can be easily generalized to the multidimensional case:

$$\frac{\partial f}{\partial t} + v\frac{\partial f}{\partial x} + a\frac{\partial f}{\partial v} = 0 \tag{4}$$

In the previous equation, v and a represent the x-components of velocity and acceleration respectively. From one-dimensional Vlasov equation (4), we split the evolution of the distribution function f into two steps:

$$\frac{\partial f_x}{\partial t} + v\frac{\partial f_x}{\partial x} = 0 \tag{5}$$

$$\frac{\partial f_v}{\partial t} + a\frac{\partial f_v}{\partial v} = 0 \tag{6}$$

The previous equations can be written in form of hyperbolic equations of conservation law type ($\partial u/\partial t + \partial(Au)/\partial x = 0$).

Equation (5) describes the time evolution of the function f_x in physical space, while equation (6) the time evolution of f_v in velocity space. We will explain in the following how the evolution of f can be obtained from those of f_x and f_v.

The solutions for f_x and f_v at the time $t + \Delta t$ can be formally written as:

$$f_x(t + \Delta t) = \Lambda_x(\Delta t) f_x(t)$$
$$f_v(t + \Delta t) = \Lambda_v(\Delta t) f_v(t)$$

where Λ_x and Λ_v will be called "translation" operators. The problem is now to obtain an explicit form for the operators Λ_x and Λ_v. It is worth noting that equations (5) and (6) are formally the same. In the following, we will discuss in detail the solution for f_x, but all the results will be valid for the evolution of f_v, as well.

Performing a simple stability analysis on the hyperbolic model equation (5) [or equivalently on equation (6)], one realizes that an algorithm based on centered schemes in evaluating the derivatives of f is not stable numerically [3, 8], and the remedy for this problem of stability is to use a noncentered approximation for the first-order derivatives, the direction of these noncentered differences being determined by the sign of v (upwind schemes).

In particular, if the information moves towards the positive x direction (i.e. $v > 0$), we obtain the value of the function f at the point $x_i + \epsilon$ as:

$$f(x_i + \epsilon) \simeq f(x_i) + \epsilon\left(\frac{\partial f}{\partial x}\right)_{x=x_i} \tag{7}$$

otherwise ($v < 0$):

$$f(x_i + \epsilon) \simeq f(x_{i+1}) - (\Delta x - \epsilon)\left(\frac{\partial f}{\partial x}\right)_{x=x_{i+1}} \tag{8}$$

We can refer to the previous formulas as first order $upwind\ Taylor\ expansions$.

The plane (x, t) will be discretized as $x_i = (i-1)\Delta x$ (with $i = 1, N$), $t_n = n\Delta t$, where i and n are integers. The simulation box is given by $[0, L_x]$, and it is divided in N intervals $I_i = [x_i - \Delta x/2, x_i + \Delta x/2]$; then, $\Delta x = L_x/N$.

The average value of the function f over a generic interval I_i, at the time t, is defined by:

$$\overline{f}(x_i, t) = \frac{1}{\Delta x}\int_{x_i - \frac{\Delta x}{2}}^{x_i + \frac{\Delta x}{2}} f(x', t)dx' \tag{9}$$

We now integrate Eq. (5) in time over the interval $[t, t+\Delta t]$, after calculating the average value of (5) over I_i, using the definition (9):

$$\overline{f}(x_i, t+\Delta t) = \overline{f}(x_i, t) - \frac{v}{\Delta x}\int_t^{t+\Delta t} dt' \int_{x_i-\frac{\Delta x}{2}}^{x_i+\frac{\Delta x}{2}} \frac{\partial}{\partial x} f(x', t')dx' \tag{10}$$

Note that, in order to use a simpler notation, in the previous and in the next formulae, we replace f_x by f. From Eq. (10), integrating by parts over x, one obtains:

$$\overline{f}(x_i, t+\Delta t) = \overline{f}(x_i, t) - \frac{v}{\Delta x}\int_0^{\Delta t} d\tau \left[f\left(x_i + \frac{\Delta x}{2}, t+\tau\right) - f\left(x_i - \frac{\Delta x}{2}, t+\tau\right)\right] \tag{11}$$

where $t' = t + \tau$. We can use a first order Taylor expansion to calculate the value of f at the boundary points of the cell I_i, taking into account the propagation direction of the information, according to the sign of v, as discussed earlier.

After some straightforward algebra, for $v > 0$, we obtain:

$$f\left(x_i + \frac{\Delta x}{2}, t+\tau\right) \simeq f(x_i, t) + \frac{\Delta x}{2}\left(\frac{\partial f}{\partial x}\right)_{x=x_i} - \tau v \left(\frac{\partial f}{\partial x}\right)_{x=x_i}$$

and:

$$f\left(x_i - \frac{\Delta x}{2}, t+\tau\right) \simeq f(x_{i-1}, t) + \frac{\Delta x}{2}\left(\frac{\partial f}{\partial x}\right)_{x=x_{i-1}} - \tau v \left(\frac{\partial f}{\partial x}\right)_{x=x_{i-1}}$$

In the same way, one can treat the case $v < 0$:

$$f\left(x_i + \frac{\Delta x}{2}, t+\tau\right) \simeq f(x_{i+1}, t) - \frac{\Delta x}{2}\left(\frac{\partial f}{\partial x}\right)_{x=x_{i+1}} - \tau v \left(\frac{\partial f}{\partial x}\right)_{x=x_{i+1}}$$

and

$$f\left(x_i - \frac{\Delta x}{2}, t+\tau\right) \simeq f(x_i, t) - \frac{\Delta x}{2}\left(\frac{\partial f}{\partial x}\right)_{x=x_i} - \tau v \left(\frac{\partial f}{\partial x}\right)_{x=x_i}$$

It is worth noting that the time derivatives of the distribution function, obtained by Taylor expanding the square bracket in Eq. (11), have been replaced by spatial derivatives, from Eq. (5):

$$\frac{\partial f}{\partial t} = -v\frac{\partial f}{\partial x} \tag{12}$$

If one defines $\sigma_v = sign(v)$ and $\alpha = (1-\sigma_v)/2$, it is possible to re-write the upwind Taylor expansions as follows:

$$f\left(x_i + \frac{\Delta x}{2}, t+\tau\right) \simeq \overline{f}_{i+\alpha} + \sigma_v\frac{\Delta x}{2}\left(\frac{\partial \overline{f}}{\partial x}\right)_{i+\alpha} - \tau v \left(\frac{\partial \overline{f}}{\partial x}\right)_{i+\alpha} \tag{13}$$

and

$$f\left(x_i - \frac{\Delta x}{2}, t+\tau\right) \simeq \overline{f}_{i-1+\alpha} + \sigma_v\frac{\Delta x}{2}\left(\frac{\partial \overline{f}}{\partial x}\right)_{i-1+\alpha} - \tau v \left(\frac{\partial \overline{f}}{\partial x}\right)_{i-1+\alpha} \tag{14}$$

In the previous formulae we used a first order Taylor expansion for the function f around the grid point x_i to obtain:

$$\overline{f}_i = f(x_i) + o(\Delta x^2) \tag{15}$$

Moreover, the following definition has been introduced:

$$\left(\frac{\partial \overline{f}}{\partial x}\right)_i = \frac{\overline{f}_{i+1} - \overline{f}_{i-1}}{2\Delta x} \tag{16}$$

Finally, using the expansions (13)-(14) in Eq. (11), and re-writing in a more compact form the physical space translation operator Λ_x, the evolution of f can be written as follows:

$$\overline{f}_i(t + \Delta t) = \sum_{j=-2}^{1} [\delta_{0,j+\alpha} + A_j(Q)]\overline{f}_{i+j+\alpha}(t) \tag{17}$$

where $Q = v\Delta t/\Delta x$ and:

$$A_{-2} = -\frac{Q}{4}(\sigma_v - Q), \qquad A_{-1} = Q + \frac{Q}{4}(\sigma_v - Q),$$
$$A_0 = -Q + \frac{Q}{4}(\sigma_v - Q), \qquad A_1 = -\frac{Q}{4}(\sigma_v - Q).$$

As already mentioned, the results obtained for the time evolution of the function f_x in the physical space can be applied to the velocity space for the evolution of f_v. In the next section, we will discuss a method extensively adopted to compose the solution for f_x and f_v in such a way to obtain the evolution of the function f in the (x, v) phase space.

Finally, a few comments about the numerical scheme we described above are in order. Equation (17) is known as *Van Leer's scheme* [9, 10, 11, 12] and it is in the form of a conservation law; in fact it is simple to verify that:

$$\sum_{j=-2}^{1} [\delta_{0,j+\alpha} + A_j(Q)] = 1 \tag{18}$$

The approach discussed in this section is similar to the "flux balance method" proposed by Fijalkow [13, 14, 15] and the resulting algorithm is second order accurate in space mesh Δx and time step Δt [6, 16].

It is easy to show that, using an upwind scheme in solving conservation law hyperbolic equations, the algorithm is stable if $\Delta t \leq \Delta x/|v|$, that is the well known Courant-Friedrichs-Lewy condition (CFL) [3].

2.2. Splitting Method

To complete the numerical integration of the Vlasov equation in one-dimensional phase space $(1D - 1V)$, we describe the so-called splitting scheme [17], that allows to obtain the solution for the time evolution of the distribution function f, as a composition of the solutions for the functions f_x and f_v, evaluated in the previous section. In order to understand the concept underlying the splitting method, one can imagine to move from a generic point a to a generic point b, in phase space, not going straight on, but following a $staircase$ path.

Therefore, the splitting method provides a solution of Eq. (4), split into (5) – (6), by applying recursively the translation operators Λ_x and Λ_v to the initial condition, in the following order:

$$f(n\Delta t) = [\Lambda_x(\Delta t/2)\Lambda_v(\Delta t)\Lambda_x(\Delta t/2)]^n f(0) =$$
$$\Lambda_x(\Delta t/2)\Lambda_v(\Delta t)[\Lambda_x(\Delta t)\Lambda_v(\Delta t)]^{n-1}\Lambda_x(\Delta t/2)f(0)$$

The previous scheme is correct at second order in time, as discussed in [17].

The generalization of the splitting algorithm to the case of multidimensional phase space is discussed in detail in [6] (Cartesian geometry) and in [16] (cylindrical geometry in velocity space).

3. The Fluid Point of View

The kinetic approach described above allows to study effects like wave-particle interaction, collisionless dissipation and so on. However, the main problem in such very detailed approach lies in the high number of spatial dimensions necessary to properly describe the problem. This means that the resolution along each direction has to be kept low, in order to afford the problem numerically, since otherwise the total number of spatial gridpoints would easily exceed the available computer memory.

In the fluid approach, instead, one takes averages of the distribution function $f(\vec{x}, \vec{v}, t)$ along the velocity directions in the phase space, thus resorting to a system of equations (the so called *fluid equations*) in which quantities change now only along the three (or less) spatial directions and time. However, such a description lacks the possibility to study the effects involving the direct interaction of the plasma particles with the electromagnetic fields. In spite of this limitation, the lower dimensionality of the system of equations allows to study several problems with much higher resolution, which is fundamental when dealing with, for example, turbulent systems where a wide range of spatial scales is involved.

From Eq. (1), one can obtain the macroscopic description of the system directly from the distribution function, by carrying out a simple integration in the velocity space (moments of f):

$$n(\vec{x}, t) = \int f(\vec{x}, \vec{v}, t) d^3\vec{v}$$

$$n\vec{v}(\vec{x}, t) = \int \vec{v} f(\vec{x}, \vec{v}, t) d^3\vec{v}$$

$$p(\vec{x}, t) = \frac{m}{3} \int v^2 f(\vec{x}, \vec{v}, t) d^3\vec{v}$$

representing the particle density, the momentum and the scalar pressure, respectively. The integration of the Vlasov equation leads to a hierarchy of infinite relations (BBJKY hierarchy)[18], each one describing the evolution of the k-th moment of the distribution function as a function of the $k+1$-th moment. This chain of infinite equations has to be closed at some value of k. Usually thermodynamics relations are used to close the system of equations at $k = 2$, thus leading to the *fluid equations* for the density n, the velocity field

$\vec{v}$ and the internal energy ϵ:

$$\frac{\partial n}{\partial t} + \nabla \cdot (n\vec{v}) = 0 \quad (19)$$

$$\frac{\partial \vec{v}}{\partial t} + (\vec{v} \cdot \nabla)\vec{v} = -\frac{1}{n}\nabla p + \frac{1}{n}\nabla \cdot \overline{\overline{\sigma}} + \frac{q}{m}(\vec{E} + \frac{1}{c}\vec{v} \times \vec{B}) \quad (20)$$

$$\frac{\partial \epsilon}{\partial t} + (\vec{v} \cdot \nabla)\epsilon = \frac{1}{n}\overline{\overline{\sigma}} \otimes \overline{\overline{U}} - k\nabla^2 T \quad (21)$$

which are in fact equivalent to the classical equations of fluid-mechanics, except for the presence of the electromagnetic fields $\vec{E}$ and $\vec{B}$ (c being the speed of light). Also in this case, $\overline{\overline{\sigma}}$ is the *stress tensor*:

$$\sigma_{ij} = \mu \left(\frac{\partial v_i}{\partial x_j} + \frac{\partial v_j}{\partial x_i} \right) + \left(\lambda - \frac{2}{3}\mu \right) (\nabla \cdot \vec{v})\delta_{ij}$$

μ being the dynamic viscosity of the fluid, λ the second viscosity coefficient and δ_{ij} the Kronecker delta, while $\overline{\overline{U}}$ is the *rate of strain tensor*:

$$U_{ij} = \frac{1}{2}\left(\frac{\partial v_j}{\partial x_i} + \frac{\partial v_i}{\partial x_j} \right)$$

Finally, k represents the coefficient of thermal conductivity of the plasma. This set of relations has to be supplemented with an equation of state which links the internal energy to the pressure and density (or temperature).

This system of equations has to be complemented with the Maxwell's relations describing the evolution of the electromagnetic fields $\vec{E}$ and $\vec{B}$:

$$\frac{\partial \vec{E}}{\partial t} = -4\pi\vec{J} + c(\nabla \times \vec{B}) \quad (22)$$

$$\frac{\partial \vec{B}}{\partial t} = -c(\nabla \times \vec{E}) \quad (23)$$

where $\vec{J}$ is the current density, in order to get a closed system of partial differential equations.

3.1. The Magnetohydrodynamics Approximation

The set of Eq.s (19-23), can be further simplified with the assumption that the phenomena under study involve quantities slowly varying with time, in particular the frequency of time depending quantities is much smaller than the cyclotron frequency of ions (low frequency approximation). This approximation is suitable to study a quite broad range of both astrophysical and laboratory plasma problems. In this case, the displacement current in Eq. (22) can be neglected and the electric field $\vec{E}$ computed directly from the velocity and magnetic field. Therefore, $\vec{E}$ does not appear any more in the equations as an unknown, which now are just the mass density, the velocity field, the magnetic field and the internal energy (or pressure). The set of equations in the Magnetohydrodynamics (MHD) approximation reads:

$$\frac{\partial \rho}{\partial t} + \nabla \cdot (\rho \vec{v}) = 0 \tag{24}$$

$$\frac{\partial \vec{v}}{\partial t} + (\vec{v} \cdot \nabla)\vec{v} = -\frac{1}{\rho}\nabla P + \frac{1}{\rho}(\nabla \times \vec{B}) \times \vec{B} + \mu \nabla^2 \vec{v} + \left(\mu + \frac{\lambda}{3}\right) \nabla(\nabla \cdot \vec{v}) \tag{25}$$

$$\frac{\partial \vec{B}}{\partial t} = \nabla \times (\vec{v} \times \vec{B}) + \eta \nabla^2 \vec{B} \tag{26}$$

$$\frac{\partial p}{\partial t} + (\vec{v} \cdot \nabla)p + \gamma p(\nabla \cdot \vec{v}) = k\nabla^2 T + (\gamma - 1)\left[2\mu U_{ij}U_{ij} + \left(\lambda - \frac{2}{3}\mu\right)(\nabla \cdot \vec{v})^2 + \eta \vec{J}^2\right] \tag{27}$$

With the further assumption that the typical velocity of the fluid is much smaller than the sound speed: $c_s = \sqrt{\frac{\partial P}{\partial \rho}}$, namely the *Mach number* of the fluid $M = |\vec{v}|/c_s$ is much smaller than one, the density of the fluid keeps constant at all times along the streamlines of the plasma, that is:

$$\frac{d\rho}{dt} = 0 \qquad \Rightarrow \qquad \nabla \cdot \vec{v} = 0 \tag{28}$$

namely the fluid velocity is to fulfill a solenoidality condition. In this case, the plasma pressure p turns out to be independent of the thermodynamic properties of the fluid but is "slaved" by the velocity field in such a way that $\vec{v}$ satisfies the incompressibility condition (28). Using this condition, by taking the divergence of Eq. (25), one gets:

$$\nabla^2 p = \nabla \cdot [(\vec{B} \cdot \nabla)\vec{B}/\rho - (\vec{v} \cdot \nabla)\vec{v}] \tag{29}$$

where $p = (P + B^2/2)/\rho$ is the total pressure per unit mass. Namely, when the pressure satisfies Eq. (29) the velocity field turns out to be solenoidal.

Hence, at low Mach numbers, Eqs. (24)-(27) can be substituted by the system:

$$\frac{\partial \vec{v}}{\partial t} + (\vec{v} \cdot \nabla)\vec{v} = \nabla p + \frac{1}{\rho}(\vec{B} \cdot \nabla)\vec{B} + \nu \nabla^2 \vec{v} \tag{30}$$

$$\nabla^2 p = \nabla \cdot [(\vec{B} \cdot \nabla)\vec{B}/\rho - (\vec{v} \cdot \nabla)\vec{v}] \tag{31}$$

$$\frac{\partial \vec{B}}{\partial t} = \nabla \times (\vec{v} \times \vec{B}) + \eta \nabla^2 \vec{B} \tag{32}$$

which represent the MHD equations in the incompressible case. Here $\nu = \mu/\rho$ is the kinematic viscosity of the plasma.

3.2. Numerical Solution of the MHD Equations

The ensemble of relations (24-27) and (30-32) have the form of a hyperbolic system containing parabolic terms as well, whenever the dissipative effects are not negligible. Furthermore, in the incompressible case, the solution of the Poisson equation for the pressure requires the solution of an elliptic equation.

Such equations are commonly used to model a variety of systems, ranging from the coronal loops excitation through the Alfvén waves coming from the photosphere [19, 20], the evolution of hydromagnetic waves which cross the solar wind current sheet [21, 22] or the magnetospheric current sheet, the magnetic reconnection phenomenon [23, 24, 25], and many others. A common feature of all such problems is the presence of a field inhomogeneity (usually in the magnetic field) crossed by MHD waves. This implies the presence of inhomogeneous directions (let us assume it as the x direction), along which generic Dirichlet or Neumann boundary conditions can be imposed, and periodic directions (let us suppose them as the y and z directions).

Moreover, all these problems are *nonideal*, namely they take place in presence of dissipative effects (viscosity, resistivity, etc.). However, like in ordinary fluid dynamics, the dissipative coefficients in ordinary plasmas are so small that their effects become relevant only at very small scales. This means that, in order to use realistic values of the dissipative terms, one should have very high resolutions to represent the whole range of scales up to those for which the dissipation is important. This, of course, is unpractical due to the huge amount of memory necessary to store the simulation variables and to the very long CPU time needed by such large resolutions. The usual solution to this problem is to set up the spatial resolution of the problems in terms of the available CPU and memory resources and choose the dissipative terms with a value that makes them act in the final part of the numerical spectrum. By the way, this is the only method (unless specific techniques for inviscid flows are used) to ensure the numerical stability of the code.

3.3. Advection Equations

To explain in detail the numerical method employed to solve this kind of equations, let us apply it to a model equation in the form:

$$\frac{\partial f}{\partial t} + \vec{F}(f) \cdot \nabla f = H(f) \tag{33}$$

where f is a generic variable, $\vec{F}$ is a vector operator representing the (nonlinear) advection term in the equations and H represents the other terms, including the dissipative ones. Note that $\vec{F}$ and H depend only on the spatial derivatives of f. Defining: $\mathcal{G}[f] = -\vec{F}(f) \cdot \nabla f + H(f)$, and introducing a time discretization: $t = t_n = n\Delta t$, standard explicit time stepping methods can be applied for approximating the time derivative in (33). For instance, for a third order Runge-Kutta scheme, we have:

$$\begin{aligned} f(t_{n+1/3}) &= f(t_n) + \frac{\Delta t}{3}\mathcal{G}[f(t_n)] \\ f(t_{n+2/3}) &= f(t_{n+1/3}) - \frac{2\Delta t}{3}\mathcal{G}[f(t_n)] + \Delta t\mathcal{G}[f(t_{n+1/3})] \\ f(t_{n+1}) &= f(t_{n+2/3}) + \frac{\Delta t}{3}\mathcal{G}[f(t_n)] - \frac{\Delta t}{2}\mathcal{G}[f(t_{n+1/3})] + \frac{\Delta t}{2}\mathcal{G}[f(t_{n+2/3})] \end{aligned}$$

The problem is now to approximate the spatial derivatives. We suppose to represent the three spatial directions through N_x, N_y and N_z discrete gridpoints, respectively. Thus, the numerical solution f of the equation is represented by the value of the function on the generic, discrete, spatial gridpoint (x_i, y_j, z_k), with $i = 1, \ldots, N_x$, $j = 1, \ldots, N_y$ and $k = 1, \ldots, N_z$. Under the assumption that the system is inhomogeneous along the x

direction and homogeneous along y and z, a generic quantity $f(x, y, z)$ can be decomposed through a discrete Fourier expansion along the periodic directions y and z:

$$f(x_i, y_j, z_k) = \sum_{m=-N_y/2}^{+N_y/2} \sum_{l=-N_z/2}^{+N_z/2} \hat{f}^{(ml)}(x_i) \exp[i(my_j + lz_k)] \tag{34}$$

Therefore, whenever the spatial derivatives along y or z are to be calculated, this can be made by transforming the original function into the Fourier space, thus getting the Fourier coefficients $\hat{f}^{(ml)}$ of the expansion (34). Hence, the value of the derivative can be easily calculated by multiplying those coefficients by im or il, according to the direction (y or z) where to compute the derivatives. Finally, the inverse Fourier transform gives back the value of the derivative in the physical space.

There are several advantages in using this technique:

1. the possibility to use Fast Fourier Transform (FFT) algorithms, which reduce the number of operations of the computation to $N \ln_2 N$, where N is the total number of gridpoints in the physical space. Normal Fourier Transform would need N^2 operations;

2. the precision of the computation is really very high. It is possible to show that, when transforming with N gridpoints, the k-th coefficient of the expansion decreases to zero faster than any integer power of k^2[26]. Therefore, even with a small number of points it is possible to get the approximation of a derivative at the machine precision;

3. we do not need to explicitly impose the periodic boundary conditions since, by definition, the Fourier Transform applies only to periodic functions and the derivatives are periodic functions as well.

Finally, we have to specify how to compute the derivative in the inhomogeneous (x) direction. Of course we cannot use the same method as above, since boundary conditions along an inhomogeneous direction cannot be periodic, by definition. However, it is possible in principle to find other basis functions which do not use any hypothesis about periodicity at the boundaries and to expand the original function on that basis. An example is the orthogonal basis of Chebyshev polynomials which allows the use of both Dirichlet or Neumann boundary conditions by fixing the values of the last coefficients of the expansion to fulfill the boundary conditions (the so-called *Tau method*[26, 27]). This method has found large application in past years, however imposing boundary conditions with Chebyshev polynomials is a fairly complicated task, especially when boundary conditions are changing with time. This is often the case in the compressible plasma/fluid framework, as we will explain below.

An alternative to spectral methods in the case of the inhomogeneous direction is the use of finite difference methods. In the general case, the computation of a derivative for a function $f(x)$ with a finite difference scheme, can be written as:

$$\left.\frac{df}{dx}\right|_{x=x_j} \simeq \sum_{i=-j_1}^{j_2} \xi_i f(x_{j+i}) \tag{35}$$

namely the derivative is expressed as a linear combination of the values of the function f in the points about x_j. The coefficients ξ_i depend on the resolution of the spatial discretization Δx and have to be determined by matching the Taylor expansions for the function $f(x)$ about the point x_j. The precision of such schemes depends on the number of points M considered in the linear combination (35): $M = j_1 + j_2 + 1$, the higher M, the more precise the scheme. Schemes where $j_1 = j_2$ are called *centered schemes* and are used generally for the assessment of the derivative on the internal points of the domain, whilst the others are generally used on the boundary of the domain.

The main advantages of this method are:

1. boundary conditions are fairly easy to implement, even when they are time dependent;

2. finite difference methods are fast: generally the computation of a derivative requires a number of operation proportional to the number of gridpoints through a coefficient which depends only on the precision of the computation (this is much faster than $N \ln_2 N$, for large N);

3. they can be easily parallelized, for running the code on massively parallel computer, with a speed-up coefficient very close to the theoretical (linear) one.

A disadvantage of this method is that finite difference evaluation of derivatives is not as precise as spectral methods, especially concerning the treatment of *phase errors*. Such errors show up when one computes the derivative for a function which is rapidly oscillating in the computational domain. When the gradients are smooth the computation of the derivative is precise whereas the precision decreases quickly when the function has steep gradients in the computational domain. This leads to unwanted dispersive effects in the simulation and, possibly, to unphysical behaviors. However, when using high order finite differences, the importance of those problems is drastically reduced.

Schemes which are very precise and with much reduced phase errors are obtained with the so called *compact finite difference* approximation [28]. This is an extension of the concept of finite difference schemes in which one writes linear combinations of both the derivatives and the function $f(x)$ in the points about x_j. In formulae:

$$\sum_{i=-l_1}^{l_2} \zeta_i f'(x_{j+i}) = \sum_{i=-j_1}^{j_2} \xi_i f(x_{j+i}) \tag{36}$$

where $f'(x)$ represents the derivative of $f(x)$.

Again, the coefficients ζ_i and ξ_i (the latter also bears the dependence on the spatial resolution Δx, as in standard finite difference formulae) are to be determined by matching the Taylor expansion of the function f in the points about x_j. Eq. (36) can be both *centered* for the use in the interior of the computational domain, or *noncentered*, for the use on the boundaries. When such relations are written in each point of the domain, one gets a system of algebraic equations in the form:

$$\overline{\overline{D_1}}\,[\bar{f}'] = \overline{\overline{C_1}}\,[\bar{f}] \tag{37}$$

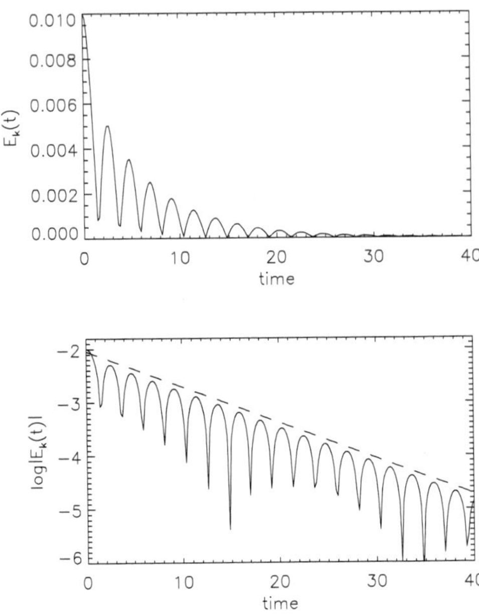

Figure 1. Time evolution of the m=1 spectral component of the electric field of a Langmuir wave (at the top). Comparison between numerical results and theoretical prediction for the linear Landau damping rate (at the bottom).

which represents the row-column product of a known square matrix $\overline{\overline{D}}_1$, built with the coefficients ζ_i, times the vector of the unknowns $\bar{f}'$, representing the values of the derivative of the function f, which are to be computed, in terms of the row-column product of the two known terms $\overline{\overline{C}}_1$ and $\bar{f}$, which are the matrix of the coefficients ξ_i and the vector representing the values of the function f in each point of the domain, respectively.

The method is powerful because, when using the same scheme for all the points (except for the boundaries, where noncentered schemes are to be used), the matrices appearing in (37) are band matrices and therefore the system can be inverted by using standard fast techniques (for instance, the LU decomposition) which require a number of operations proportional to N, the number of gridpoints, as for ordinary finite difference schemes.

Other than being very precise, compact schemes have the advantage that one can choose the values of the coefficients to fulfill specific conditions. As an example, one can match through the Taylor expansion only a limited set of the free parameters ζ_i and ξ_i and determine the others by imposing the minimization of the phase errors. Namely one can build up schemes which are highly precise but are also able to deal efficiently with phase errors. See [28] for an excellent review on the problem.

3.4. Elliptic Equations

In the incompressible case, other than the hyperbolic-parabolic equations, one has to deal also with the elliptic equation (31), coming from the need to impose the solenoidality of the velocity field. Also in this case one can exploit the periodicity in the y and z directions

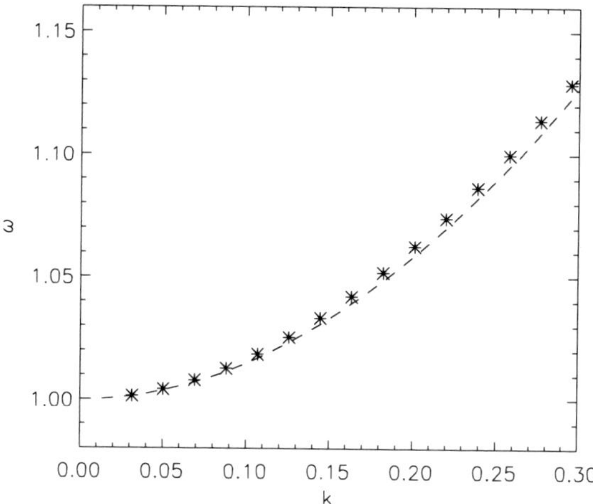

Figure 2. Bohm-Gross dispersion relation.

to rewrite the elliptic equation as a corresponding set of differential equations for each wavenumber couple (m, l) of the Fourier expansion. That is, if one is to solve the elliptic equation for the unknown $g(x, y, z)$ on the discrete set of points (x_i, y_j, z_k):

$$\nabla^2 g(x_i, y_j, z_k) = G(x_i, y_j, z_k)$$

where G is a known function, one can develop both g and G on the Fourier basis and substitute the expansion in the equation, that yields:

$$\sum_{m=-N_y/2}^{+N_y/2} \sum_{l=-N_z/2}^{+N_z/2} \left[\frac{d^2 \hat{g}^{(ml)}}{dx^2} - (m^2 + l^2)\hat{g}^{(ml)} \right] \exp\left[i(my_j + lz_k)\right] =$$

$$\sum_{m=-N_y/2}^{+N_y/2} \sum_{l=-N_z/2}^{+N_z/2} \hat{G}^{(ml)} \exp\left[i(my_j + lz_k)\right]$$

Thus, for each couple of Fourier harmonics (m, l) one has to solve the set of differential equations:

$$\frac{d^2 \hat{g}^{(ml)}}{dx^2} - (m^2 + l^2)\hat{g}^{(ml)} = \hat{G}^{(ml)} \tag{38}$$

This equation can be easily solved with standard finite difference techniques. Indeed the x second derivative operator at the point x_i can be written as a linear combination of the unknown function $\hat{g}^{(ml)}$ at the neighbor points, thus obtaining an algebraic banded system of equations solvable with standard techniques, as discussed above. Of course, all limitations concerning the precision and poor treatment of phase errors discussed above apply in this case too.

Also in this case, the use of compact differences helps in solving those problems without increasing too much the complexity of the computation. The starting point for the solution

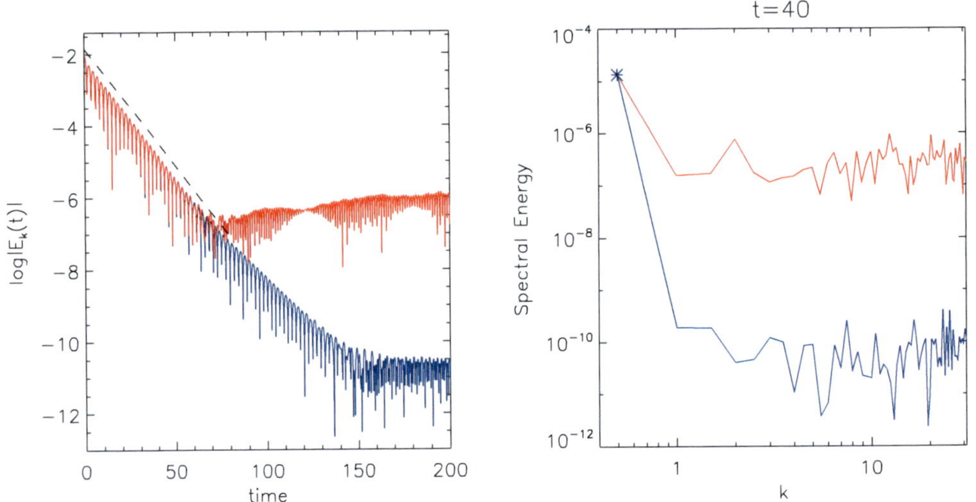

Figure 3. Left panel: linear Landau damping phenomenon numerically analyzed using an Eulerian low noise Vlasov code (blue line) and a PIC code (red line); the dashed line represents the theoretical prediction for the damping rate. Right panel: energy spectrum for the Eulerian code (blue line) and for the PIC code (red line) at $t = 40$; the blue dot in the right panel represents the energy of the mode $m = 1$, the one perturbed at $t = 0$.

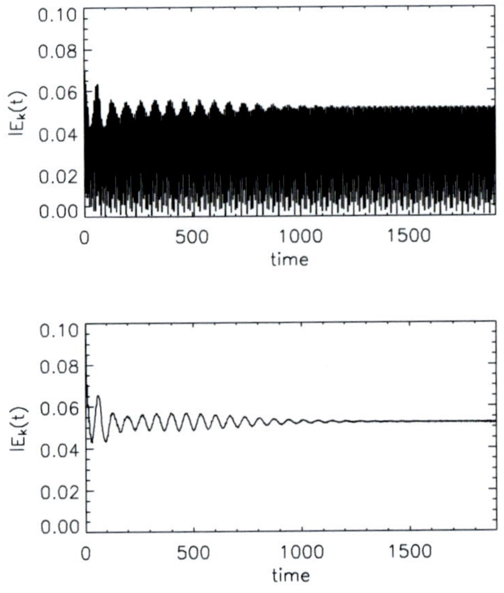

Figure 4. Time evolution of the m=1 spectral component of the electric field of a Langmuir wave (at the top), and the electric field amplitude (at the bottom). Nonlinear effects turn off Landau damping.

of equation (38) is the matrix form for the compact finite difference scheme analogous to (37) but for the second derivatives:

$$\overline{\overline{D}}_2[\bar{g}''^{(ml)}] = \overline{\overline{C}}_2[\bar{g}^{(ml)}] \tag{39}$$

where the quantities $[\bar{g}''^{(ml)}]$ and $[\bar{g}^{(ml)}]$ represent the column vector of the second derivatives of g and the function itself, respectively, on the discrete points x_i. Eq. (38), by bringing the homogeneous terms to the right hand side, can be rewritten in the form of a column vector equation:

$$[\bar{g}''^{(ml)}] = [\bar{G}^{(ml)}] + (m^2 + l^2)[\bar{g}^{(ml)}]$$

Hence, by putting this relation in (39), one gets the matrix system of equations:

$$(\overline{\overline{C_2}} - (m^2 + l^2)\overline{\overline{D_2}})[\bar{g}^{(ml)}] = \overline{\overline{D_2}}[\bar{G}^{(ml)}] \tag{40}$$

that is, the problem can be solved by inverting a new banded system, obtained from the linear combination of the matrices $\overline{\overline{C_2}}$ and $\overline{\overline{D_2}}$ which are banded matrices too. This system, can be solved with the use of standard LU decomposition techniques, for instance.

3.5. Boundary Conditions for the MHD Description

The problem of boundary conditions for the inhomogeneous direction in plasmas, as in ordinary fluids, is fairly complicated by the fact that conditions strictly depend on the problem under study. We just summarize the main issues when dealing with boundary conditions in the MHD description.

In the incompressible case, appropriate boundary conditions are found generally without problems. In this case, values at the boundaries can be fixed for the velocity and the magnetic fields (Dirichlet boundary conditions, DBC) or their x derivative, namely the derivative along the inhomogeneous direction (Neumann boundary conditions, NBC), respectively. Suitable conditions for the pressure can be imposed consequently. In Ref. [29], the authors pointed out that the choice of the boundary conditions must satisfy some compatibility relations, that is the boundary conditions for the pressure is determined by the physical boundary conditions imposed on the other variables.

With standard finite difference techniques, the value of the functions on the boundaries can be directly imposed at each time step and even during the calculation of the derivatives, in order to improve precision. The latter is less straightforward with compact finite difference schemes. When computing, for instance, the first derivative of a function on which DBC are imposed, one can directly use the value of the function in (37). For NBC, instead, one already knows the value of the derivative on the boundary, then the system of equations (37) can still be solved by simply eliminating the first and last equations, thus integrating a $(N-2) \times (N-2)$ system of equations. The method gets a little more complicated for second derivatives. However, also for second derivatives, when NBC are to be employed, one can include in the compact difference formula (36), evaluated on the boundaries, a term involving the first derivative in the form:

$$\sum_{i=-l_1}^{l_2} \zeta_i f''(x_{j+i}) = \sum_{i=-j_1}^{j_2} \xi_i f(x_{j+i}) + \gamma f'(x_j) \tag{41}$$

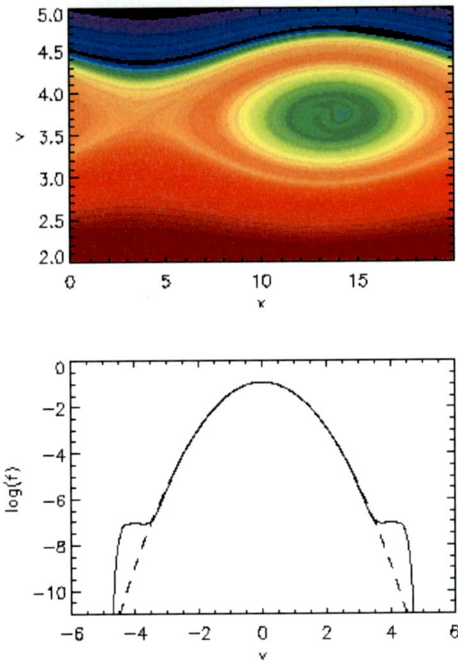

Figure 5. At the top: contour plot of the electron distribution function in phase space at $t = 1900$. At the bottom: Semilogarithmic plot of the electron distribution (averaged on x) in velocity space at $t = 1900$ (the dashed line represents a Maxwellian function).

As usual, all ζ_i, ξ_i and γ have to be computed by imposing matching conditions on the Taylor expansions of $f(x_{i+j})$. Since the value of $f'(x_j)$ is known on the boundary, this formula can be used to get the second derivative more precisely.

Incidentally, this is the only way to solve the elliptic equation for the pressure with NBC. Indeed, in this case, one must include the terms for the first derivative explicitly in (40), otherwise the matrix on the first hand side of the equation would be singular and produce numerically unstable results. See section 5 for an example of using such technique in the incompressible case.

In the compressible case, the numerical techniques are analogous to what we said till now, for general DBC or NBC. However, in this case, the boundary conditions can be imposed through a characteristic decomposition method, to overcome the difficulties of imposing standard DBC or NBC on the density, pressure and temperature. Moreover, this technique also allows the waves present in the domain to get out from the inhomogeneous boundaries [30]. The topic is too bulky to deal with here (see, for instance, [31] for a very clear and detailed review on this topic for the fluid mechanics equations) but, in principle, one has to treat boundary conditions which depend on the instantaneous values of all the quantities on the boundary. Therefore, the benefits of using finite difference schemes, which allow to impose time depending boundary conditions, are immediately understandable.

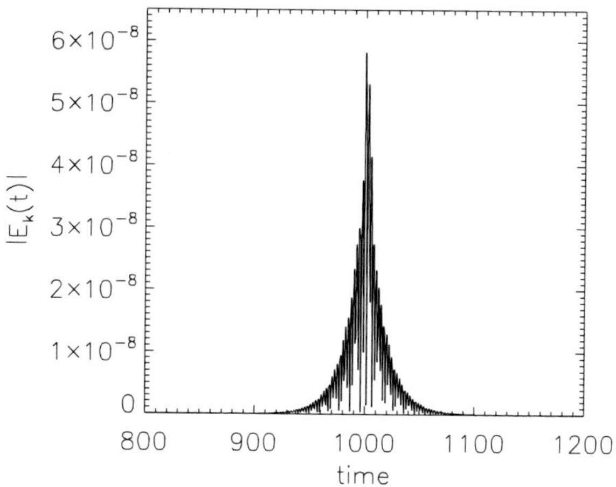

Figure 6. Time evolution of the m=1 spectral component of the electric field. The plasma wave echo.

4. Kinetic Simulations

4.1. Vlasov-Poisson Code

As an example of application of the algorithm described in sections 2.1.-2.2. for the integration of the Vlasov equation, we discuss some linear and nonlinear results, concerning the propagation of electrostatic waves in collisionless unmagnetized plasmas. We describe a numerical code that couples the Vlasov equation for the electron distribution f with the Poisson equation for the electric field:

$$\frac{\partial f}{\partial t} + v\frac{\partial f}{\partial x} - E\frac{\partial f}{\partial v} = 0 \tag{42}$$

$$\frac{\partial E}{\partial x} = 1 - \int_{-\infty}^{+\infty} f dv \tag{43}$$

In the previous equations, the ions are considered as a motionless background of neutralizing positive charge with density $n_0 = 1$ and E is the self-consistent longitudinal electric field.

In physical space we impose periodic boundary conditions, so the Poisson equation (43) is solved by using a standard FFT routine. In the Vlasov-Poisson system (42) - (43), time is normalized to the inverse of the electron plasma frequency ω_{pe} and velocity to the thermal velocity v_{th}; consequently, E is normalized to $m\omega_{pe}v_{th}/e$ (where e and m are the electron charge and mass, respectively) and lengths to the Debye length λ_D. Finally, the distribution function f is normalized to the equilibrium particle density n_0. For convenience, in this and next sections, all the quantities will be normalized with the characteristic parameters listed above.

The simulation domain in phase space is given by $D = [0, L_x] \times [-v_{max}, v_{max}]$. Outside the velocity simulation interval the distribution function is put equal to zero. Typically

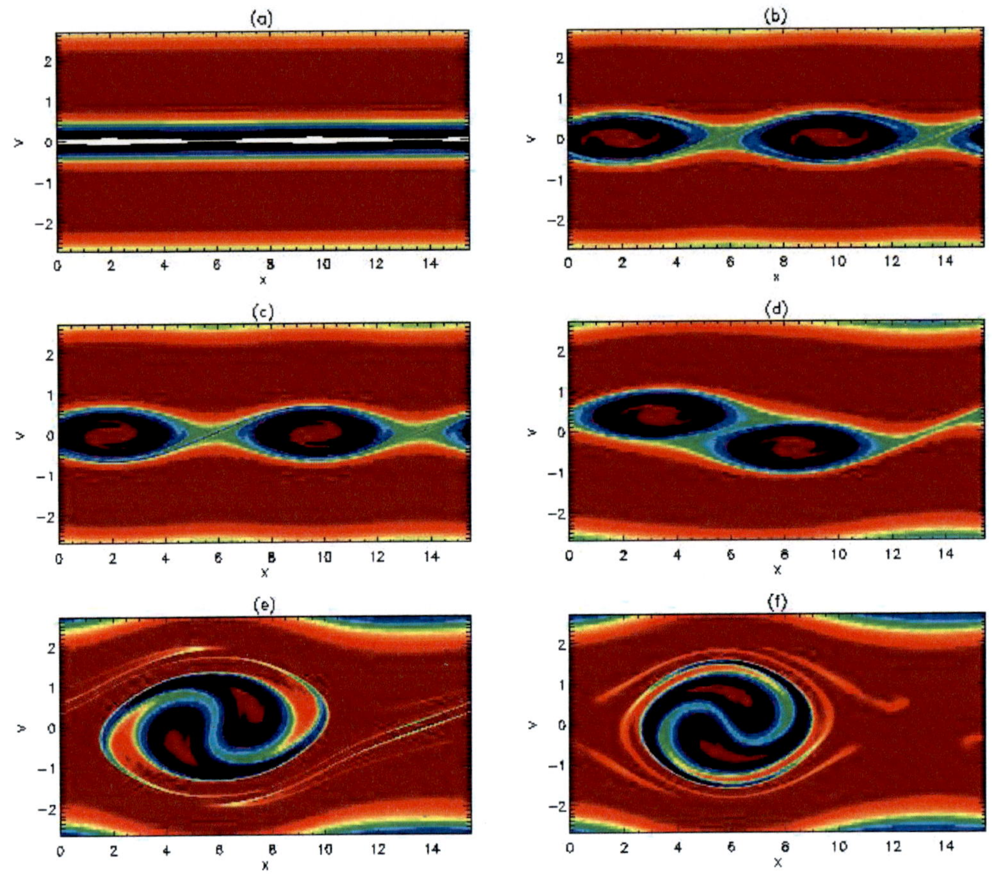

Figure 7. Coalescence and merging of two vortex structures in a two-stream plasma.

a simulation is performed using $N_x = 512$ grid points in physical space and $N_v = 2500$ grid points in velocity space. The time step $\Delta t \simeq 0.005 - 0.001$ has been chosen in such a way that the CFL condition is satisfied. An energy conservation equation has been used to control the numerical accuracy. The total energy variations remain always 10^{-2} times smaller with respect to the typical electric and kinetic energy fluctuations, all along the simulation.

4.2. Linear and Nonlinear Landau Damping

In 1946, in a seminal paper, Landau [32] showed that small amplitude electrostatic perturbations propagating in unmagnetized plasmas are damped exponentially in time, even in absence of collisions. The dissipation effect is directly related to the resonant interaction between the wave and those particles whose velocities are near the wave phase velocity v_ϕ. In the framework of linear Landau theory, under the assumption of large wavelengths (or equivalently small wavenumbers, $k^2 \ll 1$), these waves (also called Langmuir waves) propagate with frequency $\omega = \sqrt{1 + 3k^2}$ (Bohm-Gross dispersion relation) and are damped with damping rate $\gamma = -\sqrt{\pi/2}(v_\phi^2/k)e^{-v_\phi^2/2}$.

Almost twenty years later, O'Neil [33] analyzed the effect of nonlinearities on the propagation of electrostatic waves. He found that nonlinear terms, neglected by Landau in the linear theory, start playing a determinant role after a time $\tau_{trap} = 2\pi\sqrt{m/(eEk)}$. This time τ_{trap} is called *trapping time* and represents the characteristic time of oscillation of a charged particle in the trough of the wave. Nonlinear effects effectively create a population of particles trapped in the potential well; the subsequent energy exchange between the wave and these trapped particles turns off Landau damping. The wave is killed by Landau damping before the trapped particle distribution can form unless $\gamma_L \tau_{trap} < 1$; consequently, the initial amplitude of the electric perturbation must be large enough, so that the trapping time be short compared to the linear time for dissipation. In these conditions, the energy dissipation is stopped and the electric field goes on oscillating around an almost constant saturation amplitude.

The nonlinear wave-particle interaction has been subject of several analytical, numerical and experimental studies [34, 35, 36, 37, 38, 39, 40, 41, 42, 43] and is still matter of fervent discussion.

Here, we use the Vlasov-Poisson code described above to analyze numerically the propagation of linear and nonlinear Langmuir waves, and compare the numerical results with the analytical solutions provided by the kinetic theory. In our simulations, we assume for the initial distribution function a Maxwellian in velocity space, over which a modulation in physical space is superposed

$$f(x, v, t = 0) = \frac{1}{\sqrt{2\pi}} \exp\left(-v^2/2\right)[1 + A\cos\left(mkx\right)] \tag{44}$$

For the first linear simulation, the perturbation amplitude is $A = 0.01$, and $k = 2\pi/L_x$ is the wave number (we perturb the mode $m = 1$, i.e. the one with the largest wavelength that fits in the simulation domain). The length of the numerical box in physical space is $L_x = 4\pi \Rightarrow k = 0.5$, while $v_{max} = 6$.

In Fig. 1, we show the time evolution of the electric field spectral component $m = 1$, up to a time $t = 40$. At the top in the figure, the exponential decay of the electric amplitude is clearly visible, while at the bottom, the logarithmic plot shows a very good agreement between numerical results and theoretical prediction (dashed line) for the wave damping. In order to numerically reproduce the linear Bohm-Gross dispersion relation, we analyzed the time evolution of sinusoidal initial perturbations with different wavenumbers k ($k_{min} \simeq 0.05$ and $k_{max} \simeq 0.3$). The results of the simulations are reported in Fig. 2, where the dashed line represents the theoretical Bohm-Gross dispersion relation, while the dots are the numerical results. As it is easily seen from the figure, for small wavenumbers the agreement is very satisfactory. For larger wavenumbers the Bohm-Gross theory is no longer valid, since it is obtained in the approximation of large wavelengths.

To complete the discussion on the linear evolution of electrostatic waves, we shortly discuss a comparison between the numerical solution provided by an Eulerian low noise Vlasov code and a PIC code, in the problem of linear Landau damping. The PIC code considered here is a one-dimensional electrostatic code that follows the dynamics of a very large number of simulation particles ($N_p = 2 \times 10^7$). The initial condition is the same as that used in the case discussed in Fig. 1, that is the mode $m = 1$ is perturbed at $t = 0$. The system dynamics is described numerically up to a time $t = 200$. In the right panel of Fig. 3, the time evolution of the electric field spectral component $m = 1$ obtained through the Eulerian code is shown by the blue line, while the solution obtained through the PIC code is shown by the red line; the dashed line in the figure represents the theoretical prediction of the linear Landau damping rate. As it is evident from the figure, around $t = 60$ the statistical noise affects the solution obtained by the PIC code, while the Eulerian code keeps working up to $t = 150$, with a very low level of noise. The right panel in Fig. 3 shows the logarithmic plot of the spectral energy at $t = 40$ for the first 64 modes in the simulation, for the Eulerian code (blue line) and for the PIC code (red line). It is worth noting that both codes provide the correct damping rate for the mode $m = 1$ up to $t \simeq 60$. However, as it is clear from the figure, at $t = 40$ the numerical noise introduced by the PIC algorithm makes the mode $m = 2, \cdots, 64$ have a significant amount of energy, while the level of noise for the Eulerian code is sensibly lower. This comparison allows to point out that the Eulerian-like algorithms are more efficient than standard PIC algorithms in the study of low amplitude phenomena, or in the analysis of the time evolution of single mode perturbations.

In order to investigate the nonlinear evolution of the system, we increase the value of the length of the numerical box up to $L_x = 20$, so that the wavenumber gets smaller ($k = \pi/10$) and the linear damping is consequently weaker. We also increase the amplitude of the perturbation up to $A \simeq 0.063$ and run a long time simulation up to $t = 1900$. In Fig. 4, the time evolution of the mode $m = 1$ is shown. After the preliminary decaying stage, the effect of the Landau damping is stopped by nonlinearities and the electric field reaches a saturation amplitude. This effect is clearly visible in the bottom plot, where the oscillation of the electric field envelope is displayed (the high frequency oscillations have been removed).

The vortex-like structure in the phase space contour plot of the electron distribution function in Fig. 5 (top plot) represents trapped particles, and as expected these particles have a mean velocity $v \simeq 3.7$, close to the phase velocity of the wave. The velocity width of the trapped particle region is about $\Delta v_{trap} \simeq 1.6$. The theoretical expectation for the width of the trapped region is easily obtained by considering the trapping condition in the wave potential well, $1/2(v - v_\phi)^2 = E/k$. This gives a trapping width $\Delta v_{trap} = 2\sqrt{2E/k}$. Using the saturation amplitude of the electric field $E^{\rm sat} = 2E_k^{\rm sat} \simeq 0.1$ (see top graph in Fig. 4) in evaluating the theoretical trapping width, gives the value $\Delta v_{trap} = 1.595$, that is clearly in agreement with our numerical result. At the bottom in Fig. 5, the semilogarithmic plot of f, averaged on x, shows that the particle trapping effectively flattens the distribution in the region around the wave phase velocity, thus turning off Landau damping.

4.3. Plasma Waves Echoes

As discussed above, the energy of plasma waves can be dissipated in time even in absence of collisions. When a plasma wave undergoes collisionless Landau damping, the macroscopic quantities, such as electric field or charge density, are exponentially damped, according to the analytical results by Landau [32]. However, in absence of collisions, the memory of the initial perturbation cannot be lost, but it remains locked in the distribution function. For example, when an electric field of spatial dependence $\exp[-ik_1x]$ is excited in a plasma and then is damped away, it perturbs the distribution function with a modulation in the form of a

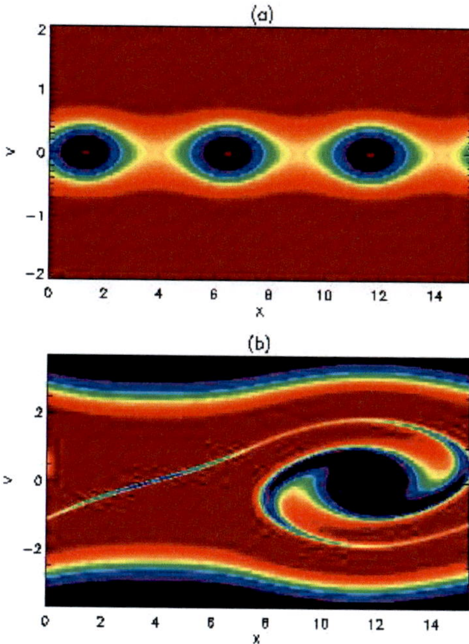

Figure 8. Coalescence and merging of three vortex structures in a two-stream plasma.

ballistic term $\exp\left[-ik_1x + ik_1vt\right]$. For large time, there is no electric field associated with this disturbance, since the integral over velocity of it will phase mix to zero. If a second wave of spatial dependence $\exp\left[-ik_1x\right]$ is lunched after a time τ, the new perturbation can interfere constructively with the old perturbation stored in the distribution function. As a result, a third wave (called echo) with wave number $k_{echo} = k_2 - k_1$ will spontaneously appear in the plasma at time $t_{echo} = \tau(k_2/k_{echo})$. This phenomenon has been first discussed analytically by Gould *et al.* in 1967 [44] and then proved experimentally in 1968 [45].

Here, we reproduce the physical phenomenology of the plasma waves echoes, using the Vlasov-Poisson code presented above. For this simulation, the initial electron distribution function has Maxwellian form. The length of the physical domain is $L_x \simeq 15.7079$ and the fundamental wavenumber is $k_0 = 2\pi/L_x \simeq 0.4$. We consider electric perturbations in the form of sinusoidal waves $\delta E_m = \epsilon\cos(k_m x)$, where $\epsilon = 0.001$ and $k_m = mk_0$. At $t = 0$ we excite the mode with the largest wavelength that fits in the simulation box $\delta E_1 = \epsilon\cos(k_1 x)$ (i. e. $m = 1$). A second wave $\delta E_2 = \epsilon\cos(k_2 x)$ is excited at $\tau = 500$ with wavenumber $k_2 = 2k_1 = 0.8$. According to the analytical results by Gould *et al.*, the plasma echo will appear at $t_{echo} = \tau(k_2/k_3) = 1000$ with wavenumber $k_{echo} = k_2 - k_1 = k_1 = 0.4$. Figure 6 shows the time evolution of the electric field spectral component $m = 1$ for $800 < t < 1200$. As predicted within the theory by Gould *et al.*, at $t = 1000$ a plasma wave echo is clearly visible with wavenumber $k_{echo} = 0.4$.

4.4. Phase Space Vortex Coalescence

The stability of vortex-like trapped particle structures has been the subject of several numerical and analytical studies, for example in the framework of the two-stream instability. In 1970, Berk *et al.* [46], using the so-called *water bag* model (see [46] and references therein) and extending the *duality principle* introduced by Dory in 1964 (see [46] and references therein), discussed the strong similarity between the Jeans instability of a one-dimensional gravitational system and the two-stream instability. Through this duality principle, they argued that the large-scale hole structures that develop in the nonlinear two-stream problem behave as gravitational bodies which attract one another and merge through a Debye-shielded Coulomb force.

More recently, in 1988 Ghizzo *et al.* [47] analyzed Bernstein-Green-Kruskal [48] (BGK) equilibria for a Vlasov plasma consisting of periodic structures exhibiting holes in phase space. A BGK mode is a nonlinear periodic stationary solution of the Vlasov-Poisson equations, characterized by the presence of trapped particle populations. Through a marginal stability analysis, the authors showed that these structures are unstable when more than one hole is present in the system. Long time numerical experiments on two-stream systems show coalescence of the vortices and stability of one-hole structures.

In 2000, Manfredi and Bertrand [49] contributed to the discussion on the vortex coalescence, analyzing both analytically and numerically the stability of the BGK modes, in the limit of small electric potential (weak inhomogeneity). Their results support the previous numerical evidences about the two-stream system; moreover, they found that one-hole BGK modes (generally thought to be stable) are marginally stable, i.e., they have a non vanishing growth rate, when velocity distributions with at least three streams are considered and some conditions are satisfied. These authors derived a simple stability criterion, which is in agreement with their numerical simulations.

In 2006 Valentini *et al.* [50] showed that the vortical structures developed in the propagation of the so-called Electron Acoustic waves, which are nonlinear electrostatic oscillations characterized by the presence of a region of trapped particles, exhibit merging phenomena, whose peculiarities are strictly related to those described earlier for the two-stream systems.

From the discussion above, it is evident that vortex merging is a common tendency of nonlinear trapped populations in plasma physics. Moreover, it is crucial to notice that the one-dimensional Vlasov plasmas are mathematically analogous to incompressible two-dimensional hydrodynamic fluids. In particular, the analogous of the Vlasov equation in fluid dynamics is the evolution equation for the vorticity $\Omega = \partial v_y/\partial x - \partial v_x/\partial y$:

$$\frac{d\Omega}{dt} = \frac{\partial\Omega}{\partial t} + v_x\frac{\partial\Omega}{\partial x} + v_y\frac{\partial\Omega}{\partial y} = 0 \tag{45}$$

The spatial coordinates in the previous equation are canonically conjugate to one another since they satisfy the equations:

$$v_x = \frac{\partial\psi}{\partial y} \quad ; \quad v_y = -\frac{\partial\psi}{\partial x} \tag{46}$$

where the Hamiltonian ψ represents the velocity stream function. The stream function is determined by the two-dimensional Poisson equation:

$$\frac{\partial^2\psi}{\partial x^2} + \frac{\partial^2\psi}{\partial y^2} = -\Omega \tag{47}$$

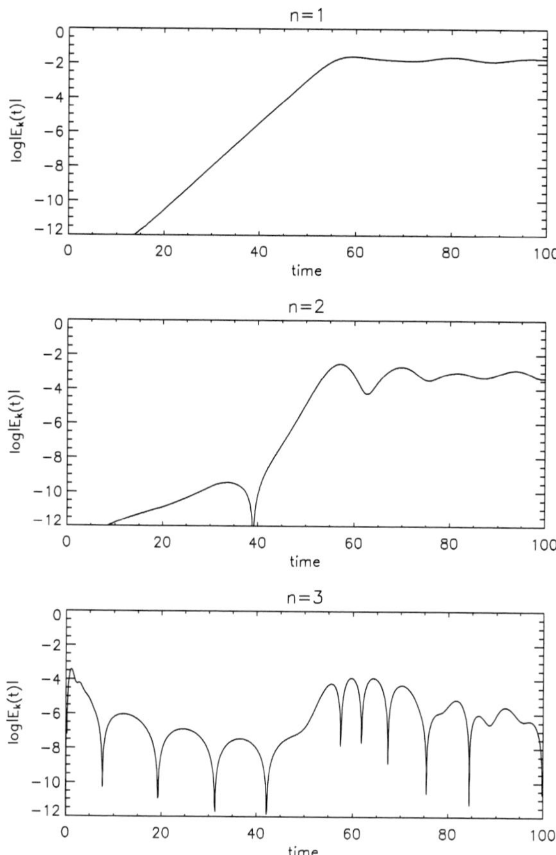

Figure 9. Time evolution of the electric field spectral components $m = 1, 2, 3$, in a two-stream plasma.

This strong mathematical link allows to apply the analysis carried on phase space fluids (Vlasov plasmas) to ordinary fluid problems where vortical structures dominate the system dynamics, such as the Kelvin- Helmotz instability.

In this section, we discuss some examples of application of the one-dimensional Vlasov Poisson code in the study of two-stream systems, in particular in the analysis of stability properties of phase space vortices. The initial condition for our simulations is a two-stream function in velocity space, perturbed in the physical space with a small amplitude modulation:

$$f(x, v) = \frac{2 - 2\xi}{3 - 2\xi}\left[1 + \frac{v^2/2 + \phi(x)}{1 - \xi}\right] \times \exp\left[-v^2/2 + \phi(x)\right]$$

The electrostatic potential perturbation is chosen to be $\phi(x) = -[A/(mk)^2] \sin{(mkx)}$.

As a first example, we study the coalescence of two phase space vortices. For this simulation, we choose the parameters $\xi = 0.99$, $L_x = 15.5 \Rightarrow k \simeq 0.405$, $A = 0.001$, $m = 2$. The time dynamics of the system is followed up to $t = 300$. The phase space contour plot of the electron distribution function is displayed in Fig. 7. At $t = 0$ the mode $m = 2$ is perturbed with a small amplitude modulation. As time goes on, the $m = 2$ mode gains energy and two vortical structures appear in phase space. This configuration remains stable for a short time; in fact, panels (d), (e), (f) in the figure show that at $t \simeq 100$ the two vortices merge and form a single large structure in phase space.

We observe the same behavior if the mode $m = 3$ is perturbed at $t = 0$ (the parameters of the simulation are those listed above). The coalescence of the three initial vortical structures is displayed in Fig. 8 in phase space. In Fig. 9, we show the time evolution of the electric field spectral components $m = 1, 2, 3$. The mode $m = 3$ excited at $t = 0$ decays and the modes $m = 1$ and $m = 2$ gain energy and grow in amplitude. At the end of the simulation, the mode $m = 1$ is dominant and remains stable in time.

The numerical evidences suggest that the mode initially perturbed remains stable, provided the wave has the largest wavelength that fits in the simulation domain, otherwise a decay instability transfers energy to the largest wavelength. In phase space, this decay to longer wavelength appears as a tendency of the vortex-like trapped particle populations to merge. The results discussed above are in agreement with those described in [47, 49].

5. Magnetohydrodynamics Simulations

We applied pseudospectral methods and compact differences to study tearing instabilities in a three-dimensional configuration. We present a comparison of the growth rates in the linear phase of the instability simulated by the numerical code with the predictions of the linear theory. We use the numerical code to describe the nonlinear evolution of the instabilities, with formation and coalescence of magnetic islands and the development of an anisotropic energy spectrum [24]. The presence of resistivity in a plasma allows the magnetic field lines to diffuse through the plasma. In this case the magnetic field lines can break and reconnect. Originally opened field lines can change their topology and form closed magnetic islands. This is not possible in the ideal limit, where the Alfvén theorem predicts that the magnetic field lines are frozen in the plasma. Magnetic reconnection takes place on resonant surfaces defined by:

$$\vec{k} \cdot \vec{B}_0 = 0 \tag{48}$$

where $\vec{B}_0$ is the equilibrium magnetic field and $\vec{k}$ is the wave vector of the perturbation.

The nonlinear evolution of resistive instabilities occurring in a current sheet can produce the development of turbulence through energy transfer from large scales to small scales, but can also be responsible of production of large scale structures by coalescence of magnetic islands. We study a configuration where a guide field is present and 3D perturbations are initially excited. The presence of the B_{0y} component has an important effect on the development of the tearing instability, namely, the resonance condition $\vec{k} \cdot \vec{B}_0(x) = 0$ is satisfied at different values of x for different wave vectors $\vec{k}$, thus unstable modes have resonant surfaces at different locations. So nonlinear interactions occur not only between modes on a single resonant surface, but also between adjoining resonant surfaces. The aspect ratios have been chosen so that energy transfer to both lower and higher wavenumbers is possible. The final state is characterized by the longest wavelength produced by coalescence in the center of the current sheet and by small scale structures on the sides. The fast 3D instability found by Dahlburg *et al.* [25, 51] is not observed in this configuration, and the time evolution of the instability displays a quasi 2D behavior, at least at large scales.

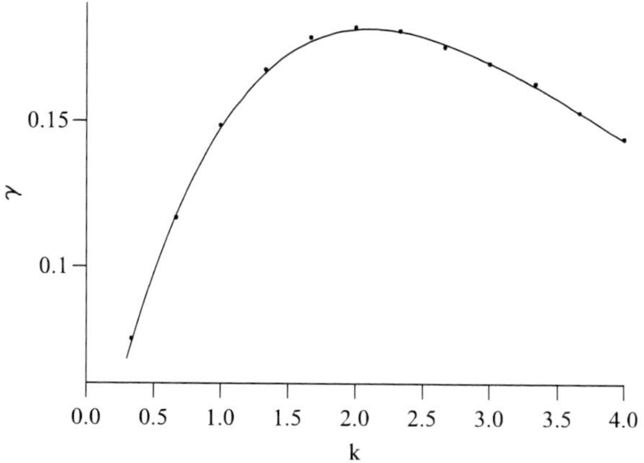

Figure 10. Growth rate for two-dimensional modes ($k_y = 0$) evaluated from the linear code (solid line) and the simulation (dots).

We solve numerically the incompressible, dissipative, magnetohydrodynamics (MHD) equations in dimensionless form:

$$\frac{\partial \vec{v}}{\partial t} + (\vec{v} \cdot \nabla)\vec{v} = -\nabla\left(P + \frac{B^2}{2}\right) + (\vec{B} \cdot \nabla)\vec{B} + \frac{1}{R_v}\nabla^2\vec{v} \tag{49}$$

$$\frac{\partial \vec{B}}{\partial t} = \nabla \times \left(\vec{v} \times \vec{B} + \frac{1}{R_M}\nabla \times \vec{B}\right) \tag{50}$$

$$\nabla \cdot \vec{v} = 0 \tag{51}$$

$$\nabla \cdot \vec{B} = 0 \tag{52}$$

where $\vec{v}$ and $\vec{B}$ are the velocity and magnetic field, respectively, P is the pressure and R_v and R_M are the kinetic and magnetic Reynolds numbers. Taking the divergence of Eq. (49) and using the condition (51) we obtain the equation for the total pressure:

$$\nabla^2 p = \nabla \cdot [(\vec{B} \cdot \nabla)\vec{B} - (\vec{v} \cdot \nabla)\vec{v}] \tag{53}$$

where $p = P + B^2/2$ is the total pressure.

We solve such equations in a three-dimensional Cartesian domain of sides: $-l_x \leq x \leq l_x$, $0 \leq y \leq 2\pi l_y$, $0 \leq z \leq 2\pi l_z$. We choose l_x as the unit length scale to nondimensionalize the equations. In such a case, the dimensions of the numerical domain become: $D = [-1, +1] \times [0, 2\pi R_y] \times [0, 2\pi R_z]$, where $R_y = l_y/l_x$ and $R_z = l_z/l_x$ are the aspect ratios in the y and z directions, respectively.

We choose as unit measure for the magnetic field a typical value B_0, which allows us to define a characteristic value of the Alfvén velocity $v_A = B_0/\sqrt{4\pi\rho}$. The quantity ρ is the mass density, uniform everywhere in the incompressible case. Hence, we express the velocity in terms of v_A and the time in terms of the typical Alfvén time: $\tau_A = l_x/v_A$. Finally, the pressure P is measured in units of ρv_A^2, while the definition of the Reynolds numbers, in terms of the kinematic viscosity ν and of the resistivity η, is: $R_v = v_A l_x/\nu$ and $R_M = v_A l_x/\eta$.

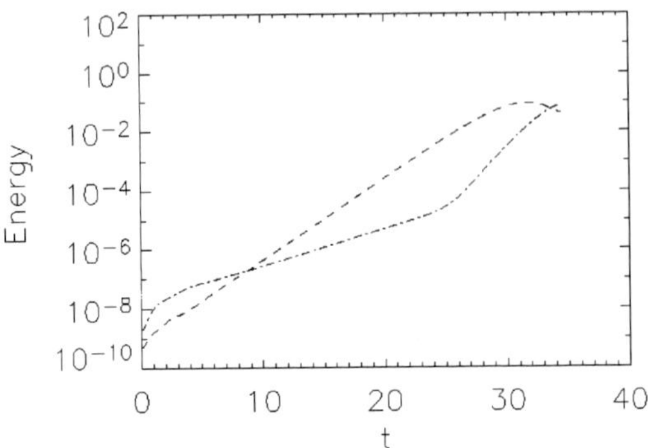

Figure 11. Time evolution of the energy for the $m = 2, n = 0$ mode (dashed line) and for $m = 1, n = 1$ mode (dot-dashed line).

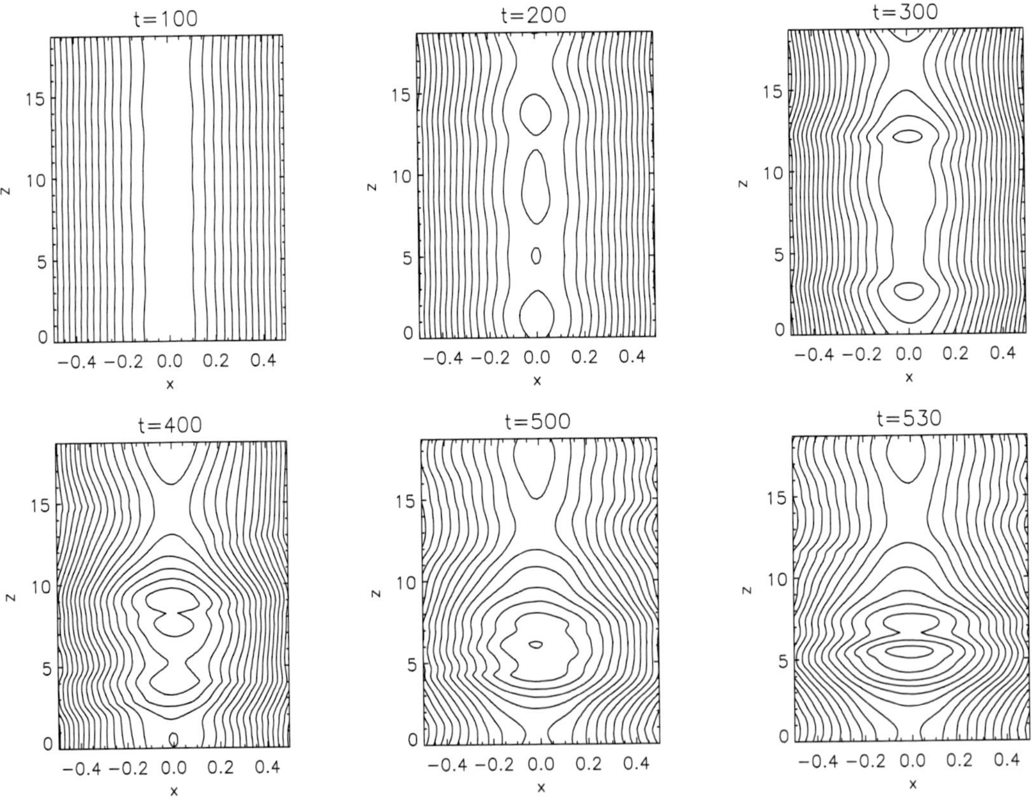

Figure 12. Magnetic field lines of the projected magnetic field $\vec{B}^{2D}(x, z) = B_x(x, 0, z)\hat{x} + B_z(x, 0, z)\hat{z}$ at different times.

We suppose that the numerical domain is bounded by two perfectly conducting walls, located at coordinates $x = \pm 1$. This corresponds to impose no diffusion of the magnetic field through the boundaries of the domain, namely that the component of $\vec{B}$ perpendicular to the conducting walls (the B_x component) must vanish at $x = \pm 1$. Moreover we suppose that the conducting walls are impenetrable and have finite viscosity, so that all the components of the velocity field are vanishing at the boundaries $x = \pm 1$. On a conducting wall the electric field has vanishing parallel components, and the same condition must hold in the plasma at the boundaries because the parallel components of the electric field are continuous crossing the walls. At the plasma boundaries $\vec{v} = 0$ so the electric field is given by $\vec{E} = \eta\vec{J}$, and the condition of vanishing parallel electric field implies $J_y = J_z = 0$. Together with the condition $B_x = 0$, this means that the x derivative of the y and z components of the magnetic field must vanish at the boundaries. Under these conditions the x component of Eq. (49), in the ideal limit, becomes $dp/dx = 0$ at the boundaries, so we impose that the gradient of the total pressure must have a vanishing $\hat{x}$ component at the boundaries to solve Eq. (53). This ensures that there is no force that could produce acceleration on the external walls and all the velocity components remain zero. Along the $\hat{y}$ and $\hat{z}$ directions, we suppose to have periodic boundary conditions.

We set up the initial condition in such a way to have a plasma that is at rest, in the frame of reference of our computational domain, permeated by a background magnetic field sheared along the $\hat{x}$ direction, with a current sheet in the middle of the simulation domain. Therefore, we set, for the background quantities:

$$\vec{v}_0 = 0$$

$$\vec{B}_0 = B_{y0}\hat{y} + B_{z0}(x)\hat{z}$$

where B_{y0} is a constant value, which has been set to 0.5, while B_{z0} is given by:

$$B_{z0}(x) = \tanh\left(\frac{x}{a}\right) - \frac{x/a}{\cosh^2\left(\frac{1}{a}\right)} \tag{54}$$

This form for the magnetic field [in particular the last term in Eq.(54)] ensures that the first derivatives along $\hat{x}$ of all the components of the magnetic field vanish at the boundaries, so that the current is completely contained in the simulation domain. The parameter a is a free parameter of our model, and represents the width of the magnetic field inhomogeneity. Under these conditions the initial total pressure is computed by solving numerically Eq. (53). In this configuration, we have a current sheet with a typical width a in the center of the simulation domain. This is an equilibrium field in the ideal limit ($1/R_M = 0$).

We perturb these equilibrium fields with three-dimensional divergenceless fluctuations of amplitude ϵ, which satisfy the boundary conditions:

$$\delta v_x = \sum_{k_y=k_{ymin}}^{k_{ymax}} \sum_{k_z=k_{zmin}}^{k_{zmax}} \epsilon \cos^2(\frac{\pi}{2}x)k^{-2}(k_y + k_z) \times \sin(k_y y + k_z z)$$

$$\delta v_y = \sum_{k_y=k_{ymin}}^{k_{ymax}} \sum_{k_z=k_{zmin}}^{k_{zmax}} \epsilon\pi \cos(\frac{\pi}{2}x)\sin(\frac{\pi}{2}x)k^{-2} \times \cos(k_y y + k_z z)$$

$$\delta v_z = \sum_{k_y=k_{ymin}}^{k_{ymax}} \sum_{k_z=k_{zmin}}^{k_{zmax}} \epsilon\pi \cos(\frac{\pi}{2}x)\sin(\frac{\pi}{2}x)k^{-2} \times \cos(k_y y + k_z z)$$

$$\delta\vec{b} = \delta\vec{v}$$

where k_y and k_z are the wave numbers along the $\hat{y}$ and $\hat{z}$ periodic directions respectively. This gives a k^{-2} spectrum for the perturbations, where k is the modulus of the wave vector: $k = \sqrt{k_y^2 + k_z^2}$.

We explicitly solve the equations (49), (50) and (53) using a compact finite difference scheme in the inhomogeneous direction (x) and a pseudo-spectral method in the periodic directions (y and z). This is realized by introducing the uniform grids:

$$x_j = -1 + j\Delta x, \quad j = 0, 1, ..., N_x, \quad \Delta x = 2/N_x$$

$$y_k = k\Delta y, \quad k = 0, 1, ..., N_y - 1, \quad \Delta y = 2\pi R_y/N_y$$

$$z_l = l\Delta z, \quad l = 0, 1, ..., N_z - 1, \quad \Delta z = 2\pi R_z/N_z$$

N_x, N_y, and N_z are the numbers of gridpoints in the x, y, and z directions respectively. For periodic coordinates (y and z) the spatial derivatives are calculated using discrete FFTs. As usual, for any variable $f(x, y, z, t)$, we assume the following decomposition:

$$f(x, y, z, t) = \sum_m \sum_n f_{m,n}(x, t) \exp\left[i\left(\frac{m}{R_y}y + \frac{n}{R_z}z\right)\right]$$

with the reality condition $f_{-m,-n} = f^*_{m,n}$. In the inhomogeneous direction a fourth-order compact difference scheme is used to compute the x derivative f' [28]:

$$f'_{j-1} + 4f'_j + f'_{j+1} = \frac{3}{\Delta x}[f_{j+1} - f_{j-1}],$$

at the interior points $j = 1, ..., N_x - 1$ and

$$2f'_{N_x-1} + f'_{N_x} = \frac{1}{2\Delta x}[5f_{N_x} - 4f_{N_x-1} - f_{N_x-2}],$$

$$2f'_1 + f'_0 = \frac{1}{2\Delta x}[-5f_0 + 4f_1 + f_2],$$

at the $j = N_x$ and $j = 0$ boundary points. For the second derivative f'' in the x direction the following fourth-order scheme is used:

$$\frac{1}{10}f''_{j-1} + f''_j + \frac{1}{10}f''_{j+1} = \frac{6}{5(\Delta x)^2}[f_{j+1} - 2f_j + f_{j-1}],$$

at the interior points $j = 1, ..., N_x - 1$ and

$$11f''_{N_x-1} + f''_{N_x} = \frac{1}{(\Delta x)^2}[13f_{N_x} - 27f_{N_x-1} + 15f_{N_x-2} - f_{N_x-3}],$$

$$11f''_1 + f''_0 = \frac{1}{(\Delta x)^2}[13f_0 - 27f_1 + 15f_2 - f_3],$$

at the $j = N_x$ and $j = 0$ boundary points.

To solve Eq. (53) we transform it in Fourier space and we obtain a second order differential equation for each Fourier harmonic:

$$\frac{d^2p^{(mn)}}{dx^2} - m^2p^{(mn)} - n^2p^{(mn)} = g^{(mn)}$$

where $g^{(mn)}(x)$ is the Fourier transform of the right hand side of Eq. (53). This equation has been solved with a compact finite difference scheme obtained by imposing the condition $dp/dx = 0$ at the boundaries. We use the following scheme at the boundaries:

$$\Big(\frac{3}{\Delta x^2} - \frac{m^2+n^2}{2}\Big)p^{(mn)}_{N_x-1} + \Big(-\frac{3}{\Delta x^2} - m^2 - n^2\Big)p^{(mn)}_{N_x} = \frac{1}{2}g^{(mn)}_{N_x-1} + g^{(mn)}_{N_x}$$
$$\Big(\frac{3}{\Delta x^2} - \frac{m^2+n^2}{2}\Big)p^{(mn)}_{1} + \Big(-\frac{3}{\Delta x^2} - m^2 - n^2\Big)p^{(mn)}_{0} = \frac{1}{2}g^{(mn)}_{1} + g^{(mn)}_{0}$$

and the following scheme at the interior points:

$$\Big(\frac{6}{5\Delta x^2} - \frac{m^2+n^2}{10}\Big)p^{(mn)}_{j-1} - \Big(\frac{12}{5\Delta x^2} + m^2 + n^2\Big)p^{(mn)}_{j} + \Big(\frac{6}{5\Delta x^2} - \frac{m^2+n^2}{10}\Big)p^{(mn)}_{j+1}$$
$$= \frac{1}{10}g^{(mn)}_{j-1} + g^{(mn)}_{j} + \frac{1}{10}g^{(mn)}_{j+1}.$$

The condition on the divergence of the magnetic field [Eq. (52)] is satisfied at $t = 0$ and taking the divergence of Eq. (50) we get:

$$\frac{d(\nabla\cdot\vec{B})}{dt} = 0.$$

Therefore Eq. (52) should be always satisfied during the simulation and we verified that the divergence of the magnetic field remains lower than the order of the numerical spatial scheme. The same condition is verified for the divergence of the velocity. Time advancement is calculated by a third-order Runge-Kutta explicit scheme. The code has been parallelized using the MPI directives. Every processor works on a section of the computational domain, which has been divided in the z direction.

The linear growth rates of the instabilities can be evaluated by solving the linearized MHD equations using a numerical method that is based on a relaxation technique. During the first stage of the simulation the energy of each Fourier harmonic grows approximately exponentially in time and it can be compared with the growth of the corresponding eigenmode, characterized by a given value of k_y and k_z. In the simulation the growth rate of a Fourier harmonic can be evaluated as $1/2$ of the energy growth rate. The linear growth rates are in agreement with the results of the simulations, as shown in Fig. 10 (dots) for two-dimensional modes obtained with $R_M = R_v = 5000$. In the simulations used for these tests we perturbed only one Fourier harmonic at a time to reduce nonlinear effects. To evaluate the growth rates from the simulations, a linear fit has been performed on the linear phase.

A new simulation has been performed with the condition that the equilibrium magnetic field is not dissipated [53] and the results have been compared with the observations of

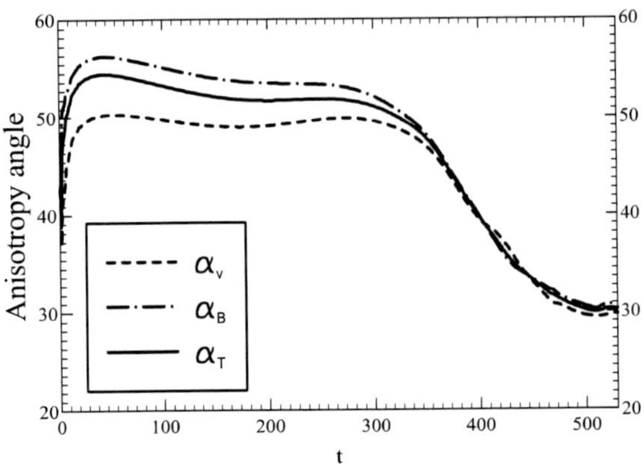

Figure 13. Anisotropy angle for kinetic energy (α_v), magnetic energy(α_B) and total energy (α_T).

secondary instabilities studied by Dahlburg *et al.* [52]. In Fig. 11 we show the evolution of the energy of the most unstable two-dimensional mode ($m = 2$, $n = 0$) and the energy of a three-dimensional mode ($m = 1$, $n = 1$). When the two-dimensional mode starts to saturate, we observe a change of slope for the energy of the $m = 1$, $n = 1$ mode, which corresponds to the growth of the secondary instability analyzed in Ref. [52].

The nonlinear evolution of the system has been studied with the following parameters:

$$N_x = 128, \quad N_y = 32, \quad N_z = 128$$
$$R_M = 5000, \quad R_v = 5000$$
$$a = 0.1, \quad R_y = 1, \quad R_z = 3$$

The equilibrium has been perturbed by exciting Fourier harmonics with $-4 < m < 4$ and $0 < n < 12$, which have resonant surfaces on both sides of the current sheet. The excited harmonics have wavelengths both shorter and longer than the most unstable mode, so nonlinear interactions can transfer energy in both directions.

We point out that the spatial resolution is not sufficient to represent a stationary fully developed turbulence. For this reason we stopped the simulation when the spectrum is formed.

At times $t > 200$ we observe a transfer of energy to smaller wavenumbers, which leads the $(0, 1)$ mode to become the most energetic one. This corresponds in the physical space to a coalescence of magnetic islands along the z direction, and the growth of the $(0, 1)$ mode can be seen as the formation of one magnetic island in the xz plane. To represent this phenomenon we consider the projection of the magnetic field onto the $y = 0$ plane: $\vec{B}^{2D}(x, z) = B_x(x, 0, z)\hat{x} + B_z(x, 0, z)\hat{z}$. In Fig. 12, the magnetic field lines of $\vec{B}^{2D}$ are shown at different times, the plots represent a narrow strip in the center of the domain. Magnetic reconnection and coalescence of magnetic islands are clearly visible.

The magnetic field structure is not fully represented in Fig. 12, since the B_y component has been neglected. However the magnetic field depends weakly on the y coordinate: if the same 2D magnetic lines are plotted in a different plane $y = const$, a similar pattern is found, except for the position of the X point, which is translated along z.

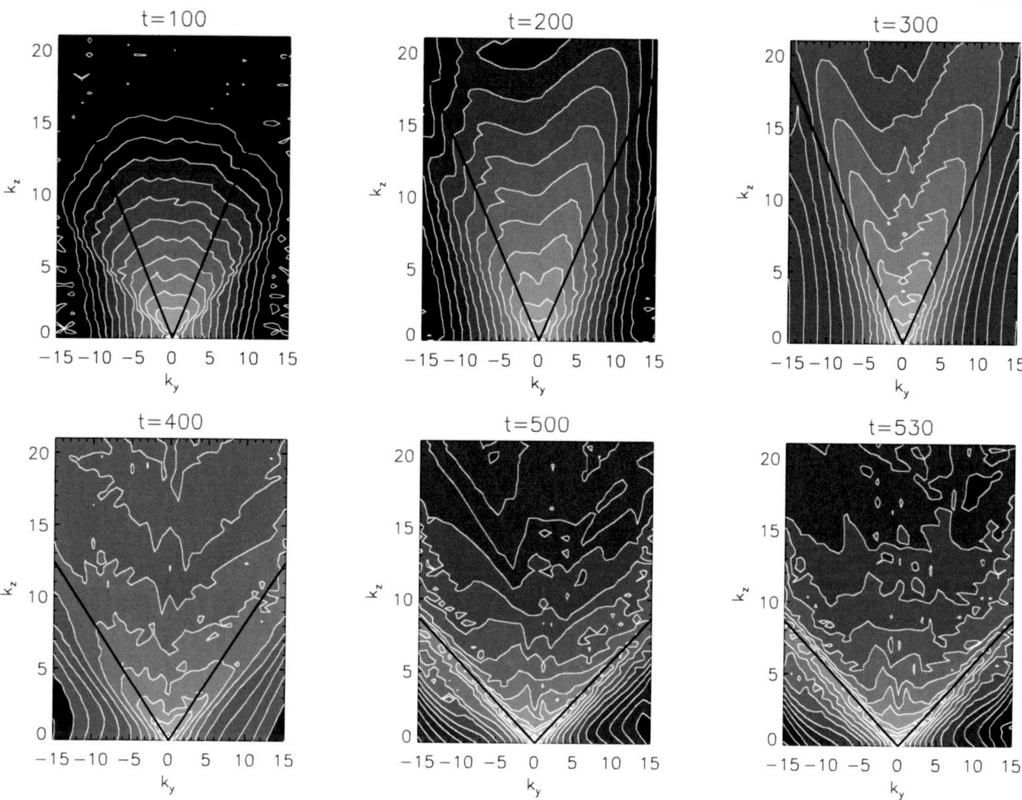

Figure 14. Contour plots of the total energy spectrum $\log(E^{tot}_{m,n})$ at different times. The straight lines have a slope corresponding to the anisotropy angle.

The energy spectrum is anisotropic, developing mainly in one specific direction in the $k_y k_z$ plane, identified by a particular value of the ratio k_y/k_z, which increases in time. This can be expressed by introducing an anisotropy angle:

$$\alpha = \tan^{-1}\sqrt{\frac{< k_z^2 >}{< k_y^2 >}}, \tag{55}$$

where $< k_z^2 >$ and $< k_y^2 >$ are the RMS wave vectors weighted by the spectral energy:

$$< k_z^2 >= \frac{\sum_{m,n}(n/R_z)^2 E_{m,n}}{\sum_{m,n} E_{m,n}}$$

$$< k_y^2 >= \frac{\sum_{m,n}(m/R_y)^2 E_{m,n}}{\sum_{m,n} E_{m,n}}.$$

and $E_{m,n}$ can be either the magnetic modal energy:

$$E^{mag}_{m,n}(t) = \frac{1}{2}\int_{-1}^{1} |\vec{B}_{m,n}(x,t)|^2 dx,$$

or the kinetic modal energy:

$$E^{kin}_{m,n}(t) = \frac{1}{2}\int_{-1}^{1} |\vec{V}_{m,n}(x,t)|^2 dx,$$

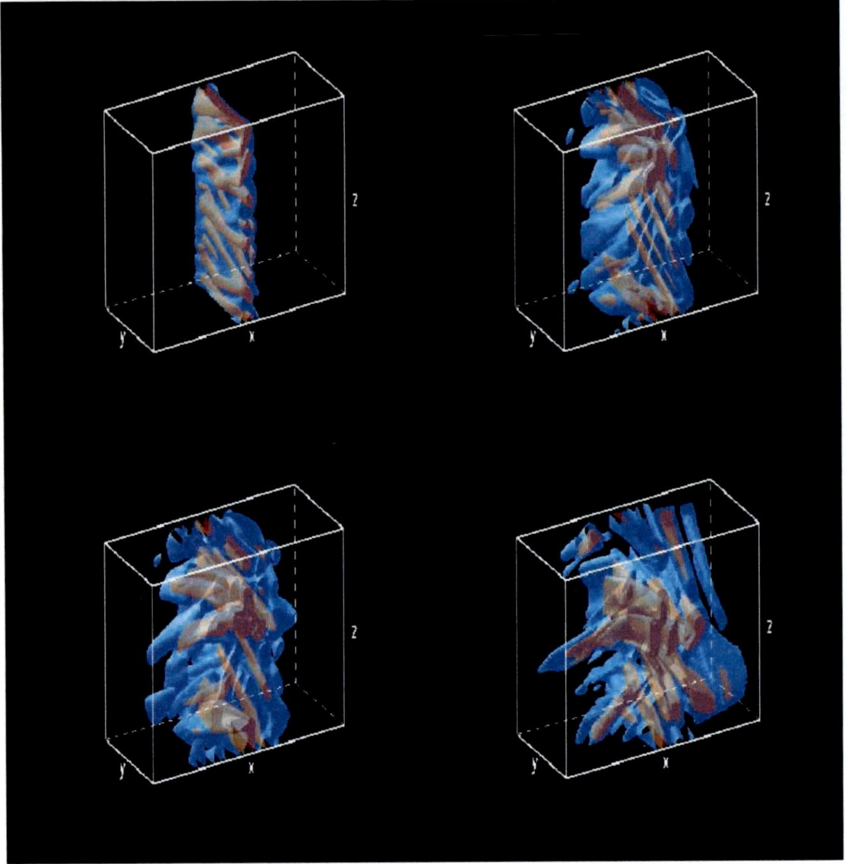

Figure 15. Current isosurfaces at $t = 50, 200, 300, 400$.

or the total energy:

$$E^{tot}_{m,n}(t) = E^{mag}_{m,n}(t) + E^{kin}_{m,n}(t),$$

In Fig. 13 the angle α is shown for kinetic energy (α_v), magnetic energy (α_B) and total energy (α_T). At the beginning of the simulation $\alpha_v = \alpha_B = \alpha_T \simeq 43$ degrees, then the anisotropy angles reach a value around 55 degrees for the magnetic energy and around 50 degrees for the kinetic energy. This indicates that the spectrum is dominated by wave vectors in the z direction, i.e., 2D modes prevail over 3D modes. After $t > 300$ they decrease rapidly and at the end of the simulation they have similar values around 30 degrees, showing that nonlinear interactions tend to reverse the initial anisotropy and in the final state wave vectors in the y direction prevail.

Fig. 14 shows the contour plots of the energy spectrum $\log(E^{tot}_{m,n})$, the straight lines have a slope corresponding to the anisotropy angle. It appears that in the latest times of the simulation the spectrum is strongly anisotropic, with most of the spectral energy aligned along the direction indicated by the anisotropy angle (straight lines). The direct energy cascade proceeds towards growing values of the ratio m/n, that is to resonant surfaces more distant from the center of the current sheet. On the contrary, at low wavenumbers energy concentrates at the central resonant surface. Therefore we observe formation of small scale structures on the lateral regions of the computational domain and coalescence in the center.

This behavior is reflected in the three-dimensional structure of the current. In Fig. 15, we show the isosurfaces of the currents at different times, with two different levels of current intensity (the red surfaces represent higher values and the blue surfaces lower values of the current). The initial equilibrium is destroyed by the formation of current filaments, with prevalence of small scale features on the lateral regions.

6. Conclusions

In this chapter we have reviewed the mixed finite difference-spectral technique as applied to the numerical solution of partial differential equations which are typically met when studying the stability properties of inhomogeneous plasma equilibria. The technique is, of course, already well known and established, and has been applied in the past to the study of several plasma configurations both in the fluid and kinetic descriptions. However, at the best of our knowledge, a unifying review of the method along with a detailed discussion of the problems one can meet when implementing the technique to the solution of fluid or kinetic equations, was still missing. The technique and the peculiarities of its implementation are reviewed in a detailed way. The advantages of the mixed finite difference-spectral approach with respect to other methods which are commonly used, for instance, in fluid mechanics, lie in the ability to deal with simple, but still rather general boundary conditions, in the case of regular geometries, while still maintaining characteristics of good precision (both for the spectral and for high order finite difference method) whenever the functions are analytical and do not have strong gradients. Last but not least, the finite difference techniques are efficiently parallelizable with respect, for instance, to fully spectral codes. This condition is crucial when dealing with problems which require very high spatial resolutions, like in the turbulent states induced by the plasma instabilities, where several decades of scales are excited. For all these reasons, the mixed finite difference-spectral approach is generally to be preferred over other techniques typical of fluid mechanics, like finite element methods or finite volume methods, which are specifically devoted to treat complicated domains or discontinuous solutions, at the price of a lower degree of precision. In the case of kinetic codes, the advantages of the mixed finite difference-spectral technique over, for instance, *Particle-in-cell* (PIC) codes is shown.

After a brief introduction to both the fluid and kinetic approaches to plasmas, we stressed the numerical difficulties and the possible effective solutions for the implementations of the mixed finite difference-spectral technique. For the fluid description of plasmas, this corresponds to the implementation of high order (compact) finite difference approximations in the direction of the plasma inhomogeneity, spectral methods (FFT) in the homogeneous directions and a variable-step time marching technique, to keep into account the large variability of characteristic time scales which can arise in plasma dynamics. For incompressible plasmas, this also involves the solution of the Poisson equation for the pressure and an appropriate choice of boundary conditions. In the kinetic case, particular care must be devoted to the choice of the time scheme, in order to satisfy (at least in an average sense) the energy conservation (*symplectic schemes*), since this is a crucial point in order to avoid unphysical effects coming from the discretization. In all cases, we explicitly give the numerical schemes which allow a correct implementation of the technique.

Finally, we have discussed in detail how such schemes have been used to build up complex numerical codes for solving several plasma instability problems in both the fluid and kinetic framework. To show the correctness of these codes and their ability to describe effectively both the linear and nonlinear phases of the instabilities, we show several examples of application to practical cases. In the fluid approach, we have shown how, for instance, the linear growth rate of the tearing instability predicted by the theory are recovered by the code. In the kinetic approach, we also show several comparisons with known theoretical results in the linear case and how the solution of the Vlasov system of equations gives improved results with respect, for instance, to a PIC code applied to the same problem.

References

[1] Ferraro V C A, Plumpton C. *An introduction to Magneto-fluid Mechanics*. Clarenford Press, 1966.

[2] Dedner A, Kemm F, Kröner D, Munz C-D, Schnitzer T, Wesenberg M. Hyperbolic Divergence Cleaning for the MHD Equations. *J. Comput. Phys.* 2002; 175: 645–673.

[3] Peyret R, Taylor T D. *Computational methods for fluid flow*. Springer-Verlag, 1983.

[4] C. K. Birdsall and A. B. Langdon, *Plasma Physics via Computer Simulation*, McGraw-Hill Book Co., Singapore (1985)

[5] D. H. Dubin, *Numerical and Analytical Methods for Scientists and Engineers Using Mathematica*, Wiley-Interscience Hoboken, New Jersey (2003)

[6] Mangeney A, Califano F, Cavazzoni C, Travnicek P. A Numerical Scheme for the Integration of the Vlasov—Maxwell System of Equations. *J. Comput. Phys.* 2002; 179: 495–538.

[7] Valentini F, Travnicek P, Califano F, Hellinger P, and Mangeney A. A hybrid-Vlasov model based on the current advance method for the simulation of collisionless magnetized plasma. *J. Comput. Phys.* 2007; 225: 753–770.

[8] Godlewski E, Raviart P A. *Numerical Approximation of Hyperbolic System of Conservation Laws*. Springer-Verlag, 1995.

[9] Harten A. High Resolution Schemes for Hyperbolic Conservation Laws. *J. Comput. Phys.* 1982; 135: 260–278.

[10] Harten A, Engquist B, Osher S, Chakravarthy S R. Uniformly High Order Accurate Essentially Non-oscillatory Schemes III. *J. Comput. Phys.* 1986; 131: 3–47.

[11] Leveque R J. High-resolution conservative algorithms for advection in incompressible flow. *Siam. J. numer. anal.* 1966; 33: 27–665.

[12] Van Leer B. Towards the Ultimate Conservative Difference Scheme III. Upstream-Centered Finite-Difference Schemes for ideal Compressible Flow. *J. Comput. Phys.* 1977; 23: 263–275.

[13] Fijalkow E. A numerical solution to the Vlasov equation. *Comput. Phys. Commun.* 1999; 116: 319–328.

[14] Fijalkow E. Numerical solution to the Vlasov equation: the 1D code. *Comput. Phys. Commun.* 1999; 116: 329–335.

[15] Fijalkow E. Numerical solution to the Vlasov equation: the 2D code. *Comput. Phys. Commun.* 1999; 116: 336–344.

[16] Valentini F, Veltri P, Mangeney A. A numerical scheme for the integration of the Vlasov—Poisson system of equations, in the magnetized case. *J. Comput. Phys.* 2005; 210: 730–751.

[17] Cheng C Z, Knorr G. The integration of the Vlasov equation in configuration space. *J. Comput. Phys.* 1976; 22: 330–351.

[18] Akhiezer A I, Akhiezer I A, Polovin R V, Sitenko A G, Stepanov K N. *Plasma electrodynamics*. Pergamon Press, 1975.

[19] Malara F, Primavera L, Veltri P. Formation of small scales via Alfven wave propagation in compressible nonuniform media. *Astrophys. Journal* 1996; 459: 347–364.

[20] Tsiklauri D, Nakariakov V M, Rowlands G. Phase mixing of a three dimensional magnetohydrodynamics pulse. *Astron. Astrophys.* 2003; 400: 1051–1055.

[21] Malara F, Primavera L, Veltri P. Compressive fluctuations generated by time evolution of Alfvenic perturbations in the solar wind current sheet. *J. Geophys. Res.* 1996; 101: 21597–21617.

[22] Goldstein M L, Roberts D A, Deane A E, Ghosh S, Wong H K. Numerical simulation of Alfvénic turbulence in the solar wind. *J. Geophys. Res.* 1999; 104: 14437–14451.

[23] Malara F, Veltri P, Carbone V. Competition among nonlinear effects in tearing instability saturation. *Phys. Fluids B* 1992; 4: 3070–3086.

[24] Onofri M, Primavera L, Malara F, Veltri P. Three-dimensional simulations of magnetic reconnection in slab geometry. *Phys. Plasmas* 2004; 11: 4837–4846.

[25] Dahlburg R B, Einaudi G. MHD unstable modes in the 3D evolution of 2D MHD structures and the diminished role of coalescence instabilities. *Phys. Lett. A* 2002; 294: 101–107.

[26] Bender C M, Orszag S A. *Advanced mathematical methods for scientists and engeneers*. International Student Edition, McGraw–Hill, 1978.

[27] Canuto C, Hussaini M Y, Quarteroni A, Zang T A. *Spectral methods in Fluid Dynamics*. Springer-Verlag, 1988.

[28] Lele S K. Compact finite difference schemes with spectral-like resolution. *J. Comput. Phys.* 1992; 103: 16–42.

[29] Sotiropoulos F, Abdallah S. The Discrete Continuity Equation in Primitive Variable Solutions of Incompressible Flow. *J. Comput. Phys.* 1991; 95: 212–227.

[30] Onofri M, Primavera L, Malara F, Londrillo P. A compressible magnetohydrodynamic numerical code with time-dependent boundary conditions in cylindrical geometry. *J. Comput. Phys.* (2007), doi: 10.1016/j.jcp.2007.06.015.

[31] Poinsot T J, Lele S K. Boundary Conditions for Direct Simulations of Compressible Viscous Flows. *J. Comput. Phys.* 1992; 101: 104–129.

[32] Landau L D. On the vibrations of the electronic plasma. *J. Phys. (Moscow)* 1946; 10: 25–34.

[33] O'Neil T. Collisionless damping of nonlinear plasma oscillations. *Phys. Fluids* 1965; 8: 2255–2262.

[34] Isichenko M B. Nonlinear Landau damping in collisionless Plasma and inviscid fuild. *Phys. Rev. Lett.* 1997; 78: 2369–2372.

[35] Manfredi G. Long-Time Behavior of Nonlinear Landau Damping. *Phys. Rev. Lett.* 1997; 79: 2815–2818.

[36] Lancellotti C, Dorning J J. Critical Initial States in Collisionless Plasmas. *Phys. Rev. Lett.* 1998; 81: 5137–5140.

[37] Brunetti M, Califano F, Pegoraro F. Asymptotic evolution of nonlinear Landau damping. *Phys. Rev. E* 2000; 62: 4109–4114.

[38] Danielson J R, Anderegg F, Driscoll C F. Measurement of Landau Damping and the Evolution to a BGK Equilibrium. *Phys. Rev. Lett.* 2004; 92: 245003-1–245003-4.

[39] Valentini F, Carbone V, Veltri P, Mangeney A. Self-consistent Lagrangian study of nonlinear Landau damping. *Phys. Rev. E* 2005; 71: 017401-1–017402-4.

[40] Valentini F, Veltri P, Mangeney A. Magnetic-field effects on nonlinear electrostatic-wave Landau damping. *Phys. Rev. E* 2005; 71: 016402-1–016402-8.

[41] Galeotti L, Califano F. Asymptotic Evolution of Weakly Collisional Vlasov-Poisson Plasmas. *Phys. Rev. Lett.* 2005; 95: 015002-1–015002-4.

[42] De Marco R, Carbone V, Veltri P. A Modified Fermi Model for Wave-Particle Interactions in Plasmas. *Phys. Rev. Lett.* 2006; 96: 125003-1–125003-4.

[43] Ivanov A I, Cairns I H. Nontrapping Arrest of Langmuir Wave Damping near the Threshold Amplitude. *Phys. Rev. Lett.* 2006; 96: 175001-1–175001-4.

[44] Gould R W, O'Neil T M, and Malmberg J H. Plasma wave echo. *Phys. Rev. Lett.* 1967; 19: 219–222.

[45] Malmberg J H, Warton C B, Gould R W, and O'Neil T M. Observation of Plasma Wave Echoes. *Phys. Fluids* 1968; 11: 1147.

[46] Berk H L, Nielsen C E, Roberts K V. Phase-space Hydrodynamics of Equivalent systems: experimental and computational observations. *Phys. Fluids* 1970; 13: 980–995.

[47] Ghizzo A, Izrar B, Bertrand P, Fijalkov E, Feix M R, Shoucri M. Stability of Bernstein Greene Kruskal plasma equilibria. Numerical experiments over a long time. *Phys. Fluids* 1988; 31: 72–82.

[48] Bernstein I B, Green J M, Kruskal M D. Exact nonlinear Plasma oscillations. *Phys. Rev. Lett.* 1957; 108: 546–550.

[49] Manfredi G, Bertrand P. Stability of Bernstein Greene Kruskal modes. *Phys. Plasmas* 2000; 7: 2425–2431.

[50] Valentini F, O'Neil T M, Dubin D H E. Excitation of nonlinear electron acoustic waves. *Phys. Plasmas* 2006; 13: 052303-1–052303-7.

[51] Dahlburg R B, Antiochos S K, Zang T A. Secondary instability in three-dimensional magnetic reconnection. *Phys. Fluids B* 1992; 4: 3902–3914.

[52] Dahlburg R B, Klimchuk J A, Antiochos S K. Magnetic Energy Release in Coronal Loops. *Astrophys. J.* 2005; 622: 1191–1201.

[53] Onofri M, Veltri P, Malara F. Development and anisotropy of three-dimensional turbulence in a current sheet. *Phys. Plasmas* 2007; 14: 062304.

In: Numerical Simulation Research Progress
Editors: Simone P. Colombo et al, pp. 141-159
ISBN 978-1-60456-783-0

Chapter 3

NUMERICAL SIMULATIONS OF THE NONLINEAR SOLITARY WAVES

***Jian-Qiang Sun*[a], *Meng-Zhao Qin*[b] *and Hua Wei*[c]**
[a]Institute of Applied Physics and Computational Mathematics,
Beijing, 100088, China
[b]Institute of Computational Mathematics, Chinese Academy of Science,
Beijing, 100080, China
[c]Department of Science, Liaoning Technical University, Fuxin, 123000, China.

Abstract

The nonlinear solitary waves arise in fluid dynamics, wave phenomena in the atmosphere and the ocean, nonlinear optics, Bose-Einstein condensation, magneto-hydrodynamics waves in warm plasma physics, etal. In the article, the symplectic and multi-symplectic methods in the structure preserving algorithms, which are well known methods in computation mathematical areas and have wide applications, are introduced. The soliton equations describing the evolution of the nonlinear solitary waves, such as the coupled nonlinear Schrodinger system, the Korteweg-de Veris equation, etal, were simulated by the symplectic and multi-symplectic methods. Numerical results showed that the symplectic and multi-symplectic methods can well simulate the evolution of the nonlinear solitary waves of the coupled nonlinear Schrodinger system, the Korteweg-de Veris equation, etal, moreover preserve the modulus square conserving property and the energy conserving property of the coupled nonlinear Schrodinger system, the Korteweg-de Veris equation, etal. the symplectic and multi-symplectic method in the structure preserving algorithms are advantage to simulating the behaviors of the nonlinear solitary waves.

1. Introduction

The soliton theory is an important branch of applied and computational mathematics. It has important applications in fluid mechanics, nonlinear optics, classical and quantum fields theories etc. It is one of the most active fields in science. Numerical simulations have played an important function in the study of the solitons. In 1965, Zabusky and Kruskal used a finite difference method, the famous Zabusky-Kruskal scheme, to show the existence of solitons which propagate with their own velocities, exerting essentially no influence on each other. They also discussed the recurrence of an initial state and guessed that the Korteweg-de Veris equation led to the recurrence [3]. Since then the interest of the soliton equations became more attractive. As the study of the soliton problems grew much complex, such as the study of the interactions of multi-solitons, quasi-solitons and the study of the existence and the interactions of the multi-dimensional solitons [2,4]. The numerical simulations of the nonlinear solitary waves in different fields have been an important topic [5-8,26]. Unfortunately, the soliton equations are nonlinear. Classical numerical methods such as the Runge-Kutta method and the multi-step method can not well simulate the behaviors of the nonlinear solitary waves with long time. As time grew, the numerical results of the soliton equations betrayed the real behaviors of the nonlinear solitary waves. It is desirable to devise good numerical schemes of the soliton equations.

It is known that Hamiltonian formalism has inherent connection with analytical mechanics, geometric optics, non-linear partial differential equations (PDEs) of first order, integrability of non-linear evolution equations. It is also under extension to infinite dimensions for various field theories, including fluid dynamics, elasticity, electrodynamics, plasma physics, etc. It is almost certain that all real physical processes with negligible dissipation can be described by Hamiltonian formalism, so the latter is becoming one of the most useful tools in the mathematical arsenal of physical and engineering sciences. It is natural that many nonlinear soliton equations can be transformed as the Hamiltonian system [1]. Feng Kang etc proposed a numerical method to solve the Hamiltonian system, which is well known as the symplectic method. It proved that the midpoint scheme of the Hamiltonian system is the symplectic scheme. The symplectic method can well simulate the Hamiltonian system with long time and preserve the inner conserving properties of the Hamiltonian system [9-11,25,29]. It is advantage to the classical numerical methods. The symplectic method has wide applications.

But there are also limitations in this approach to develop a symplectic method for PDEs. The problem with this approach is that it is global. Since we integrate over the total spatial domain,local variations in symplecticity are averaged. In fact, there may be anamalies in the spatial distribution of symplecticity that are masked by averaging. To overcome this limitation, Bridges and Reich introduced the concept of multisymplectic integrators based on a multisymplectic structure of some conservative PDEs. This approach is completely local. Thus the multisymplectic integrators have local invariant conserving properties. Reich showed that Gauss-Legendre collocation in space and time leads to multi-symplectic integrators[12-14]. Up to now, the symplectic and multisymplectic methods have become an important tool to simulate the evolution behaviors of the nonlinear solitary waves.

The purpose of this article is to show that the symplectic and multisymplectic method can well simulate the behaviors of the nonlinear solitary waves and preserve the inner characters of the soliton equations. An outline of the article is as follow. In section 2, we presented the symplectic method of the Hamiltonian system and the multisymplectic method of the multisymplectic structures of PDEs. In section 3, we simulated the evolution behaviors of the soliton equations, such as the coupled nonlinear Schrödinger system, by the symplectic method. In section 3, we simulated the evolution behaviors of the soliton equations, such as the nonlinear Klein-Gordon equation, the Kdv equation by the multisymplectic method. At last, we obtained that the symplectic and multisymplectic method are ideal methods in simulating the evolution behaviors of the nonlinear solitary waves.

2. The Symlectic and Multisymplectic Methods

We consider the canonical system in finite dimensions

$$\frac{dp_i}{dt} = -H_{q_i}, \quad \frac{dq_i}{dt} = H_{p_i}, \quad i = 1, 2, \cdots, n, \tag{1}$$

the Hamiltonian function $H(p_1, \cdots, p_n, q_1, \cdots, q_n)$. Let $z = (p_1, \cdots, p_n, q_1, \cdots, q_n)^T$, and $\nabla_z H(z) = (H_{z_1}, \cdots, H_{z_{2n}})^T$. Eq.(1) can be written as

$$\frac{dz}{dt} = J\nabla_z H(z), \qquad J = \begin{pmatrix} 0 & I_n \\ -I_n & 0 \end{pmatrix}. \tag{2}$$

Eq.(2) is defined in phase space R^{2n} with a standard symplectic structure given by the nonsingular anti-symmetric closed differential 2-form

$$\omega = \sum dp_i \wedge dq_i. \tag{3}$$

The fundamental theorem on Hamiltonian formalism says that the solution of the canonical system can be generated by a one-parameter group G_t of canonical transformations of R^{2n} (locally in t and z) such that

$$G_{t_1} G_{t_2} = G_{t_1+t_2}, \tag{4}$$

$$z(t) = G_t z(0). \tag{5}$$

A transformation $z \to \tilde{z}$ of R^{2n} is called canonical if it is a local diffeomorphism whose Jacobian $\frac{\partial \tilde{z}}{\partial z} = M$ is everywhere symplectic,i.e

$$M^T JM = J, \;\; i.e. \;\; M \in Sp(2n). \tag{6}$$

Linear canonical transformations are simply symplectic transformations.

The canonicity of G_t implies the preservation of 2-form ω, 4-from $\omega \wedge \omega, \cdots$, 2n-from $\omega \wedge \omega \wedge \cdots \wedge \omega$. They constitute the class of conservation laws of phase area of even dimensions for the Hamiltonian system. Moreover, the Hamiltonian system possesses another class of conservation laws related to the energy $H(z)$. A function $\phi(z)$ is said to be an invariant integral of Eq.(2), if it is invariant

$$\phi(z(t)) \equiv \phi(z(0))$$

which is equivalent to

$$\{\phi, H\} = 0,$$

where the Poisson bracket for two function $\phi(z)$, $\psi(z)$ are defined as

$$\{\phi, \psi\} = \phi_z^T J^{-1} \psi_z.$$

H itself is always invariant integral.

For the numerical study, we are less interested in Eq.(2) as a general system of ODE, but rather as a specific system with Hamiltonian structure. It is natural to look for those discretization system which preserve as many as possible the characteristic properties and inner symmetries of the original continuous systems. The above digressions on Hamiltonian system suggest the following guideline for difference schemes to be constructed. The transitions from the $k-th$ time step z^k to the next $(k+1)-th$ time step z^{k+1} should be canonical for all k , moreover, the invariant integers of the original system should remain invariant under these transitions.

For the Hamiltonian system (2) with Hamiltonian function H and phase flow

$$g_H^t = \exp(tJA) = I + \sum_{k=1}^{\infty} \frac{t^k}{k!}(JA)^k, \tag{7}$$

of R^{2n} (locally in t) such that the system (2) can be written as

$$\frac{d}{dt}(g_h^t z) = J(\nabla H) \circ g_z^t z, \quad \forall z \in R^{2n}. \tag{8}$$

Consider the difference for the Hamiltonian system (2), restricted mainly to the case single step scheme. Each 2-level-scheme is characterized by a transition operator relating to the old and new states by $\hat{z} = G^{\tau} z, \quad z = z^k, \quad \hat{z} = z^{k+1}$, $G^{\tau} = G_H^{\tau}$ depends on τ, H and the mode of discretization. Then true solution relating the old and new states is$z((k+1)\tau) = g_H^{\tau} z(k\tau)$.

So we are confronted with the problem of symplectic approximation to the symplectic phase flow g_H^{τ}. For the latter, there is a Lie expansion in τ

$$(g_H^{\tau} z)_i = z_i + \tau\{z_i, H\} + \frac{\tau^2}{2!}\{\{z_i, H\}, H\} + \cdots, \quad i = 1, \cdots, 2n. \tag{9}$$

Although the truncations give the legitimate approximation, but they are non-symplectic in general, so undesirable.

Consider the simplest one-legged weighted Euler scheme for Eq.(2)

$$\hat{z} = z + \tau J(\nabla H)(c\hat{z} + (1-c)z), \tag{10}$$

where c is a real constant. It is easy to show that it is symplectic if $c = \frac{1}{2}$. The only symplectic case corresponding to the time-centered Euler scheme

$$\hat{z} = z + \tau J(\nabla H)(\frac{\hat{z} + z}{2}). \tag{11}$$

High order symplectic schemes, such as symplectic Runge-Kutta method, composition symplectic method, generation function method, were founded [17-19,31].

Recently, the multisymplectic schemes, which can be viewed as the generalization of the symplectic schemes, were proposed in framework of multisymplectic geometry to solve the multisymplectic PDEs into which many soliton equations such as the KDV equation, the KP equation, the Z-K equation and the Sine-Gordon equation can be reformulated [21-23,27]. Let us consider a multi-symplectic PDEs

$$K\partial_t z + L\partial_x z = \nabla_z S(z), \tag{12}$$

with independent variable $(t, x) \in R^2$ and dependent state variable $z \in R^d, d \geq 2$. Here $K, L \in R^{d\times d}$ are two skew-symmetric matrices and $S : R^d \to R$ is a scalar-valued smooth function. ∇_z is the standard gradients in R^d. The system is multi-symplectic in the sense that K is a skew-symmetric matrix representative of the t direction and L is a skew-symmetric matrix representative of the x direction. S represents a Hamiltonian function.

Differentiating Eq.(12) and taking $dz = Z$, we can get its variational equation

$$K\partial_t Z + L\partial_x Z = S_{zz}(z)Z. \tag{13}$$

Take any two solutions U, V of the variational equation (13), then

$$\partial_t(U^T KV) + \partial_x(U^T LV) = 0, \tag{14}$$

upon introducing the two pre-symplectic form

$$\varphi(U, V) = U^T KV, \quad \text{and} \quad k(U, V) - U^T LV. \tag{15}$$

Eq.(14) is equivalent to the conservation law of multi-symplectic

$$\partial_t\varphi + \partial_x k = 0. \tag{16}$$

A more abstract approach using wedge product notation yields

$$\varphi = \frac{1}{2}dz \wedge Kdz, \quad k = \frac{1}{2}dz \wedge Ldz. \tag{17}$$

There is also an energy conservation law

$$\partial_t E(z) + \partial_x F(z) = 0, \tag{18}$$

with energy density

$$E(z) = S(z) + < z_x, Lz >, \tag{19}$$

and energy flux $F(z) = - < z_t, Lz >$, where $< .,. >$ is standard Euclidean inner product on R^d. Integrating the densities $E(z)$ over the spatial domain with suitable boundary conditions leads to global conserved quantities

$$\frac{d}{dt}\omega(z) = 0, \tag{20}$$

where $\omega(z) = \int_0^l E(z)dx$. It shows the global energy conservation.

Multi-symplectic is a geometric property of the PDEs. We naturally require a discretization to reflect this property. Based on this idea, Bridges and Reich introduced the concept of multisymplectic integrator. It has been shown that popular methods such as the center Preissman scheme are multisymplectic. Such schemes will not exactly preserve the local and global conservation of energy and momentum , however the numerical experiments show that the local and global conservation properties are preserved very well over long times. The Preissman scheme for Eq.(12) is

$$\frac{1}{\Delta t}K(z^{m+1}_{n+\frac{1}{2}} - z^{m}_{n+\frac{1}{2}}) + \frac{1}{\Delta x}L(z^{m+\frac{1}{2}}_{n+1} - z^{m+\frac{1}{2}}_{n}) = \nabla_z S(z^{m+\frac{1}{2}}_{n+\frac{1}{2}}), \tag{21}$$

where Δt is the time step, Δx is the space step, z_n^m is an approximation to $z(m\Delta t, n\Delta x)$, $z_{n+1}^{m+\frac{1}{2}} = \frac{1}{2}(z_{n+1}^{m+1} + z_{n+1}^m)$, $z_{n+\frac{1}{2}}^m = \frac{1}{2}(z_{n+1}^m + z_n^m)$, $z_{n+\frac{1}{2}}^{m+\frac{1}{2}} = \frac{1}{4}(z_n^m + z_{n+1}^m + z_n^{m+1} + z_{n+1}^{m+1})$. The discretized multisymplectic conservation is

$$\frac{\omega^1(u_n^{m+1}, v_n^{m+1}) - \omega^1(u_n^m, v_n^m)}{\Delta t} + \frac{\omega^2(u_{n+1}^m, v_{n+1}^m) - \omega^2(u_n^m, v_n^m)}{\Delta x} = 0, \tag{22}$$

where $\{u_n^m\}$ and $\{v_n^m\}$ are any two solutions of the discretized variational equation associated with

$$\frac{1}{\Delta t} K(dz_{n+\frac{1}{2}}^{m+1} - dz_{n+\frac{1}{2}}^m) + \frac{1}{\Delta x} L(dz_{n+1}^{m+\frac{1}{2}} - dz_n^{m+\frac{1}{2}}) = S_{zz}(z_{n+\frac{1}{2}}^{m+\frac{1}{2}}) dz_{n+\frac{1}{2}}^{m+\frac{1}{2}}, \tag{23}$$

$\omega^1(u,v) = < Ku, v >, \omega^2(u,v) = < Lu, v >, u, v \in R^d$, $< .,. >$ is standard Euclidean inner product on R^d. High order multisymplectic methods, such as multisymplectic Runge-Kutta methods, partitioned and splitting multisymplectic methods were founded [28,30,32].

3. Simulations of Solitary Waves by Symplectic Methods

3.1. Simulations of the Coupled Nonlinear Schrödinger System

The idea of using optical solitons as information bits in high-speed telecommunication systems was first proposed in 1973. In the following years, as fiber technology advanced, interests in optical solitons transmission grow. These techniques have spurred the interest in the theoretical modelling of pulse propagation and collision in the underlying communication systems. In an ideal fiber, optical solitons can be modelled approximately by the nonlinear Schrödinger equation, whose soliton behaviors are completely known. But the real optical fibers are birefringent, where the pulse travels at slightly different speed along the two orthogonal polarization axes. The coupled nonlinear Schrödinger (CNLS) system was derived for pulse propagation along the two polarization axes. The CNLS system can be written as

$$iu_t + u_{xx} + (|u|^2 + \beta |v|^2)u = 0, \tag{24}$$

$$iv_t + v_{xx} + (|v|^2 + \beta |u|^2)v = 0. \tag{25}$$

If we take $u(x,t) = p(x,t) + q(x,t)i$ and $v(x,t) = \mu(x,t) + \zeta(x,t)i$, Eqs.(24-25) is equivalent to

$$i(p_t + q_t i) + p_{xx} + q_{xx} i + ((p^2 + q^2) + \beta(\mu^2 + \zeta^2))(p + qi) = 0, \tag{26}$$

$$i(\mu_t + \zeta_t i) + \mu_{xx} + \zeta_{xx} i + ((\mu^2 + \zeta^2) + \beta(p^2 + q^2))(\mu + \zeta i) = 0. \tag{27}$$

Eqs.(26-27) can be rewritten as

$$p_t + q_{xx} + (p^2 + q^2 + \beta(\mu^2 + \zeta^2))q = 0, q_t - p_{xx} - (p^2 + q^2 + \beta(\mu^2 + \zeta^2))p = 0,$$
$$\mu_t + \zeta_{xx} + (\mu^2 + \zeta^2 + \beta(p^2 + q^2))\zeta = 0, \zeta_t - \mu_{xx} - (\mu^2 + \zeta^2 + \beta(p^2 + q^2))\mu = 0.$$

Thus, we can see that the CNLS system can be expressed in the Hamiltonian form

$$\frac{dz}{dt} = J \frac{\delta H(z)}{\delta z}, \quad J = \begin{pmatrix} 0 & I \\ -I & 0 \end{pmatrix}, \tag{28}$$

where $z = (p, \mu, q, \zeta)$, I is a 2×2 identity matrix, and the Hamiltonian function is

$$H(z) = \int \frac{1}{2}[p_x^2 + q_x^2 + \mu_x^2 + \zeta_x^2 - \frac{1}{2}(p^2+q^2)^2 - \frac{1}{2}(\mu^2+\zeta^2)^2 - \beta(p^2+q^2)(\mu^2+\zeta^2)]dx. \quad (29)$$

We discrete the spatial domain of Eq.(28) and expect to obtain a finite-dimensional Hamiltonian system. Denoting the $2m^{th}$ order central difference operator for $\mathcal{B} = \frac{\partial}{\partial x^2}$ by $\mathcal{B}(2m)$, we have

$$\mathcal{B}(2m) = \nabla_+ \nabla_- \sum_{j=0}^{m-1} (-1)^j \beta_j (\frac{(\Delta x)^2 \nabla_+ \nabla_-}{4})^j,$$

where $\beta_j = [(j!)^2 2^{2j}]/[(2j+1)!(j+1)]$ and ∇_+, ∇_- are forward and backward difference operators, and Δx is the spatial step length. Denoting by N the number of the spatial grid points, and letting

$$P = [p_1, p_2, \cdots, p_N, \mu_1, \mu_2, \cdots, \mu_N]^T,$$
$$Q = [q_1, q_2, \cdots, q_N, \zeta_1, \zeta_2, \cdots, \zeta_N]^T.$$

We arrive at a semi-discretization system with accuracy $O((\Delta x)^{2m})$ in the space discretization

$$\frac{d}{dt}\begin{pmatrix} P \\ Q \end{pmatrix} = \begin{pmatrix} 0 & I_{2N} \\ -I_{2N} & 0 \end{pmatrix} \begin{pmatrix} M(2m) & 0 \\ 0 & M(2m) \end{pmatrix} \begin{pmatrix} P \\ Q \end{pmatrix}, \quad (30)$$

where I_{2N} is $2N \times 2N$ identity matrix and $M(2m)$ can be written as the sum of two matrices

$$M(2m) = B(2m) + D, \quad (31)$$

where D is a diagonal matrix and $B(2m)$ is $2N \times 2N$ matrix corresponding to $\mathcal{B}(2m)$. In practice, the second order central difference approximations are usually applied

$$M(2) = B(2) + D, \quad (32)$$

where

$$B(2) = \begin{pmatrix} C(2) & 0 \\ 0 & C(2) \end{pmatrix},$$

$$C(2) = -\frac{1}{(\Delta x)^2} \begin{pmatrix} -2 & 1 & 0 & \cdots & \cdots & 0 \\ 1 & -2 & 1 & \cdots & \cdots & 0 \\ & \ddots & \ddots & \ddots & & \\ & \ddots & \ddots & \ddots & & \\ 0 & 0 & \cdots & \cdots & -2 & 1 \\ 0 & 0 & \cdots & \cdots & 1 & -2 \end{pmatrix},$$

where $C(2)$ is an $N \times N$ matrix.

$$D = \begin{pmatrix} D_1 & 0 \\ 0 & D_2 \end{pmatrix},$$

$$D_1 = \begin{pmatrix} \rho_1 & 0 & \cdots & 0 \\ 0 & \rho_2 & \cdots & 0 \\ & 0 & \ddots & 0 \\ 0 & 0 & \cdots & \rho_N \end{pmatrix}, \qquad D_2 = \begin{pmatrix} \tau_1 & 0 & \cdots & 0 \\ 0 & \tau_2 & \cdots & 0 \\ 0 & 0 & \ddots & 0 \\ 0 & 0 & \cdots & \tau_N \end{pmatrix},$$

where $\rho_k = -((p_k^2+q_k^2)+\beta(\mu_k^2+\zeta_k^2))$, $\tau_k = -((\mu_k^2+\zeta_k^2)+\beta(p_k^2+q_k^2))$, $k = 1, 2, \cdots, N$, and both D_1 and D_2 are the $N \times N$ diagonal matrices. In fact, Eq.(30) is a Hamiltonian system. In fact, let

$$Z = (P^T, Q^T), \qquad J = \begin{pmatrix} 0 & I_{2N} \\ -I_{2N} & 0 \end{pmatrix}.$$

Eq.(30) can be rewritten as

$$\frac{dZ}{dt} = J\nabla_z H(Z), \tag{33}$$

with the Hamiltonian function

$$H(P,Q) = \frac{1}{2}[P^T B(2m)P + Q^T B(2m)Q] - \sum_{l=1}^{N}[\frac{1}{4}((p_l^2+q_l^2)^2 + (\mu_l^2+\zeta_l^2)^2) + \frac{\beta}{2}(p_l^2+q_l^2)(\mu_l^2+\zeta_l^2)]. \tag{34}$$

The mid-point scheme constructs the symplectic integrator of Eq.(30). We get the symplectic scheme

$$Z^{n+1} - Z^n = \Delta t J\nabla_z H(\frac{Z^{n+1}+Z^n}{2}). \tag{35}$$

Eq.(35) can be written as

$$\frac{p_l^{n+1}-p_l^n}{\Delta t} + \frac{q_{l-1}^{n+\frac{1}{2}} - 2q_l^{n+\frac{1}{2}} + q_{l+1}^{n+\frac{1}{2}}}{(\Delta x)^2} + (|u_l^{n+\frac{1}{2}}|^2 + \beta|v_l^{n+\frac{1}{2}}|^2)q_l^{n+\frac{1}{2}} = 0, \tag{36}$$

$$\frac{\mu_l^{n+1}-\mu_l^n}{\Delta t} + \frac{\zeta_{l-1}^{n+\frac{1}{2}} - 2\zeta_l^{n+\frac{1}{2}} + \zeta_{l+1}^{n+\frac{1}{2}}}{(\Delta x)^2} + (|v_l^{n+\frac{1}{2}}|^2 + \beta|u_l^{n+\frac{1}{2}}|^2)\zeta_l^{n+\frac{1}{2}} = 0, \tag{37}$$

$$\frac{q_l^{n+1}-q_l^n}{\Delta t} - \frac{p_{l-1}^{n+\frac{1}{2}} - 2p_l^{n+\frac{1}{2}} + p_{l+1}^{n+\frac{1}{2}}}{(\Delta x)^2} - (|u_l^{n+\frac{1}{2}}|^2 + \beta|v_l^{n+\frac{1}{2}}|^2)p_l^{n+\frac{1}{2}} = 0, \tag{38}$$

$$\frac{\zeta_l^{n+1}-\zeta_l^n}{\Delta t} - \frac{\mu_{l-1}^{n+\frac{1}{2}} - 2\mu_l^{n+\frac{1}{2}} + \mu_{l+1}^{n+\frac{1}{2}}}{(\Delta x)^2} - (|v_l^{n+\frac{1}{2}}|^2 + \beta|u_l^{n+\frac{1}{2}}|^2)\mu_l^{n+\frac{1}{2}} = 0, \tag{39}$$

where $l = 1, 2, \cdots, N$.

Multiplying Eq.(36) with i, we have

$$i\frac{p_l^{n+1}-p_l^n}{\Delta t} + \frac{q_{l-1}^{n+\frac{1}{2}} - 2q_l^{n+\frac{1}{2}} + q_{l+1}^{n+\frac{1}{2}}}{(\Delta x)^2}i + (|u_l^{n+\frac{1}{2}}|^2 + \beta|v_l^{n+\frac{1}{2}}|^2)q_l^{n+\frac{1}{2}}i = 0. \tag{40}$$

Subtracting Eq.(38) from Eq.(40), we obtain

$$i\frac{u_l^{n+1}-u_l^n}{\Delta t}+\frac{u_{l+1}^{n+1}+u_{l+1}^n-2u_l^{n+1}-2u_l^n+u_{l-1}^{n+1}+u_{l-1}^n}{2(\Delta x)^2}+$$
$$(|u_l^{n+\frac{1}{2}}|^2+\beta|v_l^{n+\frac{1}{2}}|^2)\frac{u_l^{n+1}+u_l^n}{2}=0. \tag{41}$$

In the same way, we have

$$i\frac{v_l^{n+1}-v_l^n}{\Delta t}+\frac{v_{l+1}^{n+1}+v_{l+1}^n-2v_l^{n+1}-2v_l^n+v_{l-1}^{n+1}+v_{l-1}^n}{2(\Delta x)^2}+$$
$$(|v_l^{n+\frac{1}{2}}|^2+\beta|u_l^{n+\frac{1}{2}}|^2)\frac{v_l^{n+1}+v_l^n}{2}=0. \tag{42}$$

Thus, the new six-point difference scheme (41-42), which is equivalent to the symplectic integrator, is derived. The new six-point scheme of the CNLS system can preserve the modulus conserving property of the CNLS system.

THEOREM 1 The CNLS system (24-25) preserves the square conservation

$$\int_{c_1}^{c_2}|u(x,t)|^2dx=a, \tag{43}$$
$$\int_{c_1}^{c_2}|v(x,t)|^2dx=b, \tag{44}$$

where a and b are positive constants. From Theorem 1, we obtain that two soliton waves $u(x,t)$ and $v(x,t)$ in the vector case are square conserved in the interval. It is desirable that the discrete scheme would reflect the square conservation of the CNLS system in numerical simulations. However, the previous discrete schemes, generally speaking, do not reflect the square conservation, so that they can not simulate the system very well. Here, we are going to prove that the six-point scheme preserves the square conservation.

THEOREM 2 The new six-point scheme (41-42) preserves the square conservation:

$$\sum_{l=1}^{N}|u_l^n|^2=\sum_{l=1}^{N}|u_l^{n+1}|^2, \tag{45}$$
$$\sum_{l=1}^{N}|v_l^n|^2=\sum_{l=1}^{N}|v_l^{n+1}|^2. \tag{46}$$

Therefore, the new six-point scheme preserves the square conservation, which matches the square conservation of CNLS system very well [20].

We simulated the behaviors of the solitary waves of the CNLS system by the six-point scheme. When the initial condition of the coupled Schrödinger equations is as follows:

$$u(x,0)=\sqrt{2}r_1sech(r_1x+\frac{1}{2}D_0)e^{iV_0x/4}$$
$$v(x,0)=\sqrt{2}r_2sech(r_2x+\frac{1}{2}D_0)e^{iV_0x/4} \tag{47}$$

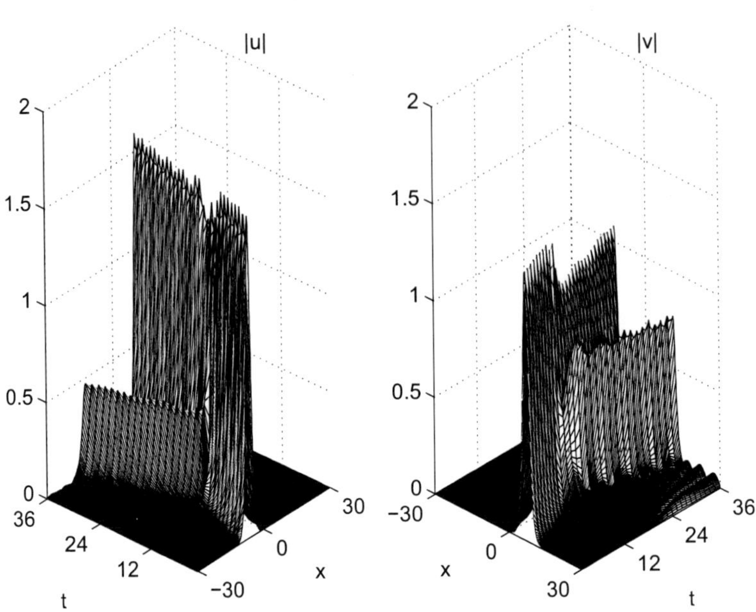

Figure 1. Numerical simulations of the two solitary waves $u(x,t)$,$v(x,t)$ in the vector case with $\beta = \frac{4}{5}$, $V_0 = 1.21$, $D_0 = 16$.

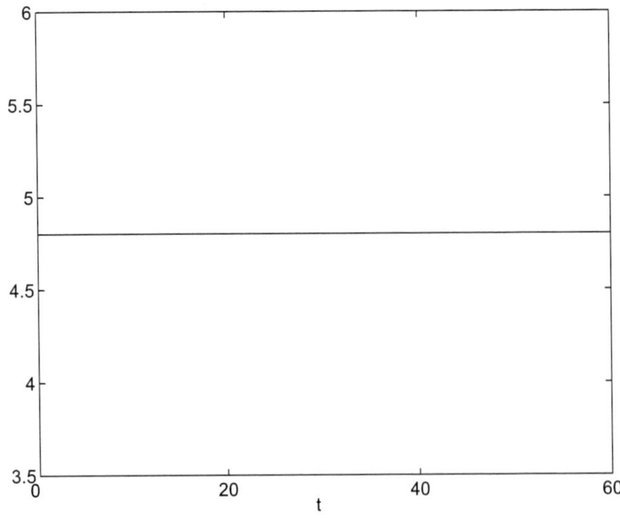

Figure 2. The modulus square sums of the solitary wave $u(x,t)$ at $t \in [0, 60]$ by the scheme (41-42)

Consider the interaction of the two solitary waves $u(x,t)$, $v(x,t)$ in the vectors case where $-30 < x < 30$, $\beta = \frac{4}{5}$, $V_0 = 1.21$, $D_0 = 16$, with a space step $\Delta x = 0.2$, a time step $\Delta t = 0.04$, and $r_1 = 1.2$, $r_2 = 1$. $|u(x,t)|$ show the modulus of the soliton waves $u(x,t)$, the same is for $v(x,t)$. The modulus square sums of the solitary wave $u(x,t)$ are denoted as $sum(t) = (\sum_{l=1}^{N} |u_l^n|^2)\Delta x$ at $t = n\Delta t$. Figure.1 show the collision behaviors of the solitary waves of the CNLS system at $t \in [0, 36]$. From Figure.1, we can see that after the collisions of the two solitary waves $u(x,t)$, $v(x,t)$ in the vectors cases, two new solitary waves produced. They travel deviating from the directions of the original two solitary waves. Figure.2 showed the modulus square sums of the solitary wave $u(x,t)$ at different times with $\beta = \frac{4}{5}$, $V_0 = 1.21$, $D_0 = 16$. From Figure.2, we can see that the symplectic scheme (41-42) can preserve the modulus square conserving property of the CNLS system. We can conclude that the symplectic scheme (41-42) can well simulate the evolution behaviors of the nonlinear solitary waves of the CNLS system.

3.2. Simulations of the Nonlinear Rossby Wave Packets

The nonlinear Rossby wave packets can also be described by the coupled nonlinear Schrödinger (CNLS) equations

$$iA_t + \alpha_1 A_{xx} + (\sigma_1|A|^2 + v_{12}|B|^2)A = 0 \tag{48}$$

$$iB_t + \alpha_2 B_{xx} + (\sigma_2|B|^2 + v_{21}|A|^2)B = 0 \tag{49}$$

where A,B are the complex amplitudes or 'envelopes' of the two wave packets respectively. In the language of the method of multiple scale, t is the slow time and x is the slow space variable moving at the linear group velocity. The coefficients α_1 and α_2 are so-called 'dispersion coefficient', σ_1 and σ_2 are the Landau constants describing the self-modulation of the wave packets, and v_{12} and v_{21} are the coupling constant of the cross-modulation between the two wave packets.

When $v_{12} = v_{21} = v$, the coupled Schrödinger equations have the symplectic structure[24]. The symplectic midpoint scheme was applied to Eqs.(48-49). We obtained the following symplectic scheme

$$i\frac{A_l^{n+1} - A_l^n}{\Delta t} + \frac{\alpha_1 \tilde{\Delta} A_l^{n+\frac{1}{2}}}{12(\Delta x)^2} + (\sigma_1|A_l^{n+\frac{1}{2}}|^2 + v|B_l^{n+\frac{1}{2}}|^2)A_l^{n+\frac{1}{2}} = 0 \tag{50}$$

$$i\frac{B_l^{n+1} - B_l^n}{\Delta t} + \frac{\alpha_1 \tilde{\Delta} B_l^{n+\frac{1}{2}}}{12(\Delta x)^2} + (\sigma_2|B_l^{n+\frac{1}{2}}|^2 + v|A_l^{n+\frac{1}{2}}|^2)B_l^{n+\frac{1}{2}} = 0 \tag{51}$$

where $\tilde{\Delta} f_l^{n+\frac{1}{2}} = (-f_{l-2}^{n+\frac{1}{2}} + 16f_{l-1}^{n+\frac{1}{2}} - 30f_l^{n+\frac{1}{2}} + 16f_{l+1}^{n+\frac{1}{2}} - f_{l+2}^{n+\frac{1}{2}})$, $f = A, B$.

The solitary behaviors of the nonlinear Rossby wave packets were simulated by the symplectic scheme (50-51). When the initial condition of the coupled Schrödinger system is as follows:

$$A(x,0) = \mu\sqrt{2\frac{\alpha_1\sigma_2 - \alpha_2 v}{\sigma_1\sigma_2 - v^2}}\{sech(\mu x)e^{i\frac{V_1}{2\alpha_1}x} + sech(\mu(x+x_0))e^{i\frac{V_2}{2\alpha_1}(x+x_0)+i\phi_1}\}$$

$$B(x,0) = \mu\sqrt{2\frac{\alpha_2\sigma_1 - \alpha_1 v}{\sigma_1\sigma_2 - v^2}}\{sech(\mu x)e^{i\frac{V_1}{2\alpha_2}x} + sech(\mu(x+x_0))e^{i\frac{V_2}{2\alpha_2}(x+x_0)+i\phi_2}\} \tag{52}$$

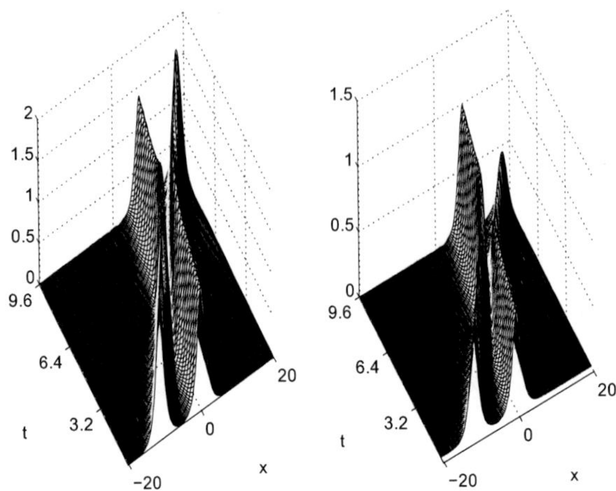

Figure 3. Numerical simulations of the collision interaction of the two envelope Rossby solitons with $\alpha_1 = 1.24$, $\alpha_2 = 1.28$, $\sigma_1 = 0.88$, $\sigma_2 = 1.15$.

We take $\Delta t = 0.001$,$\Delta x = 0.2$, $-20 < x < 20$. Figure.3 showed the the collision behaviors of the two envelope Rossby solitary waves $A(x,t)$,$B(x,t)$ with $\alpha_1 = 1.24$, $\alpha_2 = 1.28$, $\sigma_1 = 0.88$, $\sigma_2 = 1.15$. $v = 5/6$, $V_1 = 1.2$, $V_2 = 3$ by the scheme (50-51). The left of Figure.3 is the collision results of the solitary waves $A(x,t)$. The right of Figure.3 is the collision results of the solitary waves $B(x,t)$. From Figure.3, we can see that the symplectic scheme (50-51) can well simulate the collision behaviors of the two envelope Rossby solitary waves $A(x,t)$,$B(x,t)$.

4. Simulations of Solitary Waves by Multi-Symplectic Methods

4.1. Simulations of the Nonlinear Klein-Gordon Equation

The nonlinear Klein-Gordon equation

$$u_{tt} - u_{xx} + f(u) = 0, \quad (x,t) \in \Omega \subset R^2, \tag{53}$$

where $f(u) : R \to R$ is a nonlinear smooth function [15,16]. Its application can be found in many areas of physics, including nonlinear optics and electronic. Jimenez etal simulated the solitary waves of the Klein-Gordon equation by different numerical schemes [15]. But the schemes neglected the conserving properties of the Klein-Gordon equation. The Klein-Grodon equation can be transformed into the multi-symplectic structure. Set $v = u_x$, $w = u_x$, then the nonlinear Klein-Gordon equation can be written as

$$Mz_t + Kz_x = \nabla_z S(z), \tag{54}$$

where

$$M = \begin{pmatrix} 0 & -1 & 0 \\ -1 & 0 & 0 \\ 0 & 0 & 0 \end{pmatrix}, \quad K = \begin{pmatrix} 0 & 0 & 1 \\ 0 & 0 & 0 \\ -1 & 0 & 0 \end{pmatrix}, \quad z = \begin{pmatrix} u \\ v \\ w \end{pmatrix},$$

and $S(z) = \frac{1}{2}v^2 - \frac{1}{2}w^2 + F(u)$. To this PDEs, Bridges obtained the multisymplectic conservation law

$$\partial_t(dz \wedge Mdz) + \partial_x(dz \wedge Kdz) = 0. \tag{55}$$

i.e

$$\partial_t(du \wedge du_t) - \partial_x(du \wedge du_x) = 0.$$

The Preissman box scheme of Eq.(54) is equivalent to

$$\frac{1}{\Delta}M(z^{m+1}_{n+\frac{1}{2}} - z^{m}_{n+\frac{1}{2}}) + \frac{1}{\Delta x}K(z^{m+\frac{1}{2}}_{n+1} - z^{m+\frac{1}{2}}_{n}) = f(z^{m+\frac{1}{2}}_{n+\frac{1}{2}}). \tag{56}$$

where $z^{m+\frac{1}{2}}_{n+\frac{1}{2}} = \frac{1}{2}(z^{m+1}_{n+1} + z^{m+1}_{n}$, $z^{m+\frac{1}{2}}_{n} = \frac{1}{2}(z^{m+1}_{n} + z^{m}_{n})$, $z^{m+\frac{1}{2}}_{n+\frac{1}{2}} = \frac{1}{4}(z^{m}_{n} + z^{m}_{n+1} + z^{m+1}_{n} + z^{m+1}_{n+1})$, $z = (u, v, w)^T$,etc.

The Preissman scheme is multisymplectic, it involves more efforts to compute the auxiliary variables w and v. So we eliminate w and v by a trivial computation and obtain the following multisymplectic nine point scheme

$$\frac{1}{4}[\partial_t^2(u^m_{n-1} + 2u^m_n + u^m_{n+1})] - \frac{1}{4}[\partial_x^2(u^m_{n-1} + 2u^m_n + u^m_{n+1})] = \frac{1}{4}\tilde{S}(z) \tag{57}$$

where $\tilde{S}(z) = f(u^{m-\frac{1}{2}}_{n-\frac{1}{2}}) + f(u^{m-\frac{1}{2}}_{n+\frac{1}{2}}) + f(u^{m+\frac{1}{2}}_{n-\frac{1}{2}}) + f(u^{m+\frac{1}{2}}_{n+\frac{1}{2}})$, $\partial_x^2 u^j_i = (u^j_{i+1} - 2u^j_i + u^j_{i-1})/(\Delta x)^2$, $\partial_t^2 u^j_i = (u^{j+1}_i - 2u^j_i + u^{j-1}_i)/(\Delta t)^2$, $u^{j+\frac{1}{2}}_{i+\frac{1}{2}} = \frac{1}{4}(u^j_i + u^j_{i+1} + u^{j+1}_i + u^{j+1}_{i+1})$, $i = n-1, n$,$j = m-1, m$.

The energy-conserving scheme has a discrete conserved energy

$$E^n = \sum_n \frac{\Delta x}{2}[(\frac{u^{m+1}_n - u^m_n}{\Delta t})^2 + \frac{u^{m+1}_{n+1} - u^m_n}{\Delta x} \cdot \frac{u^m_{n+1} - u^m_n}{\Delta x} + F(u^{m+1}_n) + F(u^m_n)]. \tag{58}$$

which approximates the following very important first integral of with approximate boundary condition

$$E(t) = \int (\frac{1}{2}u_t^2 + \frac{1}{2}u_x^2 + F(u))dx. \tag{59}$$

In order to test the energy conservation of the multi-symplectic integrators, we define the energy as

$$Err(t) = E^n - E^0, \tag{60}$$

where E^0 is the initial energy with $t = 0$, E^n is the energy at $t = n\Delta t$ time by the multi-symplectic integrators.

We simulated the evolution behaviors of the nonlinear Klein-Gordon equation by the scheme (57). When $f(u) = sin(u)$, the initial and boundary conditions employed for the collision of two soliton waves are

$$u(x, 0) = 4atan(a \cdot e^{(-x+x_0)}) + 4atan(a \cdot e^{(x+x_0)}), \tag{61}$$

We take $-25 < x < 25$, $a = 80$, $x_0 = 10$. The time step is $\Delta t = 0.15$ and the space step is $\Delta x = 0.25$. The computation results are as follow:

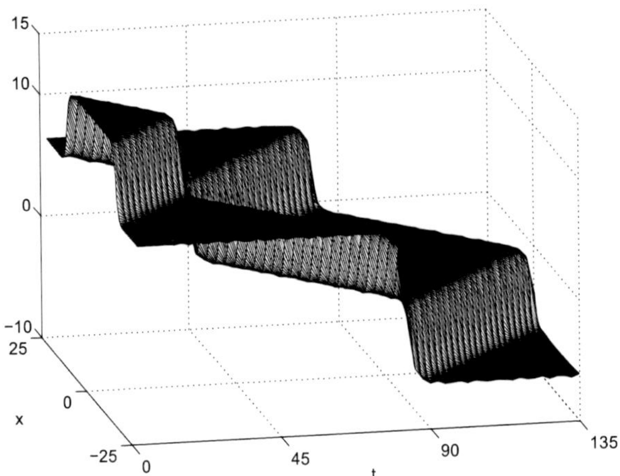

Figure 4. Numerical simulations for solitary waves of Sine Gordon equation at$t \in [0, 135]$ by the scheme (57).

Figure.4 show the numerical waveforms of the Sine-Gordon equation at$t \in [0, 135]$. From Figure.4, we can see the solitary wave is smooth. The multisymplectic scheme (57) can well simulate the evolution behaviors of the nonlinear Klein-Gordon equation with long time. Figure.5 show the energy error of the Sine-Gordon equation at$t \in [0, 45]$ by the scheme (57). From Figure.5, we can see that the multisymplectic scheme (57) can well preserve the energy conserving property of the Sine-Gordon equation except there is a big change at some time. It is obvious that the multisymplectic method can well simulate the evolution behaviors of the nonlinear solitary waves of the Sine-Gordon equation.

4.2. Simulations of the Kdv Equation

The Korteweg-de Veris equation has been found to describe various kinds of phenomena such as acoustic waves in an anharmonic crystal, waves in bubble-liquid mixtures, magnetohydrodynamics waves in warm plasma, and ion acoustic waves [26,27]. It has two fascinating and significant features. One is the existence of permanent waves solutions, including solitary wave solutions and the other is the recurrence of the initial state of the wave form. Various kinds of methods including the finite different method and the Fourier expansion have been to proposed to solve the KDV equation . But unfortunately the difference solution often exhibit nonlinear instability when a long time integration is made. We will proposed to solve the KDV equation by the multisymplectic method. The general form of the KDV equation is

$$\frac{\partial u}{\partial t} + c_1 u \frac{\partial u}{\partial x} + c_2^2 \frac{\partial^2 u}{\partial x^3} = 0, t > 0, x \in [a, b], \tag{62}$$

with the initial value $u(t = 0, x) = u_0(x)$ and the periodic boundary condition$u(t, x+a) = u(t, x + b)$, where c and δ are two real numbers.

Introduce the potential $\phi_x = u$, momenta $\nu = c_2 u_x$ and $\omega = \frac{1}{2}\phi_x + c_2 \nu_x + V'(u)$, $V(u) = c_1 u^3/6$, then the KDV equation can be written as the following multisymplectic

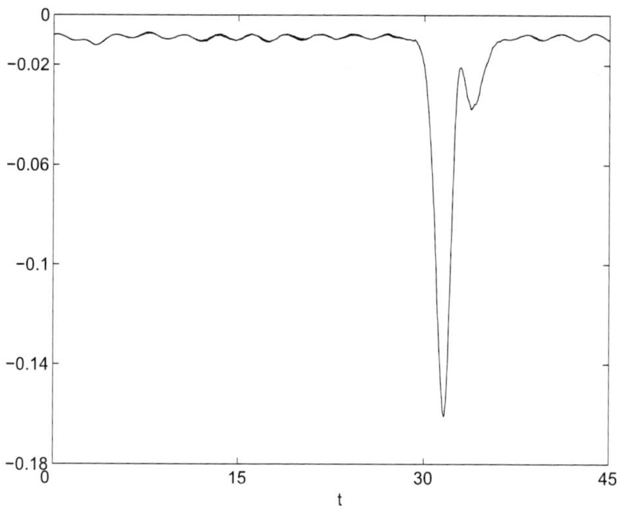

Figure 5. The energy errors of the Sine Gordon equation at$t \in [0, 45]$ by the scheme (57).

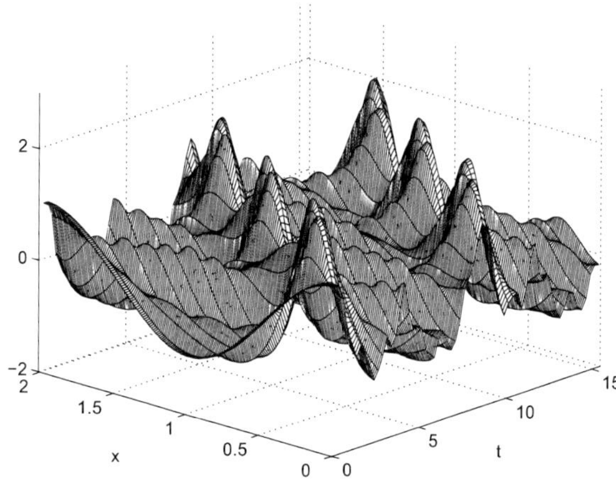

Figure 6. Numerical simulations of interactions of multi-solitons of the KDV equation at $t \in [0, 15]$.

Hamiltonian PDEs.

$$Mz_t + Kz_x = \nabla_z S(z), \tag{63}$$

where

$$M = \begin{pmatrix} 0 & \frac{1}{2} & 0 & 0 \\ -\frac{1}{2} & 0 & 0 & 0 \\ 0 & 0 & 0 & 0 \\ 0 & 0 & 0 & 0 \end{pmatrix}, \quad K = \begin{pmatrix} 0 & 0 & 0 & 1 \\ 0 & 0 & -c_2 & 0 \\ 0 & c_2 & 0 & 0 \\ -1 & 0 & \cdots & 0 \end{pmatrix}, \quad z = \begin{pmatrix} \phi \\ u \\ \nu \\ \omega \end{pmatrix},$$

and $S(z) = \frac{1}{2}\nu^2 - u\omega + V(u)$, Each of the two skew-symmetric matrices M and K can be identified with a closed two form.

$$\omega^1 =< M\mathbf{u}, \mathbf{v} >, \quad \omega^{\mathbf{2}} =< \mathbf{K}\mathbf{u}, \mathbf{v} >, \tag{64}$$

where **u**,**v** are any vectors on R^4 and $< \cdot, \cdot >$ is the standard Eulidean inner product on R^4. The multisymplectic Hamiltonian equation (63) satisfies the important multisymplectic conservation

$$\partial_t[dz \wedge Mdz] + \partial_x[dz \wedge Kdz] = 0, \tag{65}$$

which, for the KdV equation, is equivalent to

$$\partial_t[d\phi \wedge du] + 2\partial_x[d\phi \wedge d\omega + d\nu \wedge du] = 0 \tag{66}$$

where $\wedge$ is the standard exterior product operator of the differential forms.

Though the Preissman scheme is multisymplectic, it involves more effort to compute the auxiliary w, v, ϕ, so we eliminate w, v, ϕ by a trivial computation and obtain the following multisymplectic twelve-points scheme

$$\begin{aligned}\frac{1}{16\Delta t}(\delta_t u_{i+1}^j + 3\delta_t u_i^j + 3\delta_t u_{i-1}^j + \delta_t u_{i-2}^j) + \frac{1}{4\Delta x^3}(2\delta_x^3 u_i^j + \delta_x^3 u_i^{j+1} + \delta_x^3 u_i^{j-1} \\ + \frac{1}{4\Delta x}(V'(\bar{u}_i^{j-1} - V'(\bar{u}_{i-2}^{j-1}) + V'(\bar{(u)}_i^j) - V'(\bar{u}_{i-2}^j)) = 0,\end{aligned} \tag{67}$$

where $\delta_t u_i^j = u_i^{j+1} - u_i^{j-1}$, $\delta_x^3 u_i^j = u_{i+1}^j - 3u_i^j + 3u_{i-1}^j - u_{i-2}^j$, $\bar{u}_i^j = \frac{1}{4}(u_i^j + u_i^{j+1} + u_{i+1}^j + u_{i+1}^{j+1})$, and $u_m^n \approx u(m\Delta x, n\Delta t)$.

The evolution behaviors of the solitary wave of the Kdv equation were simulated by the multisymplectic twelve-points scheme (64). The parameters in the Kdv equation are taken as $c_1 = 1$, $c_2 = 0.08$. The space interval is $[0, 2]$ and the initial conditions is

$$u(x, 0) = cos(\pi x). \tag{68}$$

Figure.6 showed the evolution behaviors of the multi-solitary waves of the Kdv equation. The interactions of the multi-solitons of the Kdv equation can well be simulated by the multisymplectic twelve-points scheme (67). It is obvious that the multisymplectic method is an ideal method for simulating the evolution behaviors of the nonlinear solitary waves.

5. Conclusion

In the article, the symplectic method of the Hamilton system and the multi-symplectic method of the multi-symplectic structure of PDEs were introduced. The nonlinear solitary wave equations, such as the CNLS system, the Klein-Gordon equation, the Kdv equation, were simulated by the symplectic and multi-symplectic method. Numerical results showed that the symplectic and multisymplectic method can well simulate the behaviors of the nonlinear solitary waves and preserve the inner characters of the soliton equations. The symplectic and multisymplectic methods have wide applications in simulating the collision behaviors of the nonlinear solitary waves.

Acknowledgements

This work was supported by the National Natural Science Foundation of China (No.10401033).

References

[1] P.J.Olver, *Applications of Lie group to differential equations,* Springer, New York, (1986).

[2] R.Radhakrishnan, M.Lakshmanan, and J.Hietarinta, Inelastic collision and switching of coupled bright solitons in optical fibers,*Phys.Lett.E*, **56**,2213-2216, (1997).

[3] N.J.Zabusky, M.D.Kruskal, Interaction of soliton in a collision plasma and recurrence of initial states, *Phys.Rev.Lett. B*, **15,240-243**,(1965).

[4] Lee,S, and I.Held, Baroclinic wave packets in models and observations,*J.Atmos.Sci.*, **50**,1413-1428, (1993).

[5] S.Jimenez and L.Vazquez, Analysis of four numerical schemes for a nonlinear Klein Gordon equation, *Applied Mathematics and Computations*, **31**,61-94,(1990).

[6] Hua Wei, Liu Xue-Shen, Ding Pei-zhu, Dynamic study for numerical solutions of the Gross-Pitaevskii equation, *Chinese Journal of Computational Physics*, **23(4)**,483-488,(2006).

[7] Kanji Abe and Osamu Inoue, Fourier expansion solution of the Kortewegde Vries equation, *J.Comp.Phys*, **34(2)**,202-210, (1980).

[8] S.B.Winberg,J.F.Mcgrath, E.F.Gabl,etc., Implicit spectral methods for wave propagation problems, *J.Comp.Phys.* , **97**,311-336, (1991).

[9] R.I.McLachlan, Symplectic integration of Hamiltonian wave equations, *Numerical Mathematics,* **66**,465-470, (1994).

[10] R.I.McLachlan,Matthew Perlmutter, G.R.W.Quispel, On the nonlinear stabiltiy of symplectic integrators, *BIT Numerical Mathematics*, **44**,99-117,(2004).

[11] E.Hair, C.Lubich, On the life-span of backward error analysis.*Numer.Math.*, **76**,411 462, (1997).

[12] T.J.Bridges, Multi-symplectic structure and wave propagation, *Math. Proc. Camb. Phil. Soc.*, **121**, 147-190, (1997).

[13] S.Reich, Multi-symplectic Runge-Kutta collocation methods for Hamiltonian wave equations, *Journal of Computational Physics*, **156**:1-27, 59 (1999).

[14] J.E.Marsden, G.P.Patrick, S.Shkoller, Multisymplectic geometry variational integrators and nonlinear PDEs, *Comm.Math.Phys*, **199**, 351-359, (1999).

[15] S.Jimenez, L.Vazquez, Analysis of four numerical schemes for a nonlinear Klein Gordon equation, *Applied Mathematics and Computations*, **35**:61-94, (1990).

[16] S.Li, L.Vu-Quoc, Finite difference calculus invariant structure of a class algorithms for the nonlinear Klein Gordon equation, *SIAM J.Numer.Anal.*, **32(6)**, 1839-1875,(1995).

[17] P.J.Channell, J.C.Scovel, Symplectic integration of Hamiltonian systems, *Nonlinearity*, **3(2)**, 231-259, (1990).

[18] Y.F.Tang, A note on construction of high-order-symplectic schemes from lower-order one via formal energies, *J.Comput.Math*, **17(6)**, 561-568, (1999).

[19] Y.F.Tang, Yong-Hong Long, Formal energy of symplectic scheme for Hamiltonian systems and its applications(II), *Computer Math.Applic.*, **27(12)**,31-39, (1994).

[20] Jian-Qiang Sun, Xiao-Yan Gu, Zhong-Qi Ma, Numerical study of the soliton waves of the coupled nonlinear Schrödinger system, *Physica D*, **196**, 311-328,(2004).

[21] Jian-Qiang Sun, Meng-Zhao Qin, Multi-symplectic methods for the coupled 1D nonlinear Schrödinger system, *Computer Physics Communication*, **155**, 221-235,(2004).

[22] Jian-Qiang Sun, Xiao-Yan Gu, Zhong Qi Ma, Meng-Zhao Qin, RK-Cayley Fehlberg method on homogeneous manifolds, *Communciation in nonlinear science and numerical simulation*, **12**, 966-975, (2007).

[23] Jian-Qiang Sun,Meng-Zhao Meng, and Ting-Ting Liu, Total variation and multisymplectic structure for CNLS system, *Commun.Theor.Phys*, **46**,28-32,(2006).

[24] Jian-Qiang Sun, Zhong-Qi Ma, Meng-Zhao Qin, Simulation of envelope Rossby solitons in a pair of cublic Schrödinger equations, *Applied Mathematics and Computation*, **183**, 945-952, (2006).

[25] K.Feng, M.Z.Qin, Hamiltonian algorithms for hamiltonian systems and a comparative numerical study, *Comput.Phys.Commun*, **65**, 173-187, (1991).

[26] P.Y.Kuo, H.M.Wu, Numerical solution of K.D.V Equation, *J.Math.Anal.and Applic.*, **82**,334-345, (1981).

[27] P.F.Zhao, M.Z.Qin, Multisymplectic geometrity and multisymplectic schemes, *J.Phys.A:Math.Gen.*, **33**, 3612-3626, (2000).

[28] Y.J.Sun, M.Z.Qin, Construction of multisymplectic schemes of any finite order of modified wave equation, *J.M.P*, **41**,(11), 7854-7868, (2000).

[29] K.Feng, and S.Zai-Jiu, Volume-preserving algorithms for source-free dynamical systems, *Numer.Math*, **71**,451-463, (1995).

[30] Hong.J,Liu.Y and Sun.G, The multi-symplectic of partitional of Runge-Kutta method for Hamiltonian PDEs, *Math. Comput*, **75**,167-181, (1977).

[31] J.M.Sanz-Serna, Runge-Kutta schemes for Hamiltonian system, *BIT*, **28**,877-883, (1988).

[32] Brett N.Ryland, Robert.I.Mclachlan and Jason Frank, On multisymplectic of partitioned Runge-Kutta and splitting methods,*International Journal of Computer mathematics*, **70**, 59, (1977).

In: Numerical Simulation Research Progress
Editors: Simone P. Colombo et al, pp. 161-207
ISBN 978-1-60456-783-0

Chapter 4

SYMMETRY IN TURBULENCE SIMULATION

Dina Razafindralandy, Aziz Hamdouni and Marx Chhay
Université de La Rochelle
Avenue Michel Crépeau
17042, La Rochelle, France

Abstract

Large-eddy simulation (LES) appears to be a good alternative between DNS (direct numerical simulation) and RANS (Reynolds-Averaged Navier-Stokes) in turbulence simulation. However, the modelisation of the effect of the small scales of the flow on the large ones, which is the cornerstone of this technique, remains a challenge in the domain. In this chapter, a recent and advanced method which aims to derive models, in isothermal and non-isothermal cases, respecting the physical properties and the mathematical structure of the equations is presented. This method is based on the symmetry group of the equations, which plays a fundamental role in the understanding of physical phenomena (existence of conservation laws, ...). A large part of the chapter is devoted to expose some interesting (classical and new) applications of the symmetry group theory. A discretisation method preserving the symmetry group is also presented. Such a discretisation method is necessary to avoid the violation of physical properties of the flow at the discrete scale. The method is grounded on the theory of moving frames which will be introduced in a didactical way and applied on Burgers'equation.

1. Introduction

Most fluid flows in industrial applications are turbulent. As such, these flows contain a wide range of time and space scales, which makes a direct numerical simulation (DNS) a high consumer of time and computer-space, and even an impossibility in many cases. Indeed, a DNS must mimic the behaviour of every scale of the flow in order to be accurate. A technique which considerably reduces the cost of turbulent flow simulation is to apply an (ensemble- or time-) averaging to the equations. This method, called Reynolds-Averaged Navier–Stokes (RANS) method, is efficient in many cases but the averaging process removes too much information and, consequently, does not allow a fine description of the flow. A good compromise between DNS and RANS is the large-eddy simulation (LES) method [27]. It consists in describing only the large scales of the flow, the small scales being lost. The separation of large and small scale is done by filtering.

As well in RANS as in LES, the effect of the quantities lost during the averaging or the filtering operation must be modeled by adding a turbulence model term in the equations. This modelisation is done using various physical or mathematical assumptions [27, 28]. One of the most recent tools used to derive turbulence models in LES is the symmetry method. This method [7–9, 17] permits the construction of turbulence models respecting the physics and the mathematical structures of the equations. Oberlack was the first to suggest associating the symmetry method to turbulence modeling after stating that most classical turbulence models violate the symmetry properties of the Navier–Stokes equations [14]. This idea was followed by Razafindralandy and Hamdouni, who gave a concrete example of a class of symmetry-preserving and thermodynamically consistent LES turbulence models [25, 26] and extended the work to the non-isothermal case [23, 24].

Symmetries traduce fundamental physical properties. Indeed, thanks to Nœther's theorem, it is known that to each symmetry of a Lagrangian system corresponds a first integral ([13]). This theorem can be extended to non-Lagrangian systems ([10]). Many important properties of a flow, such as scaling laws, Kolmogorov -5/3 law, the existence of special, self-similar solutions, are also direct consequences of the scaling symmetries of the Navier–Stokes equations [6, 15, 16, 30]. Preserving the symmetry properties of the equations is then essential in LES turbulence modeling if we want the approximate solution given by the model to have the same physical qualities as the actual solution.

At the discrete scale, *i.e.* after a discretisation of the domain and the equations, the symmetry invariances should also not be broken. A way to preserve the symmetries of an equation at the discrete scale is to use an (symmetry-)invariantized scheme. The invariantization process, which utilizes the theory of moving frames [3], will be described through the example of the Burgers' equation.

Actually, invariantized schemes belong to the class of geometric integrators. Some of the most popular geometric integrators are symplectic integrators for Hamiltonian systems. Symplectic integrators are numerical schemes which preserve the physical properties of Hamiltonian systems, through the conservation of the symplectic structure. They are stable and provide numerical solutions which are accurate over a long integration time. Note that classical schemes such as the Runge-Kutta method introduce numerical (non-Hamiltonian) dissipation when applied to a Hamiltonian system and have a wrong long-term behaviour. The symplectic integrator method can however be applied only to a Hamiltonian system. The invariantization process enables one to circumvent this limitation.

This chapter aims to expose in a didactical way the power of the symmetry theory, not only in turbulence simulation but also in other fields. In section 2, some classical and new applications of the symmetry theory in mechanics are presented. Section 2 serves also to introduce progressively the basic concepts of the theory. This introduction is completed in section 3 where the systematic way of calculation of symmetries, developed by Lie, is synthetically exposed. The symmetry approach is used in section 4 for the derivation of turbulence models, in the isothermal and non-isothermal cases. Section 5 is devoted to the invariantization process of numerical schemes.

2. Panorama of the Application of Symmetries

We first begin with some definitions.

2.1. Basic Definitions

For a geometrical figure $\mathcal{F} \subset \mathbb{R}^M$, defined by an algebraic equation,

$$A(\boldsymbol{x}) = 0 \tag{1}$$

with $\boldsymbol{x} \in \mathbb{R}^M$, a symmetry transformation is a mapping of $\mathbb{R}^M$ onto $\mathbb{R}^M$ which leaves the figure $\mathcal{F}$ globally unchanged, or, equivalently, which leaves equation (1) unchanged. For example, in $\mathbb{R}^2$, rotations of angle $a = 2k\pi/3$, $k \in \{0, 1, 2\}$, are symmetries of an equilateral triangle. The set of these symmetry transformations has a group structure (closure, existence of neutral element, reversibility of each element, associativity). This group is called a symmetry group of the triangle. Reflections about the medians constitute another symmetry group of the equilateral triangle. Rotations of arbitrary angle $a \in \mathbb{R}$ constitute a symmetry group of a circle. Its elements are called continuous symmetry because they depend continuously on the parameter a (in opposition to discrete symmetries like the rotation symmetries of the triangle).

The concept of symmetry can be extended to differential equations. Consider, a differential equation

$$E(\boldsymbol{x}, \boldsymbol{u}) = 0 \tag{2}$$

where $\boldsymbol{u}$ is a function of $\boldsymbol{x}$ and E is a differential operator. A transformation g:

$$(\boldsymbol{x}, \boldsymbol{u}) \mapsto g(\boldsymbol{x}, \boldsymbol{u}) = (\widehat{\boldsymbol{x}}, \widehat{\boldsymbol{u}}) \tag{3}$$

is called symmetry tranformation of (2) if g leaves equation (2) unchanged, that is

$$E(\boldsymbol{x}, \boldsymbol{u}) = 0 \qquad \Longrightarrow \qquad E(\widehat{\boldsymbol{x}}, \widehat{\boldsymbol{u}}) = 0. \tag{4}$$

A symmetry transformation maps a solution of (2) to another solution. When (4) holds, equation (2) is said invariant under g. For example, consider the linear heat equation

$$\frac{\partial u}{\partial t} - \frac{\partial^2 u}{\partial x^2} = 0. \tag{5}$$

with $\boldsymbol{x} = (t, x)$ and $\boldsymbol{u} = u$. Under the scaling transformation

$$(t, x, u) \mapsto (\widehat{t} = \lambda^2 t, \widehat{x} = \lambda x, \widehat{u} = u), \tag{6}$$

where λ is a real parameter, equation (5) becomes

$$\frac{\partial \widehat{u}}{\partial \widehat{t}} - \frac{\partial^2 \widehat{u}}{\partial \widehat{x}^2} = 0. \tag{7}$$

which is the same equation (5). Transformation (6) is then a symmetry of (5).

If a set of transformations has a group structure, it is called a transformation group. When, moreover, each transformation of the group is a symmetry of a given equation, this group is called a symmetry group of the equation.

Note that a symmetry of a differential equation can be introduced as a symmetry in the sens of geometrical figures. However, doing this requires the introduction of more advanced mathematical tools that we wish to skip (see [17]).

In conjunction with the notion of invariance of an equation, the notion of invariance of a function can be defined. A function F is said invariant under a transformation group G if

$$F(g(\boldsymbol{x},\boldsymbol{u})) = F(\boldsymbol{x},\boldsymbol{u}) \qquad \forall\, g \in G. \tag{8}$$

This condition implies that G is a symmetry group of the equation

$$F(\boldsymbol{x},\boldsymbol{u}) = 0,$$

but the inverse does not hold.

In what follows, we take a particular interest in one-parameter transformations, *i.e.* transformations of the form

$$g_a : \ (\boldsymbol{x},\boldsymbol{u}) \mapsto \big(\widehat{\boldsymbol{x}}(\boldsymbol{x},\boldsymbol{u},a), \widehat{\boldsymbol{u}}(\boldsymbol{x},\boldsymbol{u},a)\big), \tag{9}$$

which depend continuously on a parameter a. Note that the expression of the elements g_a of a one-parameter transformation group can always be arranged such that the group is additive ($g_a \circ g_{a'} = g_{a+a'}$) and that the neutral element corresponds to $a = 0$. For example, transformations (6) can be written:

$$(t,x,u) \mapsto (\widehat{t} = \mathrm{e}^{2a}\, t, \widehat{x} = \mathrm{e}^{a}\, x, \widehat{u} = u).$$

A one-parameter transformation group is a Lie group when the composition and the inversion operators are differentiable applications.

The knowledge of symmetries of an equation gives precious information on the equation. For example, symmetries may be used to lower the order of the equation. They also permit to find self-similar solutions. More generally, from one known solution, symmetries enable to construct other solutions.

Some applications of the symmetry theory in the resolution of differential equations are shown in the next subsections.

2.2. Resolution of a Riccati Equation

Consider the Riccati equation:

$$\frac{\mathrm{d}u}{\mathrm{d}x} = xu^2 - 2\frac{u}{x} - \frac{1}{x^3}. \tag{10}$$

Under a change of variables

$$s = \ln x, \qquad v = x^2 u \tag{11}$$

equation (10) simplifies into

$$\frac{\partial v}{\partial s} = v^2 - 1. \tag{12}$$

This equation is autonomous in s and can be solved by separation of the variables. The solution is:

$$v(s) = -\tanh(s + C).$$

The question here is how to "guess" the suitable change of variables (11). In order to get an autonomous equation, we must transform equation (10) into an equivalent one, with new variables (s, v), which is invariant under the translation

$$(s, v) \mapsto (s + a, v). \tag{13}$$

To this aim, we notice that any transformation

$$g_a : (x, u) \mapsto (\mathrm{e}^a x, \mathrm{e}^{-2a} u) \tag{14}$$

is a symmetry of (10). It is easy to check that

$$g_a(\ln x, x^2 u) = (\ln x + a, x^2 u). \tag{15}$$

So, a suitable change of variable which ensures that the new equation is autonomous is

$$s = \ln x, \qquad v = x^2 u. \tag{16}$$

In summary, the symmetry approach permitted to find the adequate change of variables to solve a Riccati equation. In the next example, it is used to find an integrating factor.

2.3. Integrating Factor

Consider a first order differential equation, expressed in the form:

$$Q(x, u)\,\mathrm{d}x + P(x, u)\,\mathrm{d}u = 0. \tag{17}$$

There exists a function μ of x and u, called integrating factor, such that equation (17) multiplied by μ can be written in a form:

$$\mathrm{d}w = 0 \tag{18}$$

where w is a function of x an u. The solution of the equation is then:

$$w = C$$

where C is a constant. For example, an integrating factor of the equation

$$(xu + u^2) + (x^2 - xu)\frac{\mathrm{d}u}{\mathrm{d}x} = 0 \tag{19}$$

is

$$\mu = \frac{1}{(xu + u^2)x + (x^2 - xu)u} = \frac{1}{2x^2 u}. \tag{20}$$

Indeed, a multiplication of (19) by μ yields:

$$\frac{x+u}{2x^2}\,\mathrm{d}x + \frac{x-u}{2xu}\,\mathrm{d}u = 0$$

or

$$\mathrm{d}\left[\ln x + \ln u - \frac{u}{x} + C\right] = 0.$$

where C is a constant.

The role of an integrating factor is to facilitate the solving of an ODE. The symmetry group theory provides a way to find some by using the following result (see [17]). If equation (17) is invariant under a one-parameter symmetry group of elements

$$g_a : (x, u) \mapsto (\widehat{x}, \widehat{u})$$

then the function

$$\mu = \frac{1}{\xi Q + \eta P} \tag{21}$$

is an integrating factor, where

$$\xi = \left.\frac{\mathrm{d}\widehat{x}}{\mathrm{d}a}\right|_{a=0}, \qquad \eta = \left.\frac{\mathrm{d}\widehat{u}}{\mathrm{d}a}\right|_{a=0}.$$

In the case of an homogeneous equation like (19), the transformation group consituted by scalings:

$$(x, u) \mapsto (\mathrm{e}^a\, x, \mathrm{e}^a\, u)$$

is a symmetry group, and the corresponding integrating factor is

$$\mu = \frac{1}{xQ + uP}. \tag{22}$$

Applied to equation (19), it gives the integrating factor (20).

In these two examples, symmetries permitted to solve completely ODE's. In the following subsections, we show how to find self-similar solutions and to reduce a partial differential equation into an ODE using symmetry properties.

2.4. Reduction of a Partial Differential Equation

Consider again the example of the heat equation (5). As shown, this equation is invariant under the group of scaling transformations

$$(t, x, u) \mapsto (\mathrm{e}^{2a}\, t, \mathrm{e}^a\, x, u). \tag{23}$$

A self-similar solution of (5) under (23) is a solution $u(t, x)$ such that

$$\widehat{u}(\widehat{t}, \widehat{x}) = u(t, x)$$

or

$$u(\mathrm{e}^{2a}\, t, \mathrm{e}^{a}\, x) = u(t, x). \tag{24}$$

Deriving this equation with respect to a and taking $a = 0$ lead to

$$2t\frac{\partial u}{\partial t} + x\frac{\partial u}{\partial x} = 0. \tag{25}$$

The integration of this equation by the characteristic method shows that

$$u(t, x) = w\left(\frac{x}{\sqrt{t}}\right) \tag{26}$$

where w is an arbitrary function. Self-similar solutions of the heat equation (5) under scalings (23) have then the form (26). This result can be used to reduce the equation.

Let $s = x/\sqrt{t}$. Using (26), we get:

$$\frac{\partial u}{\partial t} = -\frac{1}{2}\frac{\mathrm{d}w}{\mathrm{d}s}\frac{x}{t^{3/2}}, \qquad \frac{\partial^2 u}{\partial x^2} = \frac{\mathrm{d}^2 w}{\mathrm{d}s^2}\frac{1}{t}. \tag{27}$$

Equation (5) becomes:

$$\frac{s}{2}\frac{\mathrm{d}w}{\mathrm{d}s} + \frac{\mathrm{d}^2 w}{\mathrm{d}s^2} = 0. \tag{28}$$

In opposition to the initial one, this equation has only one independant variable. It can be solved as a linear equation to give

$$w(s) = C_1 + C_2\,\mathrm{erf}(s/2)$$

where C_1 and C_2 are arbitrary constants and erf is the Gaussian error function. We deduce that self-similar solutions of (71) are:

$$u(t, x) = C_1 + C_2\,\mathrm{erf}\left(\frac{x}{2\sqrt{t}}\right). \tag{29}$$

Throughout these examples, we noticed the particular importance of scaling symmetries. In the next subsection, we recall the elementary way to find such symmetries, and use them further for the reduction of the equations of thin shear layer flows.

2.5. 2D Laminar Thin Shear Layer Flows

Consider the general equations of steady and laminar incompressible 2D thin layer flows without pressure gradient:

$$\begin{cases} \dfrac{\partial u}{\partial x} + \dfrac{\partial v}{\partial y} = 0, \\[2ex] u\dfrac{\partial u}{\partial x} + v\dfrac{\partial u}{\partial y} = \nu\dfrac{\partial^2 u}{\partial y^2}, \end{cases} \tag{30}$$

where $\boldsymbol{x} = (x, y)$ and $\boldsymbol{u} = (u, v)$ are respectively the space and velocity variables and ν the kinematic viscosity. Associated to boundary conditions, these equations modelise the motion of various shear layer flows (boundary layer flow, jet, wake, ...). Rather than to separate these kinds of flows (as it is the usual way in literature, [29]) for the seeking of self-similar solutions, we give a unified method to study them.

2.5.1. Scaling Symmetries and Self-similar Solutions

A scaling symmetry of (30) is a transformation

$$(x, y, u, v) \mapsto (\widehat{x} = a^{n_1} x, \widehat{y} = a^{n_2} y,\ \widehat{u} = a^{n_3} u,\ \widehat{v} = a^{n_4} v) \tag{31}$$

under which equations (30) remains unchanged. For simplicity, we write the scaling transformation in a multiplicative way rather than additive as above (*i.e.* the coefficients are of the form a^n instead of e^{na}). Equations (30) are unchanged if and only if:

$$\begin{cases} n_3 - n_1 = n_4 - n_2 \\ \\ 2n_3 - n_1 = n_4 + n_3 - n_2 = n_3 - 2n_2. \end{cases}$$

The solution of this set of equations is

$$n_3 = n_1 - 2n_2, \qquad n_4 = -n_2. \tag{32}$$

Consequently, for any values of n_1 and n_2, the following scaling transformation is a symmetry of (30):

$$(x, y, u, v) \mapsto (a^{n_1} x,\ a^{n_2} y,\ a^{n_1 - 2n_2} u,\ a^{-n_2} v). \tag{33}$$

Self-similar solutions of (30) are defined by

$$\widehat{u}(\widehat{x}, \widehat{y}) = u(x, y), \qquad \widehat{v}(\widehat{x}, \widehat{y}) = v(x, y),$$

that is

$$a^{n_1 - 2n_2} u(a^{n_1} x, a^{n_2} y) = u(x, y), \qquad a^{-n_2} v(a^{n_1} x, a^{n_2} y) = v(x, y).$$

Deriving this equation with respect to a and taking $a = 1$, it follows:

$$(n_1 - 2n_2)u + n_1 x \frac{\partial u}{\partial x} + n_2 y \frac{\partial u}{\partial y} = 0, \qquad -n_2 v + n_1 x \frac{\partial v}{\partial x} + n_2 y \frac{\partial v}{\partial y} = 0. \tag{34}$$

If $n_1 \neq 0$ and $n_2 \neq 0$, the characteristic method leads to:

$$\begin{aligned} u(x, y) &= x^{(n_1 - 2n_2)/n_1} U\left(\frac{y}{x^{n_2/n_1}}\right), \\ v(x, y) &= x^{-n_2/n_1} V\left(\frac{y}{x^{n_2/n_1}}\right), \end{aligned} \tag{35}$$

U and V are functions such that (35) is a solution of (30). Self-similar solutions are then:

$$\begin{cases} u(x,y) = x^{1-2\alpha} U\left(\dfrac{y}{x^{\alpha}}\right), \\ \\ v(x,y) = x^{-\alpha} V\left(\dfrac{y}{x^{\alpha}}\right), \end{cases} \tag{36}$$

where $\alpha = \dfrac{n_2}{n_1}$ is an arbitrary constant. The choice of α depends on the problem, as we will see later.

The extension to the cases where n_1 or n_2 vanishes is skipped. In the next subsection, the reduced equation is deduced.

2.5.2. Reduction of the Equations

Using the similarity variable $s = y/x^{\alpha}$ and relations (36), the incompressibility condition becomes:

$$(1-2\alpha)U(s) - \alpha z\dot{U}(s) + \dot{V}(s) = 0 \tag{37}$$

and the momentum equation:

$$U(s)\big[(1-\alpha)U(s) - \alpha s\dot{U}(s)\big] + \dot{U}V(s) = \nu\ddot{U}(s) \tag{38}$$

where the superscript dot symbol ($\dot{}$) stands for derivative. Equation (37) can be integrated, and, if h is a primitive of U such that, for simplicity, $h(0) = V(0) = 0$, the momentum equation gives the reduced equation:

$$(1-2\alpha)\dot{h}^2(s) - (1-\alpha)h(s)\ddot{h}(s) = \nu\dddot{h}(s). \tag{39}$$

Our approach enables to give a general form of the similariy variable for any kind of thin shear layer flows, and hence, to have a general form of the reduced equation. A particular flow can afterwards be specified by fixing the value of α as shown in the following examples.

2.5.3. Examples of Values of α

As a first example, we consider a free jet flow. For a jet flow, $u \longrightarrow 0$ when $y \to \pm\infty$. It can be shown that, consequently to this condition, the momentum flux is constant ([29]):

$$\int_{-\infty}^{+\infty} u^2(x,y)\,\mathrm{d}y = \text{constant}. \tag{40}$$

Since

$$\int_{-\infty}^{+\infty} u^2(x,y)\,\mathrm{d}y = x^{2-3\alpha}\int_{-\infty}^{+\infty} U^2(s)\,\mathrm{d}s,$$

condition (40) is satisfied if and only if

$$\alpha = \frac{2}{3}. \tag{41}$$

With this value of α, the function h verifies:

$$\dot{h}^2 + h\ddot{h} + 3\nu\dddot{h} = 0 \tag{42}$$

In the case of a boundary layer flow over a flat plane, boundary conditions leads to

$$\alpha = \frac{1}{2},$$

and h satisfies the Blasius equation

$$h(s)\ddot{h}(s) + 2\nu\dddot{h}(s) = 0. \tag{43}$$

The above analysis can be extended to the non-isothermal case.

2.6. Non-isothermal Laminar Thin Shear Layer Flows

In the non-isothermal case, the set of equations are

$$\begin{cases} \dfrac{\partial u}{\partial x} + \dfrac{\partial v}{\partial y} = 0, \\[2ex] u\dfrac{\partial u}{\partial x} + v\dfrac{\partial u}{\partial y} = \nu\dfrac{\partial^2 u}{\partial y^2} + \beta\theta, \\[2ex] u\dfrac{\partial \theta}{\partial x} + v\dfrac{\partial \theta}{\partial y} = \kappa\dfrac{\partial^2 \theta}{\partial y^2}. \end{cases} \tag{44}$$

where β is the product of the thermal expansion coefficient by the gravity acceleration and κ the thermal diffusivity. θ is the temperature, or, more precisely, the difference between the actual temperature and a reference temperature. We assume that $\beta \neq 0$.

Using similar arguments as above, it can be deduced that the scaling symmetries of (44) are

$$(x, y, u, v) \mapsto (a^{n_1}x,\ a^{n_2}y,\ a^{n_1-2n_2}u,\ a^{-n_2}v,\ a^{n_1-4n_2}\theta). \tag{45}$$

When $n_1 \neq 0$, the corresponding self-similar solutions are

$$\begin{cases} u(x,y) = x^{1-2\alpha}U\left(\dfrac{y}{x^\alpha}\right), \\[2ex] v(x,y) = x^{-\alpha}V\left(\dfrac{y}{x^\alpha}\right), \\[2ex] \theta(x,y) = x^{1-4\alpha}\Theta\left(\dfrac{y}{x^\alpha}\right). \end{cases} \tag{46}$$

where α is an arbitrary constant. These solutions reduce the initial equations (44) into:

$$\begin{cases} (1-2\alpha)\dot{h}^2 - (1-\alpha)h\ddot{h} = \nu\dddot{h} + \beta\Theta, \\ \\ (1-4\alpha)\dot{h}\Theta - (1-\alpha)h\dot{\Theta} = \kappa\ddot{\Theta}. \end{cases} \tag{47}$$

The value of α is determined by the boundary conditions. For example, for a thermal plume, boundary conditions induces that the temperature flux

$$\int_0^{+\infty} u\theta \, \mathrm{d}y = x^{2-5\alpha} \int_0^{+\infty} U(s)\,\Theta(s)\,\mathrm{d}s \tag{48}$$

must be constant. This leads to

$$\alpha = \frac{2}{5}.$$

We continue in exposing some interesting results in mechanics, and especially in turbulence, obtained with the symmetry theory. In the next subsection, we show, thanks to symmetry analysis that Burger's vortex and shear layer are self-similar solution of the Navier–Stokes equations (Grassi *et al.* [6]).

2.7. Burger's Vortex and Shear Layer Solutions of the Navier–Stokes Equations

Consider an isothermal Newtonian fluid with kinematic viscosity ν, and, to simplify, density 1. If t denotes the time variable, $\boldsymbol{x} = (x, y, z)$ the space variable, $\boldsymbol{u} = (u, v, w)$ the velocity field and p the pressure, the motion of this fluid is governed by the Navier–Stokes equations (NS):

$$\begin{cases} \dfrac{\partial \boldsymbol{u}}{\partial t} + \operatorname{div}(\boldsymbol{u} \otimes \boldsymbol{u}) + \nabla p - \operatorname{div} \mathsf{T} = 0 \\ \\ \operatorname{div} \boldsymbol{u} = 0 \end{cases} \tag{49}$$

In these expressions, $\mathsf{T} = 2\nu\mathsf{S}$ is the viscous constraint tensor and S is the strain rate tensor.

NS are invariant under space translations. In particular, the one-directional translation

$$(t, \boldsymbol{x}, \boldsymbol{u}, p) \mapsto (t, \widehat{\boldsymbol{x}}, \boldsymbol{u}, p) \qquad \text{with} \qquad \widehat{x} = x + a,\ \widehat{y} = y,\ \widehat{z} = z, \tag{50}$$

where a is the parameter, is a symmetry of (49). Equations (49) also invariant under time translations:

$$(t, \boldsymbol{x}, \boldsymbol{u}, p) \mapsto (t + a, \widehat{\boldsymbol{x}}, \boldsymbol{u}, p). \tag{51}$$

Self-similar solutions under both (50) and (51) are:

$$\begin{cases} u(t, x, y, z) = U(y, z), \\ v(t, x, y, z) = V(y, z), \\ w(t, x, y, z) = W(y, z), \\ p(t, x, y, z) = P(y, z), \end{cases} \tag{52}$$

where the U, V, W and P are arbitrary functions. Reduction of NS under (50) and (51) yields:

$$\begin{cases} V\dfrac{\partial U}{\partial y} + W\dfrac{\partial U}{\partial z} - \nu\left(\dfrac{\partial^2 U}{\partial y^2} + \dfrac{\partial^2 U}{\partial z^2}\right) = 0, \\[2ex] V\dfrac{\partial V}{\partial y} + W\dfrac{\partial V}{\partial z} + \dfrac{\partial P}{\partial y} - \nu\left(\dfrac{\partial^2 V}{\partial y^2} + \dfrac{\partial^2 V}{\partial z^2}\right) = 0, \\[2ex] V\dfrac{\partial W}{\partial y} + W\dfrac{\partial W}{\partial z} + \dfrac{\partial P}{\partial z} - \nu\left(\dfrac{\partial^2 W}{\partial y^2} + \dfrac{\partial^2 W}{\partial z^2}\right) = 0, \\[2ex] \dfrac{\partial V}{\partial y} + \dfrac{\partial W}{\partial z} = 0. \end{cases} \tag{53}$$

Hence,

$$\begin{cases} V = -\gamma y, \qquad W = \gamma z, \\[2ex] P = -\dfrac{\gamma^2}{2}[y^2 + z^2], \\[2ex] -\gamma y\dfrac{\partial U}{\partial y} + \gamma z\dfrac{\partial U}{\partial z} - \nu\left(\dfrac{\partial^2 U}{\partial y^2} + \dfrac{\partial^2 U}{\partial z^2}\right) = 0. \end{cases} \tag{54}$$

A particular solution of these equations, in which U does not depend on z and with $\gamma = \nu$, is the Burgers'shear layer solution:

$$\boldsymbol{u} = (\Gamma\,\mathrm{erf}(y/2),\ -\nu y,\ \nu z)^T \tag{55}$$

where Γ is a constant.

Other interesting symetry transformations of the Navier–Stokes equations are rotations with constant angles. For example, rotations in z direction

$$(t, \boldsymbol{x}, \boldsymbol{u}, p) \mapsto (t, \widehat{\boldsymbol{x}}, \widehat{\boldsymbol{u}}, p) \tag{56}$$

with

$$\begin{aligned} &\widehat{x} = x\cos a + y\sin a, \qquad \widehat{y} = -x\sin a + y\cos a, \qquad \widehat{z} = z \\ &\widehat{u} = u\cos a + v\sin a, \qquad \widehat{v} = -u\sin a + v\cos a, \qquad \widehat{w} = w \end{aligned} \tag{57}$$

are symmetries of (49). Self-similar solutions under these transformations are solutions depending only on t, z and $r = (x^2 + y^2)^{1/2}$:

$$\begin{cases} u(t, x, y, z) = U(t, r, z), \\ v(t, x, y, z) = V(t, r, z), \\ w(t, x, y, z) = W(t, r, z), \\ p(t, x, y, z) = P(t, r, z). \end{cases} \tag{58}$$

As shown in [6], a special solution of the corresponding reduced equations is the Burgers'vortex solution:

$$U = \frac{-\nu r}{2}, \qquad V = \frac{1 - \mathrm{e}^{-r^2/4}}{r}, \qquad W = \nu z. \tag{59}$$

In all of these examples, we saw that some particular symmetries can lead to physically interesting solutions, and, in some cases, to the complete resolution of the equation. However, we took only some special, and sometimes trivial, symmetries of the considered equations. Though, other symmetries may also be interesting. For example, Ünal showed ([30]) that scaling symmetries of the Navier–Stokes yield self-similar solutions for which the energy dissipation rate satisfies the Kolmogorov hypothesis. Scaling symmetries also parmitted Oberlack ([15]) to deduce scaling laws which predict well numerical results ([16]). In the next section, we expose briefly the method of Lie of computing all one- and infinite-parameter symmetries of an equation. One can refer to [7, 17] for complementary information.

3. Computation of One-Parameter Symmetries

Consider a group G of one-parameter transformations:

$$g_a : \ (\boldsymbol{x}, \boldsymbol{u}) \mapsto \big(\widehat{\boldsymbol{x}}(\boldsymbol{x}, \boldsymbol{u}, a), \widehat{\boldsymbol{u}}(\boldsymbol{x}, \boldsymbol{u}, a)\big), \tag{60}$$

where $\boldsymbol{x} = (x^i)_{i=1,\dots,M}$ and $\boldsymbol{u} = (u^i)_{i=1,\dots,Q}$.

Let the vectors $\boldsymbol{\xi}$ and $\boldsymbol{\eta}$ denote the infinitesimal variations of $\boldsymbol{x}$ and $\boldsymbol{u}$ under a group G of one-parameter transformations of the form (9) at $a = 0$:

$$\boldsymbol{\xi} = \left.\frac{\mathrm{d}\widehat{\boldsymbol{x}}}{\mathrm{d}a}\right|_{a=0}, \qquad \boldsymbol{\eta} = \left.\frac{\mathrm{d}\widehat{\boldsymbol{u}}}{\mathrm{d}a}\right|_{a=0} \tag{61}$$

which was already introduced in section 2.3.. The variation of a C^∞ function $F = F(\boldsymbol{x}, \boldsymbol{u})$ under trnasformation (60) at $a = 0$ is represented by

$$\boldsymbol{X}.F = \boldsymbol{\xi} \cdot \frac{\partial F}{\partial \boldsymbol{x}} + \boldsymbol{\eta} \cdot \frac{\partial F}{\partial \boldsymbol{u}}, \tag{62}$$

with

$$\boldsymbol{\xi} \cdot \frac{\partial F}{\partial \boldsymbol{x}} = \sum_{m=1}^{M} \xi_m \frac{\partial F}{\partial x_m} \qquad \text{and} \qquad \boldsymbol{\eta} \cdot \frac{\partial F}{\partial \boldsymbol{x}} = \sum_{q=1}^{Q} \eta_q \frac{\partial F}{\partial u_q}.$$

The vector field $\boldsymbol{X} = (\boldsymbol{\xi}, \boldsymbol{\eta})$, which can be written as follow:

$$\boldsymbol{X} = \boldsymbol{\xi} \cdot \frac{\partial}{\partial \boldsymbol{x}} + \boldsymbol{\eta} \cdot \frac{\partial}{\partial \boldsymbol{u}}, \tag{63}$$

is called infinitesimal generator of the group. For example, the infinitesimal generator of the group of rotations

$$(x^1, x^2, u^1, u^2) \mapsto$$

$$(x^1 \cos a - x^2 \sin a,\ x^1 \sin a + x^2 \cos a, u^1 \cos a - u^2 \sin a,\ u^1 \sin a + u^2 \cos a) \quad (64)$$

is

$$\boldsymbol{X} = -x^2 \frac{\partial}{\partial x^1} + x^1 \frac{\partial}{\partial x^2} - u^2 \frac{\partial}{\partial u^1} + u^1 \frac{\partial}{\partial u^2}. \quad (65)$$

The infinitesimal generator characterises the transformation group. Indeed, thanks to the group structure, the expression of $(\widehat{\boldsymbol{x}}, \widehat{\boldsymbol{u}})$ can be determined from $\boldsymbol{X}$ using:

$$\begin{cases} \dfrac{\mathrm{d}\widehat{\boldsymbol{x}}}{\mathrm{d}a} = \boldsymbol{\xi}(\widehat{\boldsymbol{x}}, \widehat{\boldsymbol{u}}), & \dfrac{\mathrm{d}\widehat{\boldsymbol{u}}}{\mathrm{d}a} = \boldsymbol{\eta}(\widehat{\boldsymbol{x}}, \widehat{\boldsymbol{u}}), \\ \\ \widehat{\boldsymbol{x}}(a=0) = \boldsymbol{x}, & \widehat{\boldsymbol{u}}(a=0) = \boldsymbol{u}. \end{cases} \quad (66)$$

Formula (62) is valid for a function F of $\boldsymbol{x}$ and $\boldsymbol{u}$. If we want to calculate the variation of an operator E which depends not only on $\boldsymbol{x}$ and $\boldsymbol{u}$ but also on derivatives of $\boldsymbol{u}$ up to ordre K, the vector field $\boldsymbol{X}$ must be prolonged into another vector field $\boldsymbol{X}^{(K)}$:

$$\boldsymbol{X}^{(K)} = \boldsymbol{\xi} \cdot \frac{\partial}{\partial \boldsymbol{x}} + \boldsymbol{\eta} \cdot \frac{\partial}{\partial \boldsymbol{u}} + \boldsymbol{\eta}^1 \cdot \frac{\partial}{\partial \boldsymbol{u}^{(1)}} + \cdots + \boldsymbol{\eta}^K \cdot \frac{\partial}{\partial \boldsymbol{u}^{(K)}} \quad (67)$$

wich takes into account the infinitesimal variation of these derivatives at $a = 0$. In this expression, $\boldsymbol{\eta}^k$ is defined for $k = 1, ..., K$ by:

$$\widehat{\boldsymbol{u}}^{(k)} = \boldsymbol{u}^{(k)} + a\boldsymbol{\eta}^k + O(a^2)$$

where $\widehat{\boldsymbol{u}}^{(k)}$ represents the derivatives of order k of $\widehat{\boldsymbol{u}}$ with respect to $\widehat{\boldsymbol{x}}$. We have:

$$\boldsymbol{\xi} = \boldsymbol{X}^{(K)}.\boldsymbol{x}, \qquad \boldsymbol{\eta} = \boldsymbol{X}^{(K)}.\boldsymbol{u}, \qquad \boldsymbol{\eta}^k = \boldsymbol{X}^{(K)}.\boldsymbol{u}^{(k)}.$$

The components of $\boldsymbol{\eta}^1 = \boldsymbol{X}^{(K)}\boldsymbol{u}^{(1)}$ are calculated using the formula:

$$\boldsymbol{X}^{(K)} \left(\frac{\partial u^q}{\partial x^m} \right) = D_{x^m} \boldsymbol{X}^{(K)} u^q - \sum_{i=1}^{M} \frac{\partial u^q}{\partial x^i} D_{x^m} \boldsymbol{X}^{(K)} x^i \quad (68)$$

where D_{x^m} represents the total derivation with respect to x^m. This formula is applied recursively to obtain the variation of higher derivatives.

Recall that our goal in this section is to show how to compute the symmetries of an equation. In fact, the symmetry condition (4) can be written in terms of the infinitesimal generator. Indeed, consider a differential equation of order K:

$$E(\boldsymbol{x}, \boldsymbol{u}) = 0. \quad (69)$$

Then, a group G of one-parameter transformations, generated by a vector field $\boldsymbol{X}$, is a symmetry group of (69) if and only if

$$E(\boldsymbol{x}, \boldsymbol{u}) = 0 \qquad \Longrightarrow \qquad \boldsymbol{X}^{(K)}.E = 0 \tag{70}$$

This condition determines $\boldsymbol{\xi}$ and $\boldsymbol{\eta}$ and, further, the symmetry groups of the equation.

To illustrate the procedure of calculation of symmetry groups, we take the example of the heat equation:

$$u_t - u_{xx} = 0. \tag{71}$$

In this example, $\boldsymbol{x} = (t, x)$ and $K = 2$. We denote $\boldsymbol{\xi} = (\tau, \xi)$ and $f_z = \dfrac{\partial f}{\partial z}$ for any function f and $z = t$, x or u. Using (68), we have:

$$\boldsymbol{X}^{(2)}(u_t) = \eta_t + u_t(\eta_u - \tau_t) - u_t^2\tau_u - u_x\xi_t + u_x u_t \xi_u. \tag{72}$$

Applying (68) twice, we get:

$$\begin{aligned} \boldsymbol{X}^{(2)}(u_{xx}) = \;& \eta_{xx} + u_{xx}(\eta_u - 2\xi_x) + u_x(\eta_{ux} - \tau_{xx} - \xi_{xx} + \eta_{xu}) - 2u_{tx}\tau_x \\ & -2u_{tx}u_x\tau_u - u_t u_{xx}\tau_u - 2u_t u_x \tau_{xu} - u_{xx}u_x(2\xi_u + \tau_u) \\ & +u_x^2(-2\xi_{ux} + \eta_{uu}) - u_t u_x^2 \tau_{uu} + u_x^3 \xi_{uu}. \end{aligned} \tag{73}$$

Symmetry condition (70) applied on (71) leads to

$$\boldsymbol{X}^{(2)}(u_t) - \boldsymbol{X}^{(2)}(u_{xx}) = 0. \tag{74}$$

The left-hand side of (74) is a polynomial in the derivatives of u. Considering that $u_t = u_{xx}$, this polynomial vanishes when

$$\begin{gathered} \tau_x = 0, \quad \tau_u = 0, \quad \xi_u = 0, \quad \eta_{uu} = 0, \\ \eta_t - \eta_{xx} = 0, \quad \tau_t - 2\xi_x = 0, \quad \xi_t - \xi_{xx} + 2\eta_{ux} = 0. \end{gathered} \tag{75}$$

The solution of (75) is

$$\begin{cases} \tau = b_1 t^2 + b_2 t + b_3, \\ \xi = b_1 t x + \dfrac{1}{2} b_2 x + b_4 t + b_5, \\ \eta = -\dfrac{1}{4} b_1(2t + x^2)u + \dfrac{1}{2} b_4 x u + b_6 u + b_7(t, x) \end{cases} \tag{76}$$

where b_i is a constant for $i = 1, ..., 6$ and $b_7(t, x)$ is an arbitrary solution of (71). The infinitesimal generator of the symmetry group of (71) is:

$$\boldsymbol{X} = \tau\frac{\partial}{\partial t} + \xi\frac{\partial}{\partial x} + \eta\frac{\partial}{\partial u} \tag{77}$$

which is a linear combination of the following vector fields:

$$\boldsymbol{X}_1 = -4t^2\frac{\partial}{\partial t} - 4tx\frac{\partial}{\partial x} + (2t + x^2)u\frac{\partial}{\partial u}, \qquad \boldsymbol{X}_2 = 2t\frac{\partial}{\partial t} + x\frac{\partial}{\partial x},$$

$$\boldsymbol{X}_3 = \frac{\partial}{\partial t}, \qquad \boldsymbol{X}_4 = 2t\frac{\partial}{\partial x} - xu\frac{\partial}{\partial u}, \qquad \boldsymbol{X}_5 = \frac{\partial}{\partial x}, \tag{78}$$

$$\boldsymbol{X}_6 = u\frac{\partial}{\partial u}, \qquad \boldsymbol{X}_7 = b_7(t,x)\frac{\partial}{\partial u}.$$

Each $\boldsymbol{X}_i$, $i = 1, ..., 6$, generates a one-parameter symmetry group of (71) and $\boldsymbol{X}_7$ generates an infinite-parameter one. A generic element of each of these groups can be calculated by solving (66). For instance, we get, with $\boldsymbol{X}_1$, the system:

$$\begin{cases} \dfrac{\mathrm{d}\widehat{t}}{\mathrm{d}a} = -4\widehat{t}^2, & \dfrac{\mathrm{d}\widehat{x}}{\mathrm{d}a} = -4\widehat{t}\widehat{x}, & \dfrac{\mathrm{d}\widehat{u}}{\mathrm{d}a} = (2\widehat{t} + \widehat{x}^2)\widehat{u}, \\ \widehat{t}(0) = t, & \widehat{x}(0) = x, & \widehat{u}(0) = u \end{cases} \tag{79}$$

which solution is:

$$\widehat{t} = \frac{t}{4ta+1}, \qquad \widehat{x} = \frac{x}{4ta+1}, \qquad \widehat{u} = u\sqrt{4at+1}\exp\left(\frac{x^2a}{4at+1}\right). \tag{80}$$

The next vector fields generate respectively the groups of the following transformations:

$$\begin{aligned} &\boldsymbol{X}_2 : (t,x,u) \mapsto (\mathrm{e}^{2a}\,t, \mathrm{e}^{a}\,x, u), && \boldsymbol{X}_3 : (t,x,u) \mapsto (t+a, x, u) \\ &\boldsymbol{X}_4 : (t,x,u) \mapsto (t, x+2ta, u\,\mathrm{e}^{-xa-ta^2}), && \boldsymbol{X}_5 : (t,x,u) \mapsto (t, x+a, u) \\ &\boldsymbol{X}_6 : (t,x,u) \mapsto (t, x, \mathrm{e}^{a}\,u), && \boldsymbol{X}_7 : (t,x,u) \mapsto (t, x, u + b_7(t,x)). \end{aligned} \tag{81}$$

Transformations (80)-(81) consitute the Lie symmetry group of (71).

In the next sections, we consider the symmetry group of the Navier–Stokes equations, with and without equation for temperature, and show how to derive symmetry-preserving turbulence models.

4. Symmetry in Turbulence Modeling

We first consider the case of an isothermal flow.

4.1. Isothermal Navier–Stokes Equations

Consider a Newtonian fluid governed by the isothermal Navier–Stokes equations (49). From this section on, we denote the components of the space variable $\boldsymbol{x}$ and the velocity field $\boldsymbol{u}$ with subscript: $\boldsymbol{x} = (x^i)_{i=1,2,3}$, $\boldsymbol{u} = (u^i)_i$. The symmetry group of (49) is generated by the following vector fields ([1,5,20]):

$$
\begin{aligned}
\boldsymbol{X}_1 &= \frac{\partial}{\partial t}, \\
\boldsymbol{X}_2 &= \zeta(t)\frac{\partial}{\partial p}, \\
\boldsymbol{X}_{i,j} &= -x^j\frac{\partial}{\partial x^i} + x^i\frac{\partial}{\partial x^j} - u^j\frac{\partial}{\partial u^i} + u^i\frac{\partial}{\partial u^j}, \qquad i = 1,2, \quad j > i, \\
\boldsymbol{X}_{5+i} &= \alpha^i(t)\frac{\partial}{\partial x^i} + \dot{\alpha}^i(t)\frac{\partial}{\partial u^i} - x^i\,\ddot{\alpha}^i(t)\frac{\partial}{\partial p}, \qquad i = 1,2,3, \\
\boldsymbol{X}_9 &= 2t\frac{\partial}{\partial t} + \sum_{j=1}^{3}\left(x^j\frac{\partial}{\partial x^j} - u^j\frac{\partial}{\partial u^j}\right) - 2p\frac{\partial}{\partial p},
\end{aligned}
\tag{82}
$$

where ζ and $\boldsymbol{\alpha} = (\alpha^i)$ are arbitrary functions. These generators correspond to symmetries which act on t, $\boldsymbol{x}$, $\boldsymbol{u}$ and p. One can also consider symmetries which act on the parameter ν of the flow:

$$(t, \boldsymbol{x}, \boldsymbol{u}, p, \nu) \mapsto (\widehat{t}, \widehat{\boldsymbol{x}}, \widehat{\boldsymbol{u}}, \widehat{p}, \widehat{\nu}). \tag{83}$$

Such a symmetry, called *equivalence transformation*, maps a solution of NS into a new solution of NS with a different value of ν. Doing so, we obtain an additional infinitesimal generator ([11]):

$$\boldsymbol{X}_{10} = \sum_{j=1}^{3} x^j\frac{\partial}{\partial x^j} + \sum_{j=1}^{3} u^j\frac{\partial}{\partial u^j} + 2p\frac{\partial}{\partial p} + 2\nu\frac{\partial}{\partial \nu}. \tag{84}$$

These vector fields generate the symmetry group of NS, which is spanned by the following multi- or infinite-parameter transformations:

- the *time translations* which correspond to $\boldsymbol{X}_1$:

$$(t, \boldsymbol{x}, \boldsymbol{u}, p) \mapsto (t + a, \boldsymbol{x}, \boldsymbol{u}, p), \tag{85}$$

- the *pressure translations* generated by $\boldsymbol{X}_2$:

$$(t, \boldsymbol{x}, \boldsymbol{u}, p) \mapsto (t, \boldsymbol{x}, \boldsymbol{u}, p + \zeta(t)), \tag{86}$$

- the *rotations* corresponding to the three $\boldsymbol{X}_{i,j}$'s:

$$(t, \boldsymbol{x}, \boldsymbol{u}, p) \mapsto (t, \mathsf{R}\boldsymbol{x}, \mathsf{R}\boldsymbol{u}, p), \tag{87}$$

- the *generalized Galilean transformations* characterized by $\boldsymbol{X}_6$, $\boldsymbol{X}_7$ and $\boldsymbol{X}_8$:

$$(t, \boldsymbol{x}, \boldsymbol{u}, p) \mapsto \left(t, \boldsymbol{x} + \boldsymbol{\alpha}(t), \boldsymbol{u} + \dot{\boldsymbol{\alpha}}(t), p - \boldsymbol{x}\cdot\ddot{\boldsymbol{\alpha}}(t) - \frac{1}{2}\,\boldsymbol{\alpha}(t).\ddot{\boldsymbol{\alpha}}(t)\right), \tag{88}$$

- the *first scaling transformations* corresponding to $\boldsymbol{X}_9$:

$$(t, \boldsymbol{x}, \boldsymbol{u}, p) \mapsto (\mathrm{e}^{2a}\, t, \mathrm{e}^{a}\, \boldsymbol{x}, \mathrm{e}^{-a}\, \boldsymbol{u}, \mathrm{e}^{-2a}\, p) \tag{89}$$

- and the *second scaling transformations* corresponding to $\boldsymbol{X}_{10}$:

$$(t, \boldsymbol{x}, \boldsymbol{u}, p, \nu) \mapsto (t, \mathrm{e}^{a}\, \boldsymbol{x}, \mathrm{e}^{a}\, \boldsymbol{u}, \mathrm{e}^{2a}\, p, \mathrm{e}^{2a}\, \nu) \tag{90}$$

In these expressions, R is a constant rotation matrix, *i.e.* $\mathsf{R}^T\mathsf{R} = \mathsf{I_d}$ and $\det \mathsf{R} = 1$, $\mathsf{I_d}$ being the identity matrix. Note that spatial translations can be obtained from (88) by choosing $\boldsymbol{\alpha}$ constant and the classical Galilean transformation by choosing $\boldsymbol{\alpha}$ linear in t. The first scaling transformations (89) show how the velocity and the pressure are affected when the spatio-temporal scale is multiplied by $(\mathrm{e}^a, \mathrm{e}^{2a})$, and the second scaling transformations (90) show the consequence of a space-scale change on the velocity and the pressure.

NS admit other known symmetries which could not be calculated by the same method. They are:

- the *reflections* which are discrete symmetries:

$$(t, \boldsymbol{x}, \boldsymbol{u}, p) \mapsto (t, \Lambda\boldsymbol{x}, \Lambda\boldsymbol{u}, p), \tag{91}$$

where

$$\Lambda = \begin{pmatrix} \iota_1 & 0 & 0 \\ 0 & \iota_2 & 0 \\ 0 & 0 & \iota_3 \end{pmatrix} \qquad \text{with} \qquad \iota_i = \pm 1, \;\; i = 1, 2, 3$$

- and the *material indifference* in the limit of a 2D flow in a simply connected domain ([2]):

$$(t, \boldsymbol{x}, \boldsymbol{u}, p) \mapsto (t, \widehat{\boldsymbol{x}}, \widehat{\boldsymbol{u}}, \widehat{p}), \tag{92}$$

with

$$\widehat{\boldsymbol{x}} = \mathsf{R}(t)\, \boldsymbol{x}, \qquad \widehat{\boldsymbol{u}} = \mathsf{R}(t)\, \boldsymbol{u} + \dot{\mathsf{R}}(t)\, \boldsymbol{x}, \qquad \widehat{p} = p - 2a\phi + \frac{a^2}{2}\|\boldsymbol{x}\|^2 \tag{93}$$

where $\mathsf{R}(t)$ is a 2D rotation matrix with angle at, ϕ the usual 2D stream function defined by:

$$\boldsymbol{u} = \mathrm{curl}(\phi \boldsymbol{e}_3), \tag{94}$$

$\boldsymbol{e}_3$ the unit vector perpendicular to the plane of the flow and $\|.\|$ the Euclidian norm.

Although transformations (93) constitute a one-parameter group, generated by:

$$-tx^2\frac{\partial}{\partial x^1} + tx^1\frac{\partial}{\partial x^2} - (x^2 + u^2 t)\frac{\partial}{\partial u^1} + (x^1 + u^1 t)\frac{\partial}{\partial u^2} - 2\phi\frac{\partial}{\partial p}, \tag{95}$$

they are not of the same kind as the previous ones because ϕ, and consequently $\widehat{p}$, cannot be formulated as local functions of t, $\boldsymbol{x}$, $\boldsymbol{u}$ and p.

As seen in introduction, symmetries have an important role in analysing equations. In some extent, they contains the physics of the equations. So, LES turbulent models have to respect them, otherwise fundamental properties of the flow may be lost during the modeling. In the next subsection, some standard turbulence models are analyzed under the symmetry view.

4.2. Turbulence Model Analysis

LES consists in filtering the equation of motion (49) and taking as an approximation of the actual solution $(\boldsymbol{u}, p)$ its filtered value $(\overline{\boldsymbol{u}}, \overline{p})$. Filtering (49) gives:

$$\begin{cases} \dfrac{\partial \overline{\boldsymbol{u}}}{\partial t} + \operatorname{div}(\overline{\boldsymbol{u}} \otimes \overline{\boldsymbol{u}}) + \nabla \overline{p} = \operatorname{div}(\overline{\mathsf{T}} + \mathsf{T}_s) \\ \\ \operatorname{div} \overline{u} = 0 \end{cases} \tag{96}$$

where T_s is the subgrid stress tensor defined by $\mathsf{T}_s = \overline{\boldsymbol{u}} \otimes \overline{\boldsymbol{u}} - \overline{\boldsymbol{u} \otimes \boldsymbol{u}}$ which must be modeled to close the equations. Currently, a very large number of models exists. Some of the most common ones are listed here:

- the Smagorinsky model:

$$\mathsf{T}_s^d = (C_S \overline{\delta})^2 |\overline{\mathsf{S}}| \overline{\mathsf{S}} \tag{97}$$

where the superscript ^{d} denotes the deviatoric part of a tensor, $C_S \simeq 0.148$ is the Smagorinsky constant, $\overline{\delta}$ the filter width and

$$|\overline{\mathsf{S}}| = \sqrt{2\ \operatorname{tr}(\overline{\mathsf{S}}^2)},$$

- the dynamic model:

$$\mathsf{T}_s^d = C_d \overline{\delta}^2 |\overline{\mathsf{S}}| \overline{\mathsf{S}}, \qquad \text{with} \qquad C_d = \frac{\operatorname{tr}(\mathsf{LM})}{\operatorname{tr} \mathsf{M}^2}. \tag{98}$$

where,

$$\mathsf{L} = \widetilde{\overline{\boldsymbol{u}}} \otimes \widetilde{\overline{\boldsymbol{u}}} - \widetilde{\overline{\boldsymbol{u}} \otimes \overline{\boldsymbol{u}}}, \qquad \mathsf{M} = \overline{\delta}^2 \widetilde{|\overline{\mathsf{S}}| \overline{\mathsf{S}}} - \widetilde{\overline{\delta}}^2 |\widetilde{\overline{\mathsf{S}}}| \widetilde{\overline{\mathsf{S}}} \tag{99}$$

and the tilde ($\tilde{\ }$) symbolizes a test filter, with a filter width $\widetilde{\overline{\delta}} > \overline{\delta}$,

- the structure function model:

$$\mathsf{T}_s^d = C_{SF} \overline{\delta} \sqrt{\overline{F}_2(\overline{\delta})}\ \overline{\mathsf{S}} \tag{100}$$

where

$$\overline{F}_2(\overline{\delta}) = \int_{\mathbb{R}^3} \overline{f}_2(\boldsymbol{x}, \overline{\delta})\, \mathrm{d}\boldsymbol{x},$$

f_2 is the structure function and C_{SF} is a constant,

- the gradient model:

$$\mathsf{T}_s = -\frac{\overline{\delta}^2}{12} \nabla\overline{\boldsymbol{u}}\ {}^T\nabla\overline{\boldsymbol{u}}, \tag{101}$$

- the Taylor model:

$$\mathsf{T}_s = -\frac{\overline{\delta}^2}{12} \nabla\overline{\boldsymbol{u}}\ {}^T\nabla\overline{\boldsymbol{u}} + C_g\overline{\delta}^2|\overline{\mathsf{S}}|\overline{\mathsf{S}} \tag{102}$$

  C_g being a constant,

- the rationnal model:

$$\mathsf{T}_s = -\frac{\overline{\delta}^2}{12} G * [\nabla\overline{\boldsymbol{u}}\ {}^T\nabla\overline{\boldsymbol{u}}] + C_g\overline{\delta}^2|\overline{\mathsf{S}}|\overline{\mathsf{S}}, \tag{103}$$

  where G is the kernel of the Gaussian filter and the symbol star ($*$) symbolizes the convolution operator,

- the similarity model:

$$\mathsf{T}_s = \widetilde{\overline{\boldsymbol{u}}} \otimes \widetilde{\overline{\boldsymbol{u}}} - \widetilde{\overline{\boldsymbol{u}} \otimes \overline{\boldsymbol{u}}}, \tag{104}$$

- the Lund–Novikov model:

$$\begin{aligned} -\mathsf{T}_s^d &= C_1\overline{\delta}^2|\overline{\mathsf{S}}|\overline{\mathsf{S}} + C_2\overline{\delta}^2(\overline{\mathsf{S}}^2)^d + C_3\overline{\delta}^2(\overline{\mathsf{W}}^2)^d \\ &+ C_4\overline{\delta}^2(\overline{\mathsf{S}}\,\overline{\mathsf{W}} - \overline{\mathsf{W}}\,\overline{\mathsf{S}}) + C_5\overline{\delta}^2\frac{1}{|\overline{\mathsf{S}}|}(\overline{\mathsf{S}}^2\overline{\mathsf{W}} - \overline{\mathsf{S}}\,\overline{\mathsf{W}}^2), \end{aligned} \tag{105}$$

  where $\overline{W}$ is the filtered vorticity tensor,

- and the Kosovic model:

$$-\mathsf{T}_s^d = (C_0\overline{\delta})^2\left[2|\overline{\mathsf{S}}|\overline{\mathsf{S}} + C_1(\overline{\mathsf{S}}^2)^d + C_2(\overline{\mathsf{S}}\,\overline{\mathsf{W}} - \overline{\mathsf{W}}\,\overline{\mathsf{S}})\right] \tag{106}$$

  which is a simplification of (105).

More details on these models may be found in [12, 27].

The approximate solution $(\overline{\boldsymbol{u}}, \overline{p})$ is expected to retrieve as much of the physical information as possible as the actual solution $(\boldsymbol{u}, p)$ contains, such as conservation laws, spectral properties, etc. We should then require that any symmetry of the equations of $(\boldsymbol{u}, p)$ is also a symmetry of equations (96) which govern $(\overline{\boldsymbol{u}}, \overline{p})$. More precisely, if a transformation

$$g\ :\ (t, \boldsymbol{x}, \boldsymbol{u}, p) \mapsto (\widehat{t}, \widehat{\boldsymbol{x}}, \widehat{\boldsymbol{u}}, \widehat{p}) \tag{107}$$

is a symmetry of NS equations, T_s should be modeled such that the same transformation, applied to the filtered quantities:

$$g\ :\ (t, \boldsymbol{x}, \overline{\boldsymbol{u}}, \overline{p}) \mapsto (\widehat{t}, \widehat{\boldsymbol{x}}, \widehat{\overline{\boldsymbol{u}}}, \widehat{\overline{p}}) \tag{108}$$

is a symmetry of (96). This requirement, that we summarize by

$$g \text{ symmetry of } (49) \implies g \text{ symmetry of } (96), \tag{109}$$

induces several conditions on T_s. In what follows, we check whether common models are compatible with this condition. When this is the case, the model is said invariant (this is not the definition (8) of an invariant function).

Condition (109) holds when g is a time or pressure translation, (85) or (86), provided that the model does not depend explicitely on t and p, which is generally the case. It is also straight forward to check that the mentionned models are generalized-Galilean invariant.

Equations (96) are invariant under rotations (87) and reflection (91) if, like in continuum mechanics,

$$\widehat{\mathsf{T}}_s = \Upsilon\, \mathsf{T}_s\, {}^T\Upsilon \tag{110}$$

where

$$\widehat{\mathsf{T}}_s = \mathsf{T}(\widehat{\boldsymbol{x}}, \widehat{\boldsymbol{u}}) = \mathsf{T}_s(\Upsilon \boldsymbol{x}, \Upsilon \boldsymbol{u})$$

and Υ is a (constant) rotation or reflection matrix. This relation is verified by the cited models.

Next, the two categories of scaling transformations (89) and (90) can be combined into two-parameter scaling transformations:

$$(t, \boldsymbol{x}, \boldsymbol{u}, p, \nu) \mapsto (\mathrm{e}^{2a}\, t,\ \mathrm{e}^{b+a}\, \boldsymbol{x},\ \mathrm{e}^{b-a}\, \boldsymbol{u},\ \mathrm{e}^{2b-2a}\, p,\ \mathrm{e}^{2b}\, \nu). \tag{111}$$

The first scaling transformation (89) corresponds to $b = 0$ and the second, (90), to $a = 0$. Equations (96) are invariant under (111) if and only if

$$\widehat{\mathsf{T}}_s = \mathrm{e}^{2b-2a}\, \mathsf{T}_s. \tag{112}$$

Among the listed models, only the dynamic model and the similarity model satisfy this condition, even when $a = 0$ or $b = 0$. Only these two models are then invariant under the scaling symmetries of the equations.

The same conclusion can be drawn for the 2D material indifference (93) under which the filtered equations (96) are invariant if and only if (details are in [14,22,24]):

$$\widehat{\mathsf{T}}_s = \mathsf{R}\mathsf{T}_s\, {}^T\mathsf{R}. \tag{113}$$

Indeed, only the dynamic model and the similarity model verify this last condition under some condition on the test filter (see [24]).

In summary, only few LES turbulence models are consistent with the symmetry structure of the Navier–Stokes equations. As consequence, many common models do not does not restore fundamental information on the flow, such as scaling laws which are linked to scaling symmetries ([15]). The incompatibility with the material indifference may also explain why many turbulence models are unable to correctly predict rotating flows.

In what follows, an example of a way to derive symmetry-preserving LES models is given.

4.3. Example of Symmetry-Preserving Turbulence Models

To sum up, a turbulence model preserves the symmetries of NS when it is autonomous in time and pressure, compatible the generalized Galilean transformation and verifies conditions (110), (112) and (113). Since the filtered strain tensor $\overline{\mathsf{S}}$ is generalized-Galilean invariant, the two first conditions are fulfilled if the model depends only on $\overline{\mathsf{S}}$ and on some scalar quantities which will be determined later:

$$\mathsf{T}_s = \mathsf{T}_s(\overline{\mathsf{S}}).$$

The Cayley–Hamilton theorem associated to the rotation invariance and material indifference requirements (110) and (113) leads then to:

$$\mathsf{T}_s^d = A(\chi,\zeta)\,\overline{\mathsf{S}} + B(\chi,\zeta)\ \mathrm{Adj}^d\,\overline{\mathsf{S}} \tag{114}$$

where

$$\chi = \operatorname{tr}\overline{\mathsf{S}}^2 \qquad \text{and} \qquad \zeta = \det\overline{\mathsf{S}}$$

are the invariants of $\overline{\mathsf{S}}$ under rotations, Adj stands for the operator defined by

$$(\,\mathrm{Adj}\,\overline{\mathsf{S}})\overline{\mathsf{S}} = (\det\overline{\mathsf{S}})\mathsf{I_d}, \tag{115}$$

($\mathrm{Adj}\,\overline{\mathsf{S}}$ is simply the comatrix of $\overline{\mathsf{S}}$) and A and B are arbitrary scalar functions.

Next, T_s is invariant under the first scaling transformation (89) if

$$\widehat{\mathsf{T}}_s = e^{-2a}\mathsf{T}_s. \tag{116}$$

A sufficient condition to verify this relation is that A and B are such that:

$$A(e^{-4a}\chi, e^{-6a}\zeta) = A(\chi,\zeta), \qquad B(e^{-4a}\chi, e^{-6a}\zeta) = e^{2a}B(\chi,\zeta), \tag{117}$$

since

$$\widehat{\overline{\mathsf{S}}} = \mathrm{e}^{-2a}\,\overline{\mathsf{S}} \qquad \text{and} \qquad \widehat{\mathrm{Adj}\,\overline{\mathsf{S}}} = \mathrm{e}^{-4a}\ \mathrm{Adj}\,\overline{\mathsf{S}}. \tag{118}$$

Differentiating according to a and taking $a = 0$, it follows:

$$-4\chi\frac{\partial A}{\partial\chi} - 6\zeta\frac{\partial A}{\partial\zeta} = 0, \qquad -4\chi\frac{\partial B}{\partial\chi} - 6\zeta\frac{\partial B}{\partial\zeta} = 2B. \tag{119}$$

To satisfy these equalities, one can take

$$A(\chi,\zeta) = A_1\!\left(\frac{\zeta}{\chi^{3/2}}\right), \qquad B(\chi,\zeta) = \frac{1}{\sqrt{\chi}}\,B_1\!\left(\frac{\zeta}{\chi^{3/2}}\right)$$

where A_1 and B_1 are arbitrary scalar functions. Finally, if

$$v = \frac{\zeta}{\chi^{3/2}}$$

then the model takes the form:

$$\mathsf{T}_s^d = A_1(v)\,\overline{\mathsf{S}} + \frac{1}{\sqrt{\chi}} B_1(v)\ \mathrm{Adj}^d\,\overline{\mathsf{S}}. \tag{120}$$

Relation (120) defines a model which is invariant under the second scaling transformation. The quantity v is invariant under the scaling transformations (89) and is dimensionless. A_1 and B_1 have the dimension of a viscosity. In order to have the correct dimensions for T_s, let us introduce the turbulent subgrid-scale energy q and the dissipation rate ε, defined by:

$$q = \frac{1}{2}\overline{\boldsymbol{u}'^2} \qquad \text{and} \qquad \varepsilon = 2\nu\ \mathrm{tr}\left[\overline{\mathsf{S}'^2}\right].$$

Since $\dfrac{q^2}{\varepsilon}$ has the dimension of a viscosity, we can choose that:

$$A_1(v) = \frac{q^2}{\varepsilon}\,A_2(v) \qquad \text{and} \qquad B_1(v) = \frac{q^2}{\varepsilon}\,B_2(v)$$

where A_2 and B_2 are dimensionless arbitrary scalar functions.

As mentionned by Oberlack ([14]), $\boldsymbol{u}'$ transforms in the same way than $\overline{\boldsymbol{u}}$ under the symmetries of the NS equations, except under the generalized Galilean transformation (88) and the 2D material transformation (93) for which, respectively:

$$\widehat{\boldsymbol{u}'} = \boldsymbol{u}',$$

$$\widehat{\boldsymbol{u}'} = \mathsf{R}\boldsymbol{u}'$$

(note that $\widehat{\overline{\boldsymbol{u}} + \boldsymbol{u}'}$ must be equal to $\widehat{\boldsymbol{u}}$). It can then be shown that each symmetry of the NS equations leaves q and ε invariant, except the two scaling transformations. However, the ratio q^2/ε is invariant under the first scaling transformations. Hence, the model defined by

$$\mathsf{T}_s^d = \frac{q^2}{\varepsilon}\left(A_2(v)\,\overline{\mathsf{S}} + \frac{1}{\sqrt{\chi}} B_2(v)\ \mathrm{Adj}^d\,\overline{\mathsf{S}}\right). \tag{121}$$

verifies condition (116) and is then invariant under the first scaling symmetry transformations.

The last requirement is the invariance of (96) under the second scaling transformations (90) which can be traduced by:

$$\widehat{\mathsf{T}}_s = \mathrm{e}^{2a}\,\mathsf{T}. \tag{122}$$

Since $\widehat{\overline{\mathsf{S}}} = \overline{\mathsf{S}}$, $\widehat{q} = \mathrm{e}^{2a}$ and $\widehat{\varepsilon} = \mathrm{e}^{2a}\,\varepsilon$ under (90), this condition is automatically fulfilled.

In summary, relation (121) defines a class of subgrid models which are invariant under all symmetry groups of the NS equations. In this example, we choosed q and ε as scalar parameters. Ohter choices are possible. Moreover, the symmetry approach allow us to consider other physical requirements through A_2 and B_2. In what follows, we will choose these functions such that the model conforms to the second law of thermodynamics. As shown in ([26]), the conformity to this principle guarantees the stability of the model.

4.4. Consequences of the Second Law of Thermodynamics

T_s represents the energy exchange between resolved and subgrid scales. It generates certain dissipation, which, however, may take a negative value (in the case of a backscatter). Nevertheless, in order to respect the second law of thermodynamics, we must ensure that the total dissipation (sum of the molecular and the subgrid dissipations) remains positive.

On molecular scale, the viscous constraint derives from a "potential":

$$\mathsf{T} = \frac{\partial \psi}{\partial \mathsf{S}}.$$

where the potential is

$$\psi = \nu \operatorname{tr} \mathsf{S}^2 = \nu \chi.$$

This "potential" form is important because the convexity and positivity of ψ ensures that the molecular dissipation is positive:

$$\Phi = \operatorname{tr}(\mathsf{TS}) \geq 0.$$

The tensor T_s can be considered as a subgrid constraint generating a dissipation

$$\Phi_s = \operatorname{tr}(\mathsf{T}_s \overline{\mathsf{S}}).$$

To remain compatible with the NS equations, we assume that T_s has the same form as T, *i.e.*:

$$\mathsf{T}_s = \frac{\partial \psi_s}{\partial \overline{\mathsf{S}}}. \tag{123}$$

where ψ_s is a function of the invariants χ and ζ of $\overline{\mathsf{S}}$. In this way, the constraint tensor of the filtered equations, $\overline{\mathsf{T}} + \mathsf{T}_s$, has the same "potential" form as the constraint tensor of the Navier–Stokes equations. This hypothesis refines class (120) in the following way.

One deduces from (123) that

$$\mathsf{T}_s^d = 2\frac{\partial \psi_s}{\partial \chi}\overline{\mathsf{S}} + \frac{\partial \psi_s}{\partial \zeta} \operatorname{Adj}^d \overline{\mathsf{S}}.$$

Comparing this with (120), we obtain:

$$\frac{q^2}{\varepsilon}\frac{1}{2}A_2(v) = \frac{\partial \psi_s}{\partial \chi}, \qquad \frac{q^2}{\varepsilon}\frac{1}{\sqrt{\chi}}B_2(v) = \frac{\partial \psi_s}{\partial \zeta}.$$

Thus, compatibility between the latest two equations requires:

$$\frac{\partial}{\partial \zeta}\left(\frac{1}{2}A_2(v)\right) = \frac{\partial}{\partial \chi}\left(\frac{1}{\sqrt{\chi}}B_2(v)\right).$$

If b is a primitive of B_2, a solution of this equation is

$$A_2(v) = 2b(v) - 3v\dot{b}(v) \qquad \text{and} \qquad B_2(v) = \dot{b}(v). \tag{124}$$

Then, hypothesis (123) induces the existence of a real-valued dimensionless scalar continued function b such that:

$$\mathsf{T}_s^d = \frac{q^2}{\varepsilon}\left[\left(2b(v) - 3v\dot{b}(v)\right)\overline{\mathsf{S}} + \frac{1}{\sqrt{\chi}}\dot{b}(v)\ \mathrm{Adj}^d\,\overline{\mathsf{S}}\right]. \tag{125}$$

Now, let Φ_T be the total dissipation. We have:

$$\Phi_T = \mathrm{tr}[(\overline{\mathsf{T}} + \mathsf{T}_s)\overline{\mathsf{S}}].$$

Using (120) and (124), it can be shown that $\Phi_T \geq 0$ if and only if

$$\nu + \frac{q^2}{\varepsilon} b(v) \geq 0. \tag{126}$$

Consequently, if b satisfies the condition (126) then the total dissipation is positive.

In summary, a model belonging to the class (125) with a real continuous dimensionless function b verifying (126) is a model respecting the symmetry group of the NS equations and is conform with the second law of thermodynamics.

5. Numerical Test

A numerical test on a symmetry-preserving and thermodynamically consistent model was carried out in ([21,25]). The main results are reproduced here.

We consider an air flow inside a ventilated room (Nielsen's cavity, figure 1) which interests us for applications in building domain. The Reynolds number, computed from the height of the intake opening is about 5000. The function b is chosen linear. The numerical scheme, based on a finite difference discretisation method, is described in [4]. A $72\times52\times26$ grid is used. The time step is 7.10^{-3}s and the calculation is led to 1200s.

Figure 2 reports the horizontal velocity profiles, along the vertical line defined by ($x_1 = 2L/3$, $x_3 = W/2$), given by the Smagorinsky model, the dynamic model and our model (called here "invariant model"). It can be observed that the invariant model gives a result in good agreement with experiments, except near the floor, but in all cases better than those provided by the two other models. Near the upper wall, it predicts particularly well the velocity profile. Note that no wall function was used.

The above analysis and derivation of turbulence model were done for an isothermal fluid. This study can be extended to the case of non-isothermal fluid, as shown in the next subsection.

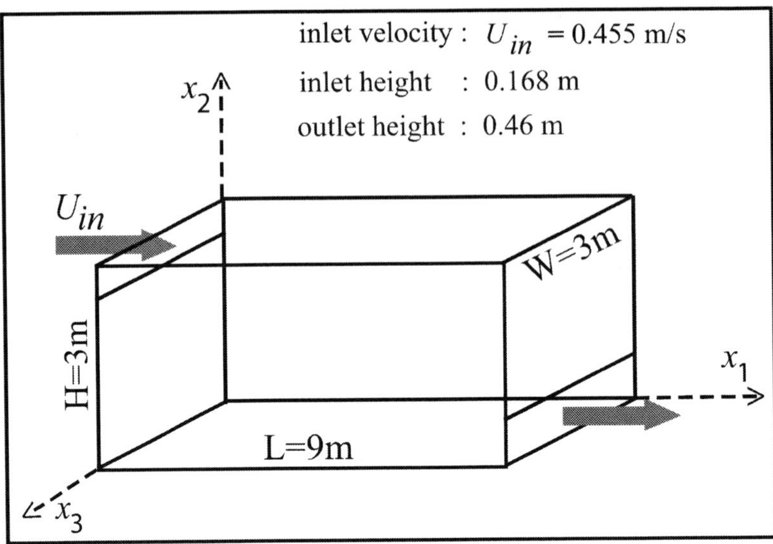

Figure 1. Geometry of the room.

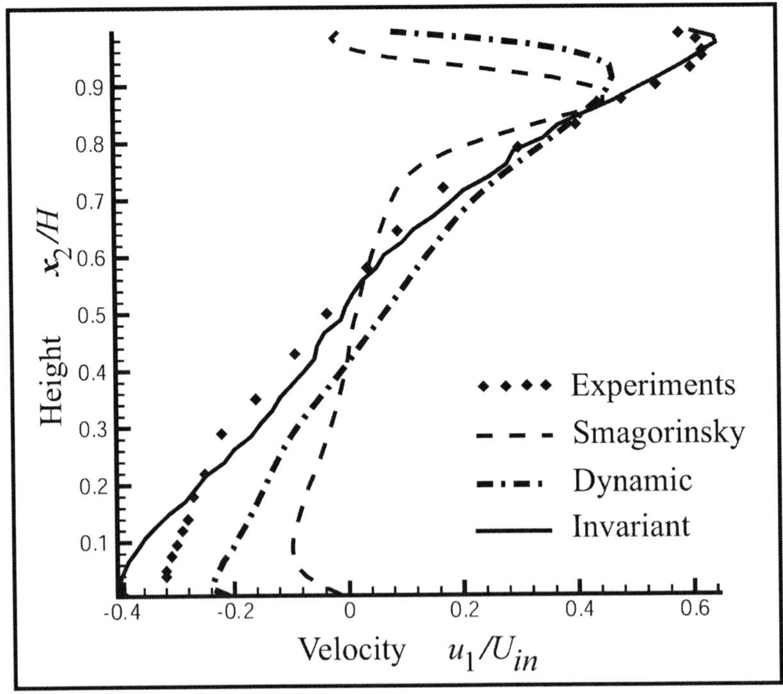

Figure 2. Mean velocity profiles at $x_1 = 2L/3$, $x_3 = W/2$.

5.1. Non-isothermal Flow

The non-isothermal Navier–Stokes equations are

$$\begin{cases} \dfrac{\partial \boldsymbol{u}}{\partial t} + \operatorname{div}(\boldsymbol{u} \otimes \boldsymbol{u}) + \dfrac{1}{\rho}\nabla p - \operatorname{div}\, \mathsf{T} - \beta\theta\, \boldsymbol{e}_3 = 0 \\ \\ \dfrac{\partial \theta}{\partial t} + \operatorname{div}(\theta \boldsymbol{u}) - \operatorname{div}\, \mathbf{h} = 0 \\ \\ \operatorname{div}\, \boldsymbol{u} = 0 \end{cases} \tag{127}$$

In these expression, $\mathbf{h} = \kappa \nabla \theta$ is the heat flux and θ, β and κ are as defined in section 2.6.. $\boldsymbol{e}_3$ is the ascending vertical unitary vector.

The symmetry group of (127) is generated by the following vector fields

$$\begin{aligned} \boldsymbol{X}_1 &= \frac{\partial}{\partial t}, \\ \boldsymbol{X}_2 &= \zeta(t)\frac{\partial}{\partial p}, \\ \boldsymbol{X}_{1,2} &= -x^2\frac{\partial}{\partial x^1} + x^1\frac{\partial}{\partial x^2} - u^2\frac{\partial}{\partial u^1} + u^1\frac{\partial}{\partial u^2}, \\ \boldsymbol{X}_{4+i} &= \alpha_i(t)\frac{\partial}{\partial x^i} + \dot{\alpha}_i(t)\frac{\partial}{\partial u^i} - x^i\ddot{\alpha}_i(t)\frac{\partial}{\partial p} \qquad i = 1, 2, 3, \\ \boldsymbol{Y}_1 &= \beta g\, x^3\frac{\partial}{\partial p} + \frac{\partial}{\partial \theta} \\ \boldsymbol{Y}_2 &= 2t\frac{\partial}{\partial t} + \sum_{j=1}^{3}\left(x^j\frac{\partial}{\partial x^j} - u^j\frac{\partial}{\partial u^j}\right) - 2p\frac{\partial}{\partial p} - 3\theta\frac{\partial}{\partial \theta} \end{aligned} \tag{128}$$

If the transformations are allowed to act on ν and κ, one obtains equivalence transformations generated by the vector field:

$$\boldsymbol{Y}_3 = \sum_{j=1}^{3}\left(x^j\frac{\partial}{\partial x^j} + u^j\frac{\partial}{\partial u^j}\right) + 2p\frac{\partial}{\partial p} + \theta\frac{\partial}{\partial \theta} + 2\nu\frac{\partial}{\partial \nu} + 2\kappa\frac{\partial}{\partial \kappa}. \tag{129}$$

Vector fields $\boldsymbol{X}_1$, $\boldsymbol{X}_2$ and $\boldsymbol{X}_{4+i}$, $i = 1, 2, 3$, generate respectively the time, pressure and generalized Gelilean translations which were described in section 4.1.. Compared to the isothermal case, the only rotation symmetries of (127) are the horizontal rotations, generated by $\boldsymbol{X}_{1,2}$. The isotropy of the equations is broken by the existence of the privileged direction $\boldsymbol{e}_3$.

Vector field $\boldsymbol{Y}_1$ generates a new group of pressure-temperature translations:

$$(t, \boldsymbol{x}, \boldsymbol{u}, p, \theta) \longmapsto (t, \boldsymbol{x}, \boldsymbol{u}, p + a\,\beta x_3, \theta + a), \tag{130}$$

whereas $\boldsymbol{Y}_2$ and $\boldsymbol{Y}_3$ are the non isothermal versions of $\boldsymbol{X}_9$ and $\boldsymbol{X}_{10}$. $\boldsymbol{Y}_2$ corresponds to the first group of scaling transformations:

$$(t, \boldsymbol{x}, \boldsymbol{u}, p, \theta) \longmapsto (\mathrm{e}^{2a}\, t, \mathrm{e}^{a}\, \boldsymbol{x}, \mathrm{e}^{-a}\, \boldsymbol{u}, \mathrm{e}^{-2a}\, p, \mathrm{e}^{-3a}\, \theta) \tag{131}$$

and $\boldsymbol{Y}_2$ to the second group of scaling transformations of the form:

$$(t, \boldsymbol{x}, \boldsymbol{u}, p, \theta, \nu, \kappa) \longmapsto (t, \mathrm{e}^{a}\, \boldsymbol{x}, \mathrm{e}^{a}\, \boldsymbol{u}, \mathrm{e}^{2a}\, p, \mathrm{e}^{a}\, \theta, \mathrm{e}^{2a}\, \nu, \mathrm{e}^{2a}\, \kappa). \tag{132}$$

Reflections in x^1 and x^2 directions which constitute a subgroup of the group of reflections (91) are also symmetries of (127). Reflections in the third direction:

$$(t, \boldsymbol{x}, \boldsymbol{u}, p, \theta) \mapsto (t, \Lambda \boldsymbol{x}, \Lambda \boldsymbol{u}, p, -\theta) \tag{133}$$

where Λ is the reflection matrix:

$$\Lambda = \begin{pmatrix} 1 & 0 & 0 \\ 0 & 1 & 0 \\ 0 & 0 & -1 \end{pmatrix} \qquad \text{with}$$

is theoretically a symmetry of (127) but is not physically admissible if θ is assumed to satisfy a condition like positiveness.

Lastly, a flow evoluating in the horizontal plane is (horizontal) 2D material indifferent.

In the next subsection, turbulence models are analysed according to their compatibility with these symetries.

5.1.1. Model Analysis

After filtering, equations (127) become

$$\begin{cases} \dfrac{\partial \overline{\boldsymbol{u}}}{\partial t} + \operatorname{div}(\overline{\boldsymbol{u}} \otimes \overline{\boldsymbol{u}}) + \nabla \overline{p} - \beta \overline{\theta}\, \boldsymbol{e}_3 = \operatorname{div}(\overline{\mathsf{T}} + \mathsf{T}_s), \\[2ex] \dfrac{\partial \overline{\theta}}{\partial t} + \operatorname{div}(\overline{\theta} \overline{\boldsymbol{u}}) = \operatorname{div}(\overline{\mathbf{h}} + \mathbf{h}_s), \\[2ex] \operatorname{div} \overline{\boldsymbol{u}} = 0. \end{cases} \tag{134}$$

where $\mathbf{h}_s = \overline{\boldsymbol{u}}\overline{\theta} - \overline{\theta \boldsymbol{u}}$ is the subgrid flux which much be modeled with T_s in order to close the equations. Some of the most commonly used models are:

- the Smagorinsky model:

$$\mathsf{T}_s^d = C_S\overline{\delta}^2|\overline{\mathsf{S}}|\overline{\mathsf{S}}, \qquad \mathbf{h}_s = \frac{C_S\overline{\delta}^2}{\mathrm{Pr}_s}|\overline{\mathsf{S}}|\overline{\mathsf{S}} \tag{135}$$

where Pr_s is a subgrid Prandtl constant,

- the dynamic model:

$$\mathsf{T}_s^d = \frac{\mathrm{tr}(\mathsf{LM})}{\mathrm{tr}(\mathsf{M}^2)}\overline{\delta}^2|\overline{\mathsf{S}}|\overline{\mathsf{S}}, \qquad \mathbf{h}_s = \frac{\mathrm{tr}(\mathcal{LM})}{\mathrm{tr}(\mathcal{M}^2)}\overline{\delta}^2|\overline{\mathsf{S}}|\nabla\overline{\theta}, \tag{136}$$

with L and M defined in (99) and

$$\mathsf{L} = \widetilde{\overline{\boldsymbol{u}}}\widetilde{\overline{\theta}} - \widetilde{\overline{\boldsymbol{u}}\overline{\theta}}, \qquad \mathsf{M} = \widetilde{\overline{\delta}}^2|\widetilde{\overline{\mathsf{S}}}|\nabla\widetilde{\overline{\theta}} - \overline{\delta}^2\widetilde{|\overline{\mathsf{S}}|\nabla\overline{\theta}}, \tag{137}$$

- the Eidson model:

$$\mathsf{T}_s^d = 2\,C_E\overline{\delta}^2\sqrt{|\overline{\mathsf{S}}|^2 - \frac{\beta g}{\mathrm{Pr}_s}\frac{\partial\overline{\theta}}{\partial x_3}}\;\overline{\mathsf{S}}, \quad \mathbf{h}_s = \frac{C_E\overline{\delta}^2}{\mathrm{Pr}_s}\sqrt{|\overline{\mathsf{S}}|^2 - \frac{\beta g}{\mathrm{Pr}_s}\frac{\partial\overline{\theta}}{\partial x_3}}\;\overline{\theta}, \tag{138}$$

C_E being a constant,

- and the modified Eidson model which avoid a negative term inside the radical sign:

$$\begin{aligned}\mathsf{T}_s^d &= 2\,C_E\overline{\delta}^2\frac{1}{|\overline{\mathsf{S}}|}\left(|\overline{\mathsf{S}}|^2 - \frac{\beta g}{\mathrm{Pr}_s}\frac{\partial\overline{\theta}}{\partial x_3}\right)\overline{\mathsf{S}},\\ \mathbf{h}_s &= \frac{C_E\overline{\delta}^2}{\mathrm{Pr}_s}\frac{1}{|\overline{\mathsf{S}}|}\left(|\overline{\mathsf{S}}|^2 - \frac{\beta g}{\mathrm{Pr}_{sg}}\frac{\partial\overline{\theta}}{\partial x_3}\right)\overline{\theta}.\end{aligned} \tag{139}$$

Using similar arguments as in the isothermal case, it is straight forward to show that all these models are invariant under time-, pressure- and generalized Galilean translations, and (constant or horizontal time-dependent) rotations. Moreover, these models do not depend explicitely on the temperature but on its derivatives; consequently, their are invariant under pressure-temperature translations (130).

A model is invariant under the scaling transformations

$$(t, \boldsymbol{x}, \boldsymbol{u}, p, \theta, \nu, \kappa) \mapsto (\mathrm{e}^{2a}\,t,\ \mathrm{e}^{b+a}\,\boldsymbol{x},\ \mathrm{e}^{b-a}\,\boldsymbol{u},\ \mathrm{e}^{2b-2a}\,p,\ \mathrm{e}^{b-3a}\,\theta,\ \mathrm{e}^{2b}\,\nu,\ \mathrm{e}^{2b}\,\kappa) \tag{140}$$

which are combinations of (131) and (132), if and only if

$$\widehat{\mathsf{T}}_s = \mathrm{e}^{2b-2a}\,\mathsf{T} \qquad \text{and} \qquad \widehat{\mathbf{h}}_s = \mathrm{e}^{2b-4a}\,\mathbf{h}. \tag{141}$$

Under scalings (140), the Smagorinsky model changes as follows:

$$\widehat{\mathsf{T}}_s = \mathrm{e}^{-4a}\,\mathsf{T}_s \qquad \text{and} \qquad \widehat{\mathbf{h}}_s = \mathrm{e}^{-8a}\,\mathbf{h}_s. \tag{142}$$

This shows that this model is invariant neither under (131) nor under (132). Containing a term similar to the Smagorinsky model, the Eidson and modified Eidson models are also not invariant. In contrast, the similarity and the dynamic models fulfill conditions (141) and are then invariant under the scaling transformations (131) and (132).

In summary, as in the isothermal case, only the similarity model and the dynamic model are compatible with the symmetries of the equations, provided that the test filter inherent to these models does not destroy this property. In the next paragraph, the way exposed in section 4.3. of deriving invariant models is extended to the non-isothermal case.

5.1.2. New Symmetry-Preserving Turbulent Models

In order to have a model compatible with the translation symmetries of the equations, we assume that T_s and $\mathbf{h}_s$ are functions of $\overline{\mathsf{S}}$, $\overline{\Theta} = \nabla\theta$ and other scalar functions which will be precised later:

$$\mathsf{T}_s = \mathsf{T}_s(\overline{\mathsf{S}}, \overline{\Theta}), \qquad \mathbf{h}_s = \mathbf{h}_s(\overline{\mathsf{S}}, \overline{\Theta}). \tag{143}$$

Next, the Cayley–Hamilton theorem yields:

$$\begin{aligned} \mathsf{T}_s^d &= E_1\overline{\mathsf{S}} + E_2 \operatorname{Adj}^d \overline{\mathsf{S}} + E_3(\overline{\Theta} \otimes \overline{\Theta})^d \\ &+ E_4[\overline{\mathsf{S}}(\overline{\Theta} \otimes \overline{\Theta})]^d + E_5[\overline{\mathsf{S}}(\overline{\Theta} \otimes \overline{\Theta})\overline{\mathsf{S}}]^d \\ \mathbf{h}_s &= E_6\overline{\Theta} + \kappa E_7\overline{\mathsf{S}}\,\overline{\Theta} + E_8\overline{\mathsf{S}}^2\overline{\Theta}. \end{aligned} \tag{144}$$

The coefficients E_i are fonctions of the invariants built from $\overline{\mathsf{S}}$ et $\overline{\Theta}$ which are

$$\chi = \operatorname{tr}\overline{\mathsf{S}}^2, \quad \zeta = \det\overline{\mathsf{S}}, \quad \vartheta = \overline{\Theta}^2, \quad \omega_1 = \overline{\Theta}\cdot\overline{\mathsf{S}}\,\overline{\Theta}, \quad \omega_2 = \overline{S}\,\overline{\Theta}\cdot\overline{\mathsf{S}}\,\overline{\Theta}. \tag{145}$$

The model is invariant under the first scaling transformations (131) if

$$\widehat{\mathsf{T}}_s = \mathrm{e}^{-2a} \qquad \text{and} \qquad \widehat{\mathbf{h}}_s = \mathrm{e}^{-4a}. \tag{146}$$

Doing as in section 4.3., we deduce that condition (146) is verified if

$$\begin{aligned} \mathsf{T}_s^d &= F_1\overline{\mathsf{S}}^d + \chi^{-1/2}F_2 \operatorname{Adj}^d \overline{\mathsf{S}} + \chi^{-3/2}F_3(\overline{\Theta} \otimes \overline{\Theta})^d \\ &+ \chi^{-2}F_4[\overline{\mathsf{S}}(\overline{\Theta} \otimes \overline{\Theta})]^d + \chi^{-5/2}F_5\overline{\mathsf{S}}[(\overline{\Theta} \otimes \overline{\Theta})\overline{\mathsf{S}}]^d \\ \mathbf{h}_s &= F_6\overline{\Theta} + \chi^{-1/2}F_7\overline{\mathsf{S}}\,\overline{\Theta} + \chi^{-1}F_8\overline{\mathsf{S}}^2\overline{\Theta} \end{aligned} \tag{147}$$

where the F_i's, $i = 1, \ldots, 8$, are functions of (v, v_2, v_3, v_4), defined by:

$$v = \frac{\zeta}{\chi^{3/2}}, \qquad v_2 = \frac{\vartheta}{\chi^2}, \qquad v_3 = \frac{\omega_1}{\chi^{5/2}}, \qquad v_4 = \frac{\omega_2}{\chi^3}. \tag{148}$$

Finally, in order to have a model invariant under the second scaling symmetry, we choose the scalar quatities on which depends the model like in the isothermal case. This gives:

$$\begin{aligned}
\mathsf{T}_s^d &= \frac{q^2}{\varepsilon}\Big(H_1\overline{\mathsf{S}}^d + \chi^{-1/2}H_2 \operatorname{Adj}^d\overline{\mathsf{S}} + \chi^{-3/2}H_3(\overline{\Theta}\otimes\overline{\Theta})^d \\
&\quad + \chi^{-2}H_4[\overline{\mathsf{S}}(\overline{\Theta}\otimes\overline{\Theta})]^d + \chi^{-5/2}H_5\overline{\mathsf{S}}[(\overline{\Theta}\otimes\overline{\Theta})\overline{\mathsf{S}}]^d\Big) \qquad (149)\\
\mathbf{h}_s &= \frac{q^2}{\mathrm{Pr}_s\,\varepsilon}\Big(H_6\overline{\Theta} + \chi^{-1/2}H_7\overline{\mathsf{S}}\,\overline{\Theta} + \chi^{-1}H_8\overline{\mathsf{S}}^2\overline{\Theta}\Big),
\end{aligned}$$

where Pr_s is a subgrid Prandtl number and the H_i's, $i = 1,\ldots,8$, are functions of (v, v_2, v_3, v_4). Expressions (149) defines a general class of symmetry-preserving turbulencc models. When $\overline{\Theta} = \mathbf{0}$, (149) reduce to the isothermal model (121).

The number of arbitrary functions in (149) can be lowered by making other hypotheses. The viscous tensor T and the flux $\mathbf{h}$ can be written in a "potential" form:

$$\mathsf{T} = \frac{\partial\psi}{\partial\mathsf{S}}, \qquad \mathbf{h} = \frac{\partial\psi_h}{\partial\theta} \qquad (150)$$

where $\psi = \nu \operatorname{tr}\mathsf{S}^2$ and $\psi_h = \kappa\Theta^2/2$. We assume that the subgrid constraint tensor and the subgrid flux derive also from a "potential". This hypothesis is fulfilled if the model has the form:

$$\begin{aligned}
\mathsf{T}_s^d &= \frac{q^2}{\varepsilon}\Bigg[\left(2b - 3v\frac{\partial b}{\partial v} - 4v_2\frac{\partial b}{\partial v_2} - 5v_3\frac{\partial b}{\partial v_3} - 6v_4\frac{\partial b}{\partial v_4}\right)\overline{\mathsf{S}} + \\
&\quad + \chi^{-1/2}\frac{\partial b}{\partial v}\operatorname{Adj}^d\overline{S} + \nu\chi^{-3/2}\frac{\partial b}{\partial v_3}(\overline{\Theta}\otimes\overline{\Theta})^d + 2\chi^{-2}\frac{\partial b}{\partial v_4}[\overline{S}(\overline{\Theta}\otimes\overline{\Theta})]^d\Bigg],\\
\mathbf{h}_s &= \frac{q^2}{\mathrm{Pr}_s\,\varepsilon}\left[\frac{\partial b_t}{\partial v_2}\overline{\Theta} + \chi^{-1/2}\frac{\partial b_t}{\partial v_3}\overline{S}\,\overline{\Theta} + \chi^{-1}\frac{\partial b_t}{\partial v_4}\overline{S}^2\overline{\Theta}\right] \qquad (151)
\end{aligned}$$

where b and b_t are arbitrary functions of (v, v_2, v_3, v_4).

Relations (151) represent a strongly coupled model. Of course, this expression can be simplified. For example, if b and b_t are assumed to depend only on v and v_2 (*i.e.* on $||\overline{\mathsf{S}}||^2$, $\det\overline{\mathsf{S}}$ and $||\nabla\overline{\theta}||^2$), then the model reduces to:

$$\begin{aligned}
\mathsf{T}_s^d &= \frac{q^2}{\varepsilon}\left[\left(2b - 3v\frac{\partial b}{\partial v} - 4v_2\frac{\partial b}{\partial v_2}\right)\overline{\mathsf{S}} + \frac{\partial b}{\partial v}\frac{\operatorname{Adj}^d\overline{S}}{||\overline{\mathsf{S}}||}\right],\\
\mathbf{h}_s &= \frac{q^2}{\mathrm{Pr}_s\,\varepsilon}\,b_t'\,\nabla\overline{\theta}
\end{aligned} \qquad (152)$$

where b'_t is an arbitrary function of (v, v_2). If, moreover, the subgrid tensor T is assumed to be independent of the temperature, then:

$$\begin{aligned} \mathsf{T}_s^d &= \frac{q^2}{\varepsilon}\left[\left(2b(v) - 3v\dot{b}(v)\right)\overline{\mathsf{S}} + \dot{b}(v)\frac{\mathrm{Adj}^d\,\overline{S}}{\|\overline{\mathsf{S}}\|}\right], \\ \mathbf{h}_s &= \frac{q^2}{\mathrm{Pr}_s\,\varepsilon}\,b'_t(v, v_2)\nabla\overline{\theta}. \end{aligned} \tag{153}$$

Finally, if b and b'_t are linear, for example:

$$b(v) = C\mathring{\delta}^2 v, \qquad b'_t(v, v_2) = C_t\mathring{\delta}^2 v_2 \tag{154}$$

where $\mathring{\delta}$ is a ratio between the filter width and a characteristic length of the geometry, then

$$\begin{aligned} \mathsf{T}_s^d &= C\mathring{\delta}^2\frac{q^2}{\varepsilon}\left[-\frac{\det\overline{\mathsf{S}}}{\|\overline{\mathsf{S}}\|^3}\overline{\mathsf{S}} + \frac{\mathrm{Adj}^d\,\overline{S}}{\|\overline{\mathsf{S}}\|}\right], \\ \mathbf{h}_s &= C_t\mathring{\delta}^2\frac{q^2}{\mathrm{Pr}_s\,\varepsilon}\frac{\|\nabla\overline{\theta}\|^2}{\|\overline{\mathsf{S}}\|^4}\nabla\overline{\theta}. \end{aligned} \tag{155}$$

The model proposed here are compatible with the symmetry properties of the underlying equations (and their solutions). However, these properties may be destroyed during the numerical resolution when the numerical scheme are not invariant under the symmetry transformations. In the last paragraph, we show briefly how to make a numerical scheme invariant under the symmetry group of an equation.

6. Invariant Schemes

6.1. Basic Definitions

Consider a differential equation

$$E(\boldsymbol{x}, \boldsymbol{u}) = 0 \tag{156}$$

on a domain $\Omega \subset \mathbb{R}^M$. A discretization of Ω is a network of discrete points

$$\underline{\boldsymbol{x}} = (\boldsymbol{x}_1, ..., \boldsymbol{x}_J)$$

verifying a relation

$$\Phi(\underline{\boldsymbol{x}}) = 0. \tag{157}$$

Φ is called the mesh. Similarly, the unknown variable $\boldsymbol{u}$ is discretized into a sequence $\underline{\boldsymbol{u}} = (\boldsymbol{u}_l)_{l=1,...J}$. We denote $\boldsymbol{z}_j = (\boldsymbol{x}_j, \boldsymbol{u}_j) \in \mathbb{R}^M \times \mathbb{R}^Q$ and $\boldsymbol{z} = (\boldsymbol{z}_1, ..., \boldsymbol{z}_J)$.

A discretizetion scheme of equation (156), with an accuracy order $(q_1, ..., q_M)$, is a couple of functions (N, Φ) such that

$$N(\boldsymbol{z}) = O\left((\Delta x^1)^{q_1}, ..., (\Delta x^M)^{q_M}\right) \tag{158}$$

$$\tag{159}$$

$$\Phi(\boldsymbol{z}) = 0 \tag{160}$$

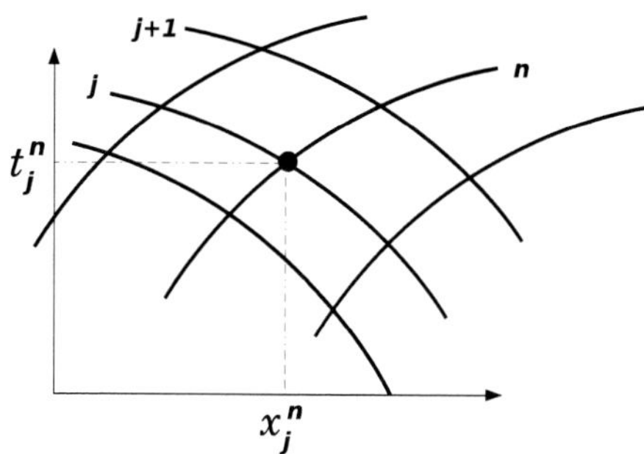

Figure 3. Grid points on a non-cartesian grid.

as soon as $\boldsymbol{u}_j = \boldsymbol{u}(\boldsymbol{x}_j)$ for all $j = 1, ..., J$. Δx^m is the step size in the m-th direction, $m = 1, ..., M$. Φ has been extended to $\boldsymbol{z}$ in (160) such that N and Φ has the same argument. This extention is also necessary when the mesh changes with the values of $\boldsymbol{u}$ (adaptative mesh, ...).

As an example, consider the heat equation (71). In this case, $\boldsymbol{x} = (t, x)$ and $\boldsymbol{u} = u$. The domain Ω is descretized into $\underline{\boldsymbol{x}} = (t_j^n, x_j^n)_{n,j}$ as shown on Figure 3. We denote $u_j^n = u(t_j^n, x_j^n)$. An orthogonal (cartesian) mesh on Ω is defined by

$$\begin{aligned} t_{j+1}^n - t_j^n &= 0, \\ x_j^{n+1} - x_j^n &= 0. \end{aligned} \tag{161}$$

This mesh is regular if

$$\begin{aligned} t_j^{n+1} - t_j^n &= t_j^n - t_j^{n-1}, \\ x_{j+1}^n - x_j^n &= x_j^n - x_{j-1}^n. \end{aligned} \tag{162}$$

Relations (161) and (162) define Φ. Note that when the mesh is orthogonal, we can denote $t_j^n = t^n$ and $x_j^n = x_j$ thanks to (161).

The Euler explicit descretization scheme of the heat equation (71) on an orthogonal and regular mesh is then defined by:

$$\frac{u_j^{n+1} - u_j^n}{k} - \frac{u_{j+1}^n - 2u_j^n + u_{j-1}^n}{h^2} = O(k, h^2) \tag{163}$$

along with (161) and (162). k and h are respectively the time and space step-sizes. N is defined by (163).

We now introduce the concept of symmetry in a numerical scheme.

A numerical scheme defined by (N, Φ) is said symmetric under a transformation group G if G is a symmetry group of both the discretized equations and the mesh equation, *i.e.*:

$$\begin{cases} N(\boldsymbol{z}) = 0 & \Rightarrow \quad N(g(\boldsymbol{z})) = 0 \\ \Phi(\boldsymbol{z}) = 0 & \Rightarrow \quad \Phi(g(\boldsymbol{z})) = 0 \end{cases} \qquad \forall g \in G. \tag{164}$$

Note that conditions (164) are satisfied if N and Φ are invariant functions under G, that is

$$N(g(\boldsymbol{z})) = N(\boldsymbol{z}) \qquad \text{and} \qquad \Phi(g(\boldsymbol{z})) = \Phi(\boldsymbol{z}), \qquad \forall g \in G, \tag{165}$$

but conditions (165) are not required.

Starting from any existing scheme (N, Φ), our aim is to derive a new scheme $(\tilde{N}, \tilde{\Phi})$ which is invariant under the symmetry group of the equation. This will be done using the concept of moving frame.

6.2. Invariantization of a Numerical Scheme

Consider a multi-parameter group G acting regularly and freely on a manifold $\mathcal{M}$. A lowered bold dot symbol (**.**) is used to indicate the action of G on $\mathcal{M}$. A (right) moving frame is a map $\rho : \mathcal{M} \mapsto G$ verifying the equivariance condition ([18, 19]):

$$\rho[g\boldsymbol{.}\boldsymbol{y}] = \rho[\boldsymbol{y}]g^{-1} \qquad \forall(\boldsymbol{y}, g) \in \mathcal{M} \times G. \tag{166}$$

This is a generalization of Cartan's moving frame (repère mobile) definition ([3]). If condition (166) holds, then

$$\rho[g\boldsymbol{.}\boldsymbol{y}]\boldsymbol{.}(g\boldsymbol{.}\boldsymbol{y}') = \rho[\boldsymbol{y}]\boldsymbol{.}\boldsymbol{y}' \qquad \forall(\boldsymbol{y}, \boldsymbol{y}', g) \in \mathcal{M}^2 \times G. \tag{167}$$

If G_t is the group of translations, $G_t = \{t_{\boldsymbol{a}},\ t_{\boldsymbol{a}}\boldsymbol{.}\boldsymbol{y} = \boldsymbol{y} + \boldsymbol{a}\}$, then the map ρ such that:

$$\rho \ : \ \boldsymbol{y} \mapsto t_{-\boldsymbol{y}} \tag{168}$$

is a moving frame. Indeed, for any $\boldsymbol{y}' \in \mathcal{M}$ and any $t_{\boldsymbol{a}} \in G_t$,

$$\rho[t_{\boldsymbol{a}}\boldsymbol{.}\boldsymbol{y}]\boldsymbol{.}\boldsymbol{y}' = \rho[\boldsymbol{y} + \boldsymbol{a}]\boldsymbol{.}\boldsymbol{y}' = t_{-(\boldsymbol{y}+\boldsymbol{a})}\boldsymbol{.}\boldsymbol{y}' = \boldsymbol{y}' - \boldsymbol{y} - \boldsymbol{a} = \rho[\boldsymbol{y}]t_{\boldsymbol{a}}^{-1}\boldsymbol{.}\boldsymbol{y}'.$$

Note that ρ is not unique since $\rho_{\boldsymbol{y}_0}[\boldsymbol{y}] = t_{-\boldsymbol{y}+\boldsymbol{y}_0}$ is a moving frame for any $\boldsymbol{y}_0 \in \mathcal{M}$. In fact, a general way of construction of moving frames, utilizing a cross section, was proposed by Olver ([18, 19]); and the constant $\boldsymbol{y}_0$ depends on the choice of cross section.

Another example is the scaling symmetry group acting of the Burgers' equation (171):

$$G_s = \{s_a : \ (t, x, u) \mapsto (a^2 t, ax, u/a)\}.$$

For any $\lambda \in \mathbb{R}$, the map:

$$\rho[(t, x, u)] = s_{\lambda/x}. \tag{169}$$

is a moving frame at (t, x, u) when $x \neq 0$, corresponding to G_s.

With a moving frame, a numerical scheme invariant under the symmetry group G of an equation can be derived as follows. Consider a numerical discretization scheme (N, Φ) of the equation and a moving frame ρ corresponding to G. A fundamental theorem ([18]) shows that the discretization scheme $(\tilde{N}, \tilde{\Phi})$ defined by

$$\tilde{N}(\boldsymbol{z}) = N(\rho[\boldsymbol{z}]\boldsymbol{.}\boldsymbol{z}), \qquad \tilde{\Phi}(\boldsymbol{z}) = \Phi(\rho[\boldsymbol{z}]\boldsymbol{.}\boldsymbol{z}) \tag{170}$$

is an invariant numerical scheme of the same order for the same equation.

We apply this theorem to the Burgers' equation.

6.3. Application to the Burgers' Equation

Consider the Burgers'equation:

$$\frac{\partial u}{\partial t} + u\frac{\partial u}{\partial x} = \nu\frac{\partial^2 u}{\partial x^2}. \tag{171}$$

The symmetry transformations of (171) are:

- time translations:

$$g_1 \; : \; (t, x, u) \longmapsto (t + a_1, x, u), \tag{172}$$

- space translations:

$$g_2 \; : \; (t, x, u) \longmapsto (t, x + a_2, u), \tag{173}$$

- scaling transformations:

$$g_3 \; : \; (t, x, u) \longmapsto (te^{2a_3}, xe^{a_3}, ue^{-a_3}), \tag{174}$$

- projections:

$$g_4 \; : \; (t, x, u) \longmapsto \left(\frac{t}{1 - a_4 t}, \frac{x}{1 - a_4 t}, (1 - a_4 t)u + a_4 x\right), \tag{175}$$

- and Galilean boosts:

$$g_5 \; : \; (t, x, u) \longmapsto (t, x + a_5 t, u + a_5). \tag{176}$$

a_i is the parameter of g_i, for each i=1,...,5. We start with the Euler forward-time centered-space (FTCS) scheme:

$$\frac{u_j^{n+1} - u_j^n}{\Delta t} + u_j^n\left(\frac{u_{j+1}^n - u_{j-1}^n}{2\Delta x}\right) = \nu\frac{u_{j+1}^n - 2u_j^n + u_{j-1}^n}{\Delta x^2} \tag{177}$$

on the orthogonal and regular mesh (161)-(162).

The scheme (N, Φ) defined by relations (177), (161) and (162) is invariant under time translation, space translation and scaling transformation groups, like most of standard schemes on Burgers' equation. We need only to apply the invariantization process under the Galilean boost and projection groups. However, for convenience, we take into account the time and space translation groups. These groups can be gathered into the four-parameter group G_0 which generic element is defined by:

$$g_0 = g_5 \circ g_4 \circ g_2 \circ g_1 \; : \; (t, x, u) \mapsto (\widehat{t}, \widehat{x}, \widehat{u}) \qquad \text{with}$$

$$\begin{aligned}
\widehat{t} &= \frac{t + a_1}{1 - a_4(t + a_1)} \\
\widehat{x} &= \frac{(x + a_2) + a_5(t + a_1)}{1 - a_4(t + a_1)} \\
\widehat{u} &= (1 - a_4(t + a_1))u + (x + a_2)a_4 + a_5.
\end{aligned} \tag{178}$$

Each transformation g_i is obtained from g_0 by setting $a_j = 0$ for all $j \neq i$.

A moving frame ρ associated to g_0 is an element of G_0. This means that $\rho[\boldsymbol{z}].\boldsymbol{z}$ is of the form (178), with particular values of a_i's, depending on $\boldsymbol{z}$. Determining $\rho[\boldsymbol{z}]$ is then equivalent to deciding the values of the a_i's.

6.3.1. Transformation of the Grid

A basic stencil for the FTCS scheme is represented by:

$$\mathsf{X}_j^n = (\boldsymbol{x}_j^{n+1}, \boldsymbol{x}_j^n, \boldsymbol{x}_{j+1}^n, \boldsymbol{x}_{j-1}^n)$$

where $\boldsymbol{x}_j^n = (t_j^n, x_j^n)$. At each point $\boldsymbol{z}_j^n$, we choose the moving frame such that

$$a_1 = -t_j^n \qquad \text{and} \qquad a_2 = -x_j^n. \tag{179}$$

In this way, the transformed scheme does not depend explicitely on x_i^m 's and t_i^m 's but on the step sizes:

$$k_j^n = t_j^{n+1} - t_j^n, \qquad h_j^n = x_{j+1}^n - x_j^n,$$
$$\sigma_j^n = x_j^{n+1} - x_j^n, \qquad t_{j-1}^n - t_j^n.$$

Indeed, if we denote $\widehat{\boldsymbol{x}}_i^m = \rho[\boldsymbol{z}_j^n].\boldsymbol{x}_i^m$, this choice of a_1 and a_2 leads to:

$$\begin{aligned}
\widehat{\boldsymbol{x}}_j^{n+1} &= \left(\frac{(x_j^{n+1} - x_j^n) + a_5(t_j^{n+1} - t_j^n)}{1 - a_4(t_j^{n+1} - t_j^n)}, \frac{t_j^{n+1} - t_j^n}{1 - a_4(t_j^{n+1} - t_j^n)}\right), \\
\widehat{\boldsymbol{x}}_j^n &= (0, 0), \\
\widehat{\boldsymbol{x}}_{j+1}^n &= \left(\frac{(x_{j+1}^n - x_j^n) + a_5(t_{j+1}^n - t_j^n)}{1 - a_4(t_{j+1}^n - t_j^n)}, \frac{t_{j+1}^n - t_j^n}{1 - a_4(t_{j+1}^n - t_j^n)}\right), \\
\widehat{\boldsymbol{x}}_{j-1}^n &= \left(\frac{(x_{j-1}^n - x_j^n) + a_5(t_{j-1}^n - t_j^n)}{1 - a_4(t_{j-1}^n - t_j^n)}, \frac{t_{j-1}^n - t_j^n}{1 - a_4(t_{j-1}^n - t_j^n)}\right).
\end{aligned} \tag{180}$$

With an orthogonal and regular mesh, defined by (161)-(162), we have:

$$\begin{gathered}
\widehat{\boldsymbol{x}}_j^{n+1} = \left(\tfrac{k}{1-a_4k}, \tfrac{a_5k}{1-a_4k}\right), \\
\widehat{\boldsymbol{x}}_j^n = (0,0), \qquad \widehat{\boldsymbol{x}}_{j+1}^n = (0,h), \qquad \widehat{\boldsymbol{x}}_{j-1}^n = (0,-h).
\end{gathered} \tag{181}$$

The transformed grid stays regular in space because

$$\widehat{x}_{j+1}^n - \widehat{x}_j^n = \widehat{x}_j^n - \widehat{x}_{j-1}^n. \tag{182}$$

In fact, $\widehat{h} = h$. Moreover, time layers stay flat as for the original mesh grid since

$$\widehat{t}_{j+1}^n - \widehat{t}_j^n = 0. \tag{183}$$

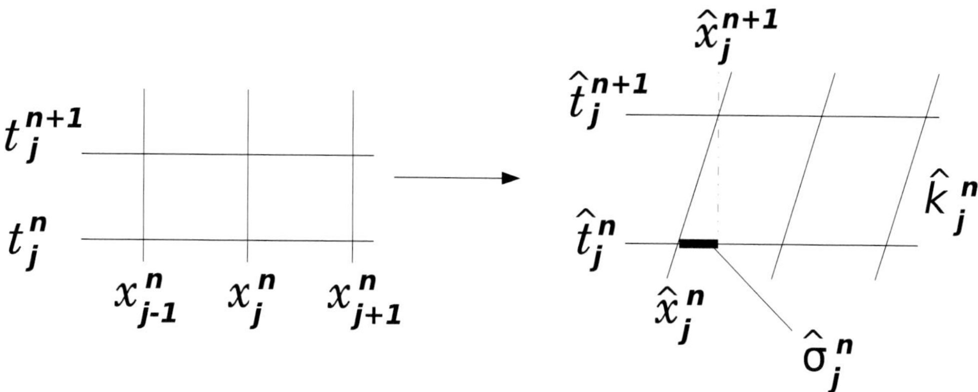

Figure 4. Transformation of the grid.

But the time step size is modified since:

$$\widehat{t}_j^{n+1} - \widehat{t}_j^n = \frac{k}{(1-a_4k)} = \widehat{k}_j^n. \tag{184}$$

This induces a translation $\widehat{\sigma}_j^n = \widehat{x}_j^{n+1} - \widehat{x}_j^n$ of the spatial layers at each time increment, with:

$$\widehat{\sigma}_j^n = \widehat{x}_j^{n+1} - \widehat{x}_j^n = a_5\frac{k}{1-a_4k}, \tag{185}$$

as seen on figure 4. The new grid $\tilde{\Phi}$ is defined by (182)-(185).

The next step is to define $\tilde{N}$.

6.3.2. Invariantization of the Scheme

To apply the invariantization process (170), we need to calculate the image $\widehat{u}_k^m = \rho[\boldsymbol{z}_j^n].u_k^m$, for all $\widehat{u}_k^m$ appearing in the stencil. We have:

$$\widehat{u} = (1 - a_4(t+a_1))u + (x+a_2)a_4 + a_5. \tag{186}$$

With the previous choice of a_1 and a_2 and the regularity of the mesh, it follows:

$$\begin{aligned}
\widehat{u}_j^{n+1} &= (1-a_4k)u_j^{n+1} + a_5, \\
\widehat{u}_j^n &= u_j^n + a_5, \\
\widehat{u}_{j+1}^n &= u_{j+1}^n + a_4h + a_5, \\
\widehat{u}_{j-1}^n &= u_{j-1}^n - a_4h + a_5.
\end{aligned} \tag{187}$$

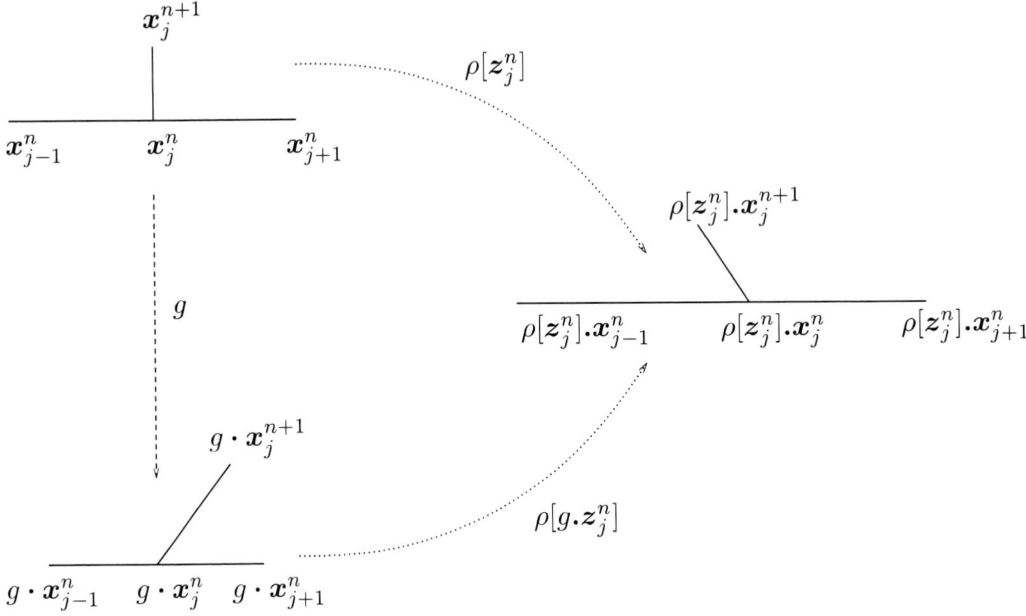

Figure 5. Equivariance condition applied to the stencil.

The new scheme $\tilde{N}$ is then defined by

$$\begin{aligned}\frac{(1-a_4k)u_j^{n+1}-u_j^n}{k}(1-a_4k)+(u_j^n+a_5)\left(\frac{u_{j+1}^n-u_{j-1}^n}{2h}+a_4\right)\\ =\nu\frac{u_{j+1}^n-2u_j^n+u_{j-1}^n}{h^2}.\end{aligned} \tag{188}$$

We now have to specify the values of a_4 and a_5 to make the transformed scheme invariant.

6.3.3. Determination of a_4 and a_5

The scheme is invariant only if a_4 and a_5 are such that equivariance condition (167) is satisfied at each descrete point. This condition reduces to:

$$\rho[\boldsymbol{z}_j^n].\boldsymbol{z}_i^m=\rho[g.\boldsymbol{z}_j^n]g.\boldsymbol{z}_i^m \tag{189}$$

for any $\boldsymbol{z}_i^m$ belonging to the stencil and any g belonging to G_0, with $a_1=-t_j^n$ and $a_2=-x_j^n$. Condition (189) is illustrated on Figure 5 (u is not represented on the figure).

The group element g involved in (189) is defined by two parameters λ and μ as follows:

$$g(t,x,u)=\left(\tfrac{t-t_j^n}{1-\mu(t-t_j^n)},\tfrac{(x-x_j^n)+\lambda(t-t_j^n)}{1-\mu(t-t_j^n)},(1-\mu(t-t_j^n))u+(x-x_j^n)\mu+\lambda\right). \tag{190}$$

A moving frame at a point $g.\boldsymbol{z}$ is determined by two parameters $\bar{a}_4$ and $\bar{a}_5$. We have, on the one hand, from (181) and (187):

$$
\begin{aligned}
\rho[\boldsymbol{z}_j^n].\boldsymbol{z}_j^n &= \big(0, 0, u_j^n + a_5\big),\\
\rho[\boldsymbol{z}_j^n].\boldsymbol{z}_j^{n+1} &= \big(\tfrac{k}{1-a_4k}, \tfrac{a_5k}{1-a_4k}, (1-a_4k)u_j^{n+1} + a_5\big),\\
\rho[\boldsymbol{z}_j^n].\boldsymbol{z}_{j\pm1}^n &= \big(0, h, u_{j\pm1}^n + a_4h + a_5\big),
\end{aligned}
\tag{191}
$$

and, on the other hand:

$$
\begin{aligned}
\rho[g.\boldsymbol{z}_j^n]g.\boldsymbol{z}_j^n &= \big(0, 0, u_j^n + (\lambda + \bar{a}_5)\big),\\
\rho[g.\boldsymbol{z}_j^{n+1}]g.\boldsymbol{z}_j^{n+1} &= \left(\tfrac{k}{1-(\mu+\bar{a}_4)k}, \tfrac{(\lambda+\bar{a}_5)k}{1-(\mu+\bar{a}_4)k}, [1-(\mu+\bar{a}_4)k]u_j^{n+1} + (\lambda+\bar{a}_5)\right),\\
\rho[g.\boldsymbol{z}_{j\pm1}^n]g.\boldsymbol{z}_j^n &= \big(0, h, u_{j\pm1}^n \pm (\mu + \bar{a}_4)h + (\lambda + \bar{a}_5)\big).
\end{aligned}
\tag{192}
$$

The discrete invariance condition (189) is verified when

$$
\begin{aligned}
\bar{a}_4 &= a_4 - \mu,\\
\bar{a}_5 &= a_5 - \lambda.
\end{aligned}
\tag{193}
$$

In order to determine a familly of moving frames allowing us to invariantize the numerical scheme, we assume an algebraic expression for a_4. To keep the explicit form of the numerical scheme, a_4 should not depend on u^{n+1}. Moreover, to keep the order of the convective term, the parameter of symmetrization must be at most of degree one. Next, the invariance of the scheme under g_4 requires that there are no term in h and k alone. Finally, as a_4 is homogeneous to the inverse of a time scale, we choose:

$$
a_4 = \frac{au_{j+1}^n + bu_j^n + cu_{j-1}^n}{h} \tag{194}
$$

where a, b, and c are real constants. Similar arguments for a_5, which is homogeneous to a velocity, allows to take a general form:

$$
a_5 = du_{j+1}^n + eu_j^n + fu_{j-1}^n \tag{195}
$$

where d, e and f are real constants. Since a_4 and a_5 must satisfy (193), these constants verify:

$$
\begin{aligned}
c - a &= 1,\\
a + b + c &= 0,\\
d - f &= 0,\\
d + e + f &= -1.
\end{aligned}
\tag{196}
$$

More information on these constants can be obtained by optimizing the order of accuracy.

6.3.4. Order of Accuracy

A fundamental result guarantees that the invariantization of a numerical scheme using the moving frame technique preserves the consistency. However, the order of consistency may change.

As the invariantization process does not affect the diffusion term, we only need to consider the unsteady term T_t and the convective T_c term:

$$T_t = \frac{u_j^{n+1} - u_j^n}{k}, \qquad T_c = u_j^n \, \frac{u_{j+1}^n - u_{j-1}^n}{2h} \tag{197}$$

corresponding to the non-viscous Burgers' equation:

$$\frac{\partial u}{\partial t} + u\frac{\partial u}{\partial x} = 0$$

The consistency condition for the symmetrized scheme is:

$$\lim_{h,k\mapsto 0} \tilde{T}_t + \tilde{T}_c = \frac{\partial u}{\partial t} + u\frac{\partial u}{\partial x} \tag{198}$$

where $\tilde{T}_t$ and $\tilde{T}_c$ are the invariantized unsteady and convection terms. A Taylor expansion gives:

$$\tilde{T}_t + \tilde{T}_c = \frac{\partial u}{\partial t} + u\frac{\partial u}{\partial x} - (a+c)\frac{h}{2}u\frac{\partial^2 u}{\partial^2 x} + \frac{dx}{dt}(\frac{\partial u}{\partial x} + ...) + O(k, h^2)$$

The invariantized scheme has the same order of consistency if

$$a + c = 0. \tag{199}$$

This condition determines a_4, which takes the expression:

$$a_3 = -\frac{u_{j+1}^n - u_{j-1}^n}{2h}, \tag{200}$$

but gives no additional constraint on a_5. In fact, with (200), the Taylor expansion of $\tilde{T}_c$ is:

$$\tilde{T}_c = 0. \tag{201}$$

This means that the invariantized convective term vanishes, and, therefore, the convective phenomenon are produced by the symmetrized unsteady term.

Notice that a_5 does not appear when the invariantized numerical scheme is expressed in the regular and orhthogonal original mesh. It is no longer the case if the mesh grid is not orthogonal, nor if the numerical solution is expressed in the transformed frame of reference.

We end up with some numerical results to show that invariantized schemes have some physical intersting properties compared to non-invariant ones.

6.4. Numerical Tests

The first test aims to check if the solutions given by various standard and invariant schemes are Galilean invariant.

Consider the Burgers'equation on $\Omega = \{(t, x) \in [0, 1] \times [-2, 2]\}$. Boundary conditions are such that the exact solution is

$$u(t, x) = \frac{-\sinh(\frac{x}{2\nu})}{\cosh(\frac{x}{2\nu}) + \exp(-\frac{t}{4\nu})} \tag{202}$$

with $\nu = 5.10^{-3}$. We consider the standard Euler FTCS, the Lax–Wendroff and the Crank–Nicolson schemes, and their invariantized versions. The space step size is $h = 2.10^{-2}$ and the CFL number is 1/2 in the original referential.

At each time step, the frame of reference is shifted by a Galilean translation

$$(t, x) \mapsto (t, x + \lambda t). \tag{203}$$

It is expected that u follows this shifting according to (176). Figure 6 shows, in the original frame of reference, how the standard schemes react to this shifting. They show clearly that, when λ is high, the standard schemes produce locally important errors. In particular, a blow-up arises with the FTCS scheme when $\lambda = 1$. On the contrary, the invariantized version of these schemes present no oscillation, as can be observed on Figure 7.

Consistency analysis shows that, as the original schemes are no longer consistent with the equation when $\lambda \neq 0$. This inconsistency introduces a numerical error which grows with λ, independently of the step sizes. As for them, the invariantized schemes respect the physical property of the equation and provide quasi-identical solutions when λ changes.

Another numerical test was carried out with FTCS scheme. The exact solution is a self-similar solution under projections (175):

$$u(t, x) = \frac{1}{t}\left(x - \tanh(\frac{x}{2\nu})\right). \tag{204}$$

It corresponds to a viscous chock. The chock tends to be entropic when ν becomes closer to 0. Figure 8 shows the numerical solutions obtained with the standard and the invariantized FTCS schemes at $t = 2s$, with $\nu = 10^{-2}$, $k = 5.10^{-2}$ and $h = 5.10^{-2}$. It shows that the solution given by the invariantized scheme remains close to the exact solution whereas the original FTCS scheme presents a poor performance close to the chock location.

Since the invariant scheme has the same invariance property as the solution under projections, it does not produce an artificial error like the non-invariant FTCS. This shows the ability of invariantized scheme to respect the physics of the equation.

7. Conclusion

In the first part of this chapter, corresponding to section 2. we showed that the symmetry theory and, more generally, the geometric approach constitute an efficient tool for the

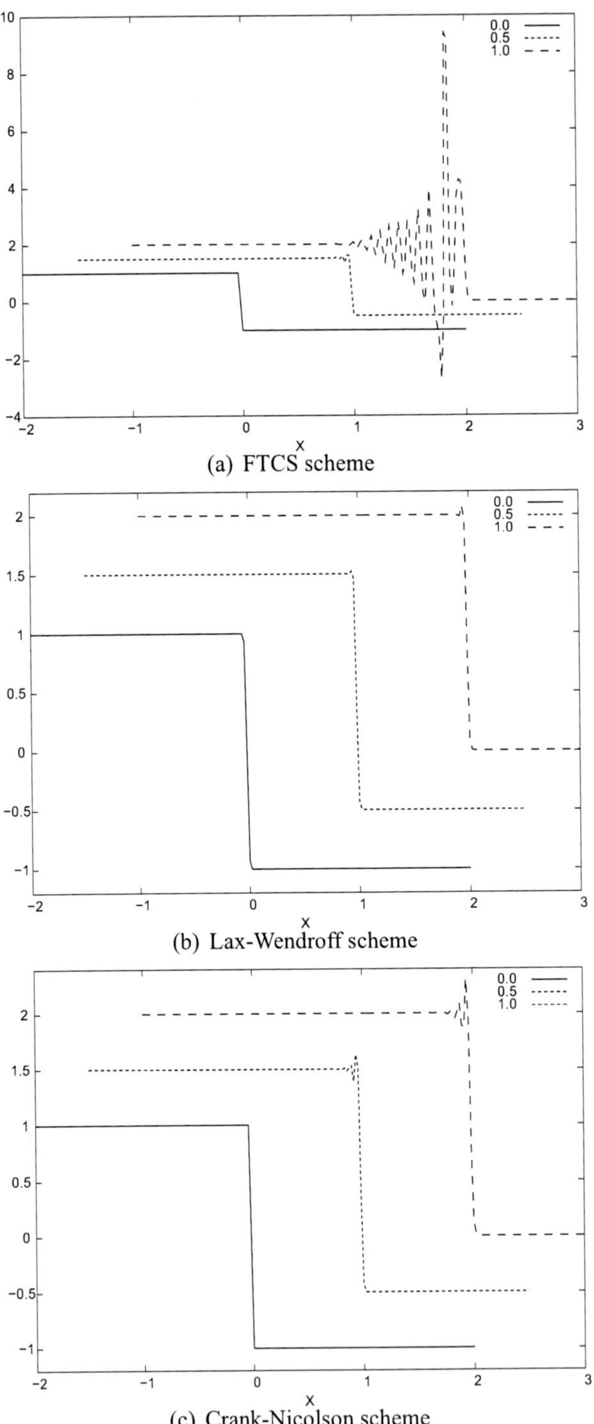

(a) FTCS scheme

(b) Lax-Wendroff scheme

(c) Crank-Nicolson scheme

Figure 6. Profiles of u versus x, with $\lambda = 0$, $\lambda = 0.5$ and $\lambda = 1$, at $t = 1$s.

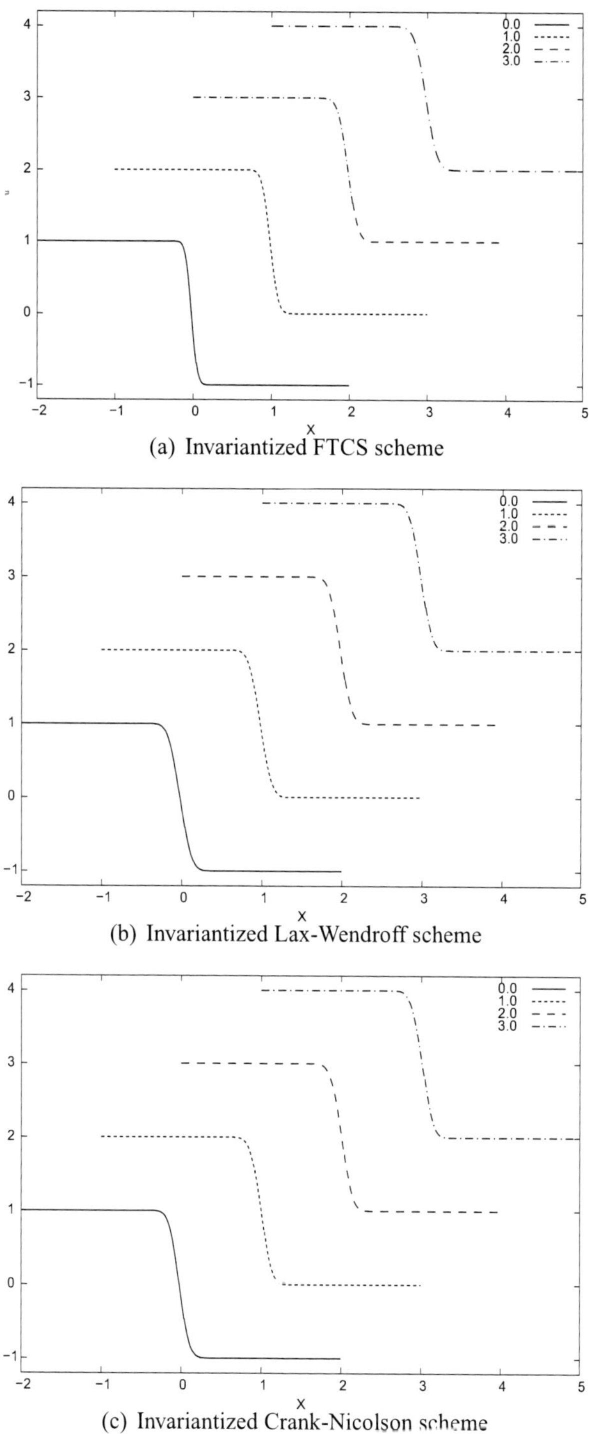

(a) Invariantized FTCS scheme

(b) Invariantized Lax-Wendroff scheme

(c) Invariantized Crank-Nicolson scheme

Figure 7. Velocity profiles with $\lambda = 0$, $\lambda = 1$, $\lambda = 2$ and $\lambda = 3$, at $t = 1$s.

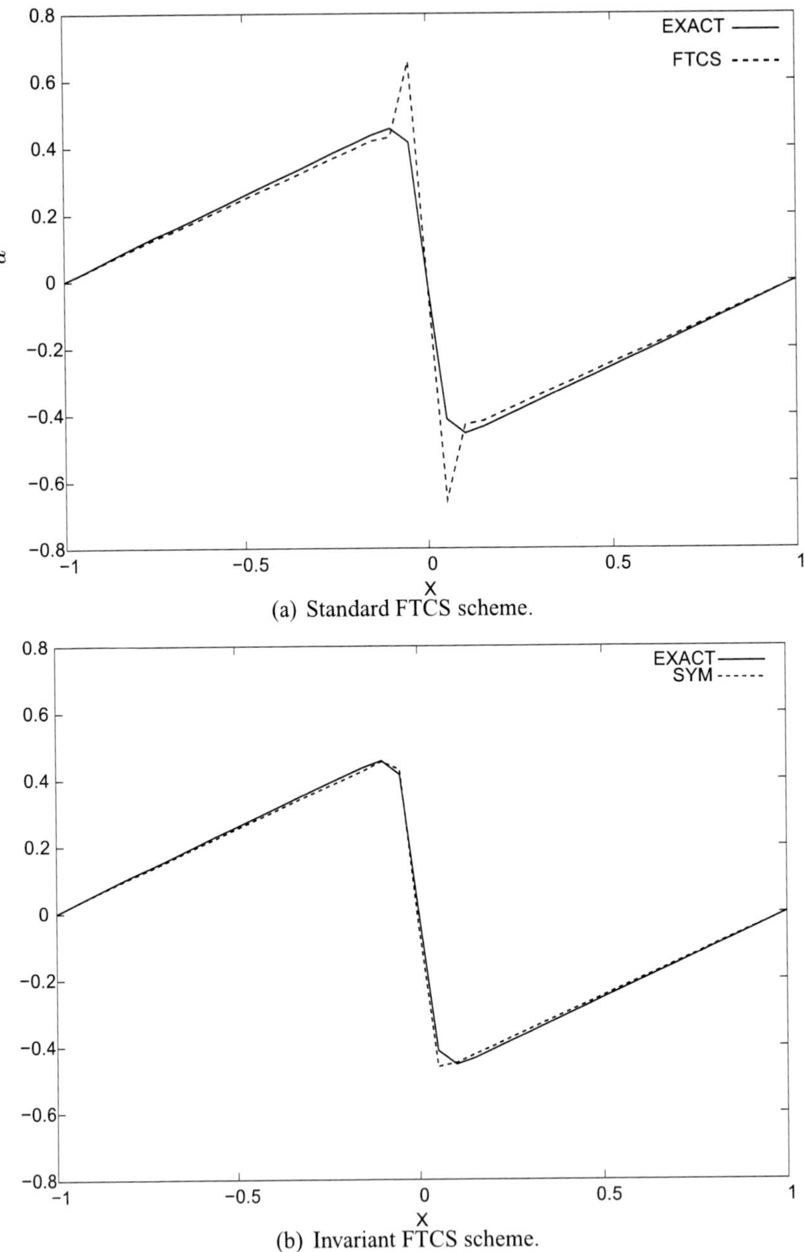

(a) Standard FTCS scheme.

(b) Invariant FTCS scheme.

Figure 8. Numerical solutions with $\nu = 10^{-2}$, $\Delta t = 5.10^{-2}$, $\Delta x = 5.10^{-2}$, $t = 2$s. Self-similar case.

understanding and the resolution of many problems involving differential equations, and especially in fluid machanics. The examples given in this part was inspired from results in the literature, except subsections 2.5. and 2.6. where original results on a unified view on thin shear layer problems was exposed.

The symmetry theory provides also a mathematical way to build the models which respect the physics of the equations. This approach allows us moreover some liberty such that other criteria can be introduced for the construction of the models. Here, thermodynamical arguments were used as additional criterium.

At last, a way to build invariant numerical schemes was given. These schemes respect the physical properties of the equation at the discrete point and have naturally good performances compared to non-invariant schemes. Considering long range, the invariantization process will be used for turbulence simulation.

References

[1] V.O. Bytev. Group-theoretical properties of the navier–stokes equations. *Numerical Methods of Continuum Mechanics [in Russian]*, **3**(3):13–17, 1972.

[2] B.J. Cantwell. Similarity transformations for the two-dimensional, unsteady, stream-function equation. *Journal of Fluid Mechanics*, **85**:257–271, 1978.

[3] E. Cartan. La méthode de repère mobile, la théorie des groupes continus et les espaces généralisés. In *Actualités Scientifiques et Industrielles*. N.**194**. 1935.

[4] Q. Chen, Y. Jiang, C. Béghein, and M. Su. Particulate dispersion and transportation in buildings with large eddy simulation. Technical report, Massachusetts Institute of Technology, 2001.

[5] Y.A. Danilov. Group properties of the maxwell and navier–stokes equations. *Preprint, Khurchatov Inst. Nucl. Energy, Acad. Sci. USSR*, 1967.

[6] V. Grassi, R.A. Leo, G. Soliani, and P. Tempesta. Vorticies and invariant surfaces generated by symmetries for the 3D Navier-Stokes equation. *Physica A*, **286**:79–108, 2000.

[7] N.H. Ibragimov. *CRC handbook of Lie group analysis of differential equations. Vol 2: Applications in engineering and physical sciences*. CRC Press, 1995.

[8] N.H. Ibragimov. *CRC handbook of Lie group analysis of differential equations. Vol 3: New trends in theorical developments and computational methods*. CRC Press, 1996.

[9] N.H. Ibragimov. *New trends in theorical developments and computational methods*. CRC handbook of Lie group analysis of differential equations, Vol 3. CRC Press, 1996.

[10] N.H. Ibragimov. A new conservation theorem. *Journal of Mathematical Analysis and Apllications*, **333**:311–328, 2007.

[11] N.H. Ibragimov and G. Ünal. Equivalence transformations of Navier-Stokes equations. *Bulletin of the Technical University of Istanbul*, **47**(1-2):203, 1994.

[12] T. Iliescu, V. John, W.J. Layton, G. Matthies, and L. Tobiska. A numerical study of a class of LES models. *International Journal of Computational Fluid Dynamics*, **17**:75 – 85, 2003.

[13] E. Nœther and A. Tavel. Invariant variation problems. *Transport Theory and Statistical Physics*, **1**(3):183–207, 1971. English traduction of the Nœther's original paper in 1918.

[14] M. Oberlack. Invariant modeling in large-eddy simulation of turbulence. In *Annual Research Briefs*. Stanford University, 1997.

[15] M. Oberlack. A unified approach for symmetries in plane parallel turbulent shear flows. *Journal of Fluid Mechanics*, **427**:299–328, 2001.

[16] M. Oberlack, W. Cabot, B. Pettersson Reif, and T. Weller. Group analysis, direct numerical simulation and modelling of a turbulent channel flow with streamwise rotation. *Journal of Fluid Mechanics*, **562**:355–381, 2006.

[17] P. Olver. *Applications of Lie groups to differential equations*. Graduate texts in mathematics. Springer-Verlag, 1986.

[18] P. Olver. Geometric foundations of numerical algorithms and symmetry. *Applicable Algebra in Engineering, Communication and Computing*, **11**(5):417–436, 2001.

[19] P. Olver. Moving frames. *Journal of Symbolic Computation*, **36**(3-4):501–512, 2003.

[20] V.V. Pukhnachev. Invariant solution of Navier-Stokes equations describing motions with free boundary. *Doklady Akademii Nauk SSSR*, pages 202–302, 1972.

[21] D. Razafindralandy and A. Hamdouni. Subgrid models preserving the symmetry group of Navier-Stokes equations. *Comptes Rendus Mécanique*, **333**:481–486, 2005.

[22] D. Razafindralandy and A. Hamdouni. Consequences of symmetries on the analysis and construction of turbulence models. *Symmetry, Integrability and Geometry: Methods and Applications*, **2**:Paper 052, 2006.

[23] D. Razafindralandy and A. Hamdouni. Analysis of subgrid models of heat convection by symmetry group theory. *Comptes Rendus Mécanique*, **(4)**:225–230, 2007.

[24] D. Razafindralandy and A. Hamdouni. Invariant subgrid modelling in large-eddy simulation of heat convection turbulence. *Theoretical and Computational Fluid Dynamics*, 2007.

[25] D. Razafindralandy, A. Hamdouni, and C. Béghein. A class of subgrid-scale models preserving the symmetry group of Navier-Stokes equations. *Communications in Nonlinear Science and Numerical Simulation*, **12**(3):243–253, 2007.

[26] D. Razafindralandy, A. Hamdouni, and M. Oberlack. Analysis and development of subgrid turbulence models preserving the symmetry properties of the Navier–Stokes equations. *European Journal of Mechanics/B*, **26**, 2007.

[27] P Sagaut. *Large eddy simulation for incompressible flows. An introduction.* Scientific Computation. Springer, 2004.

[28] R. Schiestel. *Les écoulements turbulents. Modélisation et simulation.* Mermes, Paris., 1998.

[29] H. Schlichting and K. Gersten. *Boundary layer theory.* Springer-Verlag, Germany, 2000.

[30] G. Ünal. Application of equivalence transformations to inertial subrange of turbulence. *Lie Group and Their Applications*, **1**(1):232–240, 1994.

In: Numerical Simulation Research Progress
Editors: Simone P. Colombo et al, pp. 209-248
ISBN 978-1-60456-783-0

Chapter 5

THE SHOOTING METHOD IN HYDROTHERMAL OPTIMAL CONTROL PROBLEMS

L. Bayón, J.M. Grau, M.M. Ruiz and P.M. Suárez
University of Oviedo, Department of Mathematics
and
E.U.I.T.I. Campus of Viesques, Gijón, 33203, Spain.

Abstract

This study presents an in-depth analysis of the behavior of the shooting method for the two-point boundary value problem that appears in the resolution of the short-term hydrothermal coordination problem when optimal control theory is used.

First we shall consider a problem without restrictions. We present a theorem that establishes sufficient conditions of existence of the extremals of the functional defined on the whole optimization interval. We analyze some additional hypotheses whose fulfilment guarantees that the extremals represent the strong relative minimum, and other hypotheses that ensure compatibility with any restriction on the admissible volume. We also present a result which makes it evident that the convexity of a functional is not necessary to guarantee that the extremals are solutions.

We then go on to study the problem arising in the case where the hydro- and thermal plants are subject to certain restrictions. Under these conditions, the non-emptiness of the set of admissible functions is sufficient to ensure the existence of a solution. We shall also study the conditions that guarantee non-existence of the boundary solutions. When we consider boundary solutions, the solution need not be "algebraically interior". To this end, we have developed the mathematical machinery to solve the problem in a very satisfactory way, from both the theoretical and algorithmic/computational point of view.

Finally we illustrate the performance of our work with several numerical examples and the running of the proposed algorithms using Mathematica ©.

1. Introduction

This study analyzes the behavior of the shooting method in solving the two-point boundary value problem (TPBVP) that appears in the solution to the classic short-term hydrothermal coordination problem (STHC) when optimal control theory is used. A hydrothermal system is made up of hydro- and thermal plants that must jointly satisfy a certain demand in electric power during a specific time interval. Thermal plants generate power at the expense of fuel consumption, while hydro-plants obtain power from the energy liberated by water moving through a turbin, a limited quantity of water being available during the optimization period. The short-term hydrothermal coordination (STHC) problem plays a very important role in the safety, reliability and economic operation of electric power systems whereby the generation of hydro- and thermal plants is allocated so as to minimize total operating cost in a schedule horizon of 1 day or 1 week while satisfying various constraints on plants and a certain demand in electric power.

As this large-scale nonlinear problem has been the subject of intensive investigation for several decades now, the bibliography describing different formulations and solution methodologies is vast. Dynamic programming [1,2] and mixed integer linear programming [3,4] methods have been widely used in different formulations. Promising results have been obtained using the Lagrangian relaxation technique to generate near optimal solutions [5,6]. However, the drawback of this approach lies in the primal solution, which is infeasible. As a result, certain heuristic procedures are needed to obtain a feasible primal solution. In recent years, evolutionary computational optimization techniques have constituted a tool that has shown a certain ability to solve this problem. These evolutionary algorithms can be implemented in various forms, such as genetic algorithms [7,8], evolutionary programming [9], simulated annealing [10] and evolutionary strategy [11].

The main drawback with the majority of these methods is the difficulty of treating large-scale systems, besides the fact that they require substantial simplifying assumptions to make the problem computationally tractable. In previous papers [12-15], we proposed Pontryagin's Minimum Principle (PMP) to solve the STHC problem, which is formulated within the framework of Optimal Control Theory (OCT) [16-19]. The advantage of this infinite-dimensional technique compared to previous approaches lies in the possibility of obtaining theoretical results whose implementation is feasible regardless of the size of the problem. Considering an elevated number of discretization intervals may make the use of other methods unviable, whereas our technique would be more plausible in this case, while barely increasing the computational effort.

A number of methods exist for solving the TPBVP, including shooting, collocation and finite difference methods. Among the shooting methods [20,21], the Simple Shooting Method (SSM) and the Multiple Shooting Method (MSM) appear to be the most well known and widely used methods. The Simple Shooting Method transforms a TPBVP into an initial value problem in which the initial values of selected parameters are varied to satisfy the desired end conditions. The boundary conditions are satisfied when the differential equation is integrated over the optimization interval, employing the initial condition obtained using the SSM. It should be noted that serious problems with the convergence of the SSM can arise if the starting initial condition is not close to the solution. This drawback of the SSM can be addressed by implementing what is known as the Multiple Shooting Method (MSM).

The MSM is similar to the SSM in sofar as unknown parameters are selected at the initial time; however, the differential equation is not integrated all the way to the final time. Instead, the "distance" from a corresponding point on a pre-selected reference path is checked continuously as the integration proceeds and the integration is aborted when the distance exceeds a tolerance value. Then, the integration is started again from the corresponding point on the reference path and the above step is repeated, until the system is integrated to the final time. An advantage of this approach over the SSM is that convergence can now be obtained for a larger class of TPBVPs. A serious drawback of this method is that if the differential equations are re-integrated to result in one continuous trajectory for the system, the actual final values may not be close to the desired final values. This is a common problem when solving TPBVPs resulting from optimal control, like our problem. Bearing in mind the above considerations, we decided to implement a SSM, obtaining good results.

This study, which presents an in-depth analysis of the behavior of the shooting method for the STHC problem, is organized as follows. Section 2 addresses a problem without restrictions. We consider a theorem that establishes sufficient conditions of existence of the extremals of the functional defined over the entire optimization interval. In some particular cases, the hypotheses of this theorem can be relaxed and we consider two cases of special interest. However, the existence of extremals does not guarantee the existence of a solution to the problem. We analyze some additional hypotheses whose fulfilment guarantees that the extremals represent the strong relative minimum, and other hypotheses that ensure compatibility with any restriction on the admissible volume. Although convex functionals form the ideal framework for addressing optimization problems, we also present a result which makes it evident that the convexity of a functional is not necessary to guarantee that the extremals are solutions of the problem. Finally, we present an algorithm to construct the solution which is much more versatile than the majority of classical algorithms. Its true importance becomes evident in the study of problems with restrictions.

In Section 3, we study the problem arising in the case where the hydro- and thermal plants are subject to certain restrictions. Under these conditions, the non-emptiness of the set of admissible functions is sufficient to ensure the existence of a solution. We shall also study the conditions that guarantee non-existence of the boundary solutions. When we consider boundary solutions, the problem is not reduced to solving a boundary-value problem for a second-order equation. The solution need not be "algebraically interior". With this purpose in mind, we developed the mathematical machinery to solve the problem in a very satisfactory way, from both the theoretical and algorithmic/computational points of view.

Section 4 illustrates the performance of our work with various numerical examples and the running of the proposed algorithms using Mathematica ©. Finally, Section 5 summarizes the main contributions of our research.

2. Problem without Restrictions

This study addresses the STHC problem for a hydrothermal system with one hydro-plant and m thermal power plants. In prior studies [22,23], it was proven that the problem of optimization of a hydrothermal system with m thermal power plants may be reduced to the study of a hydrothermal system made up of one single thermal power plant, called the *thermal equivalent*.

The problem consists in minimizing the cost of fuel needed to satisfy a certain power demand during the optimization interval $[0, T]$. Said cost may be represented by the functional

$$F(P) = \int_0^T \Psi(P(t))dt,$$

where Ψ is the function of thermal cost of the thermal equivalent and $P(t)$ is the power generated by said plant. Moreover, the following equilibrium equation of active power are to be fulfilled

$$P(t) + H(t, z(t), z'(t)) = P_d(t), \ \forall t \in [0, T],$$

where $P_d(t)$ is the power demand and $H(t, z(t), z'(t))$ is the power contributed to the system at the instant t by the hydro-plant, being: $z(t)$ the volume that is discharged up to the instant t (in what follows, simply volume) by the plant, and $z'(t)$ the rate of water discharge of the plant at the instant t. Taking into account the equilibrium equation, the problem reduces to calculating the minimum of the functional

$$F(z) = \int_0^T L(t, z(t), z'(t))dt = \int_0^T \Psi\left(P_d(t) - H\left(t, z(t), z'(t)\right)\right) dt.$$

If we assume that b is the volume of water that must be discharged during the entire optimization interval, the following boundary conditions will have to be fulfilled

$$z(0) = 0, \ z(T) = b.$$

Thus, we shall try to consider the functions P_d, Ψ and H as general as possible without any restrictions except for those which are natural for problems of this type. For the sake of convenience, we assume throughout the paper that they are sufficiently smooth and are subject to the following additional assumptions:

Function of thermal cost. Let us assume that the function of thermal cost

$$\Psi : \mathbb{R} \longrightarrow \mathbb{R}$$

satisfies: $\Psi'(x) > 0$, $\forall x \in \mathbb{R}$ and thus is strictly increasing. This restriction is absolutely natural: it reads more cost to more generated power. Let us assume as well that $\Psi''(x) > 0$, $\forall x \in \mathbb{R}$ and is therefore strictly convex. The traditionally employed models meet this restriction.

Function of effective hydraulic generation. Let us assume that the hydraulic generation

$$H(t, z, z') : \Omega_H = [0, T] \times \mathbb{R} \times \mathbb{R} \longrightarrow \mathbb{R}$$

is strictly increasing with respect to the rate of water discharge z', i.e. $H_{z'} > 0$. Let us also assume that $H(t, z, z')$ is concave with respect to z', i.e. $H_{z'z'} \leq 0$. The assumptions we have made guarantee the fulfilment of the following inequalities:

$$L_{z'z'}(t, z, z') > 0; \ L_{z'}(t, z, z') < 0.$$

The real models meet these two restrictions; the former means the higher the rate of water discharge, the greater the power.

The only restrictions on the admissible functions will be their inclusion in $KC^1[0, T]$ (continuous with the piecewise continuous derivatives) and that of the available volume.

Our study will therefore include the pumped-storage plants[1] once it is allowed that the function of effective hydraulic generation $H(t, z, z')$ is defined for negative values of z' (the rate of water discharge). We shall hence allow the function of the thermal cost Ψ to admit negative values of the arguments (the thermal power), which is equivalent to the assumption that the hydraulic effective power might exceed the demanded power and is able to deliver negative costs (profits[2]) which compensate for the positive thermal costs.

Therefore we may expound the mathematical problem in the following terms.

Definition 1. *We denote by $\wp_b$ the problem of minimizing the functional*

$$F(z) = \int_0^T L(t, z(t), z'(t))dt$$

over the set

$$\Theta_b = \{z \in KC^1[0, T] \mid z(0) = 0,\, z(T) = b\},$$

with L of the form

$$L(t, z(t), z'(t)) = \Psi(P_d(t) - H(t, z(t), z'(t))).$$

If z satisfies Euler's equation for the functional F, we have that, $\forall t \in [0, T]$, Euler's equation is fulfilled

$$L_z(t, z(t), z'(t)) - \frac{d}{dt}\left(L_{z'}(t, z(t), z'(t))\right) = 0.$$

Integrating we have, $\forall t \in [0, T]$, the integral form of Euler's equation, known as the Du Bois-Reymond equation

$$-L_{z'}(t, z(t), z'(t)) + \int_0^t L_z(s, z(s), z'(s))ds = -L_{z'}(0, z(0), z'(0)) = K. \qquad (1)$$

We shall call (1) the *first coordination equation* for $z(t)$, and the positive constant K will be termed the *coordination constant* of the extremal.

If we divide Euler's equation by $L_{z'}(t, z(t), z'(t)) < 0$, $\forall t$, we have that

$$\frac{L_z(t, z(t), z'(t))}{L_{z'}(t, z(t), z'(t))} - \frac{\frac{d}{dt}\left[L_{z'}(t, z(t), z'(t))\right]}{L_{z'}(t, z(t), z'(t))} = 0.$$

[1]At certain moments, the pumped-storage plants can pump the water in order to use it more efficiently later on. As the rate of water discharge is negative during pumping, this means that the generated power is likewise negative (being consumed by the pump).

[2]It is quite possible that the power plants generate an excess of power and transfer it to other systems with the resulting economic compensation. On the other hand, if instead of minimizing the fuel cost with a given quantity of water, we consider the inverse problem, which consists in minimizing the water consumed with a prescribed limited quantity of fuel, then the pumped-storage plants (which play the role of thermal plants) furnish an example in which negative costs can be produced (water recovery).

Integrating,

$$\int_0^t \frac{L_z(s,z(s),z'(s))}{L_{z'}(s,z(s),z'(s))}ds - \ln\left|L_{z'}(s,z(s),z'(s))\right|\Big|_0^t = 0$$

$$\int_0^t \frac{L_z(s,z(s),z'(s))}{L_{z'}(s,z(s),z'(s))}ds - \ln\left|L_{z'}(t,z(t),z'(t))\right| + \ln\left|L_{z'}(0,z(0),z'(0))\right| = 0$$

$$\left|L_{z'}(0,z(0),z'(0))\right| = \left|L_{z'}(t,z(t),z'(t))\right| \cdot \exp\left[-\int_0^t \frac{L_z(s,z(s),z'(s))}{L_{z'}(s,z(s),z'(s))}ds\right]$$

However, bearing in mind that $L_{z'}(t,z(t),z'(t)) < 0$ and that

$$\frac{L_z(s,z(s),z'(s))}{L_{z'}(s,z(s),z'(s))} = \frac{H_z(s,z(s),z'(s))}{H_{z'}(s,z(s),z'(s))}$$

we have that, $\forall t \in [0,\tau]$

$$-L_{z'}(t,z(t),z'(t)) \cdot \exp\left[-\int_0^t \frac{H_z(s,z(s),z'(s))}{H_{z'}(s,z(s),z'(s))}ds\right] = -L_{z'}(0,z(0),z'(0)) = K. \quad (2)$$

We shall call (2) the *second coordination equation* for $z(t)$.

For the sake of brevity in subsequent expressions, we shall denote

$$E(t,q) = \exp\left(-\int_0^t \frac{H_z(s,q(s),q'(s))}{H_{z'}(s,q(s),q'(s))}ds\right)$$

In a previous paper [24], a problem of hydrothermal optimization with pumped-storage plants was addressed without considering constraints for the admissible generated power. In this kind of problem, the derivative of H with respect to z' $(H_{z'})$ presents discontinuity at $z' = 0$, which is the border between the power generation zone (positive values of z') and the pumping zone (negative values of z').

The mathematical problem $\wp_b$ was stated in the following terms:

$$\min_{z\in\Theta_b} F(z) = \min_{z\in\Theta_b} \int_0^T L(t,z(t),z'(t))dt$$
$$\Theta_b = \{z \in KC^1[0,T] \mid z(0) = 0, z(T) = b\}$$

where $L(\cdot,\cdot,\cdot)$ and $L_z(\cdot,\cdot,\cdot)$ are the class C^0 and $L_{z'}(t,z,\cdot)$ is piecewise continuous ($L_{z'}(t,z,\cdot)$ is discontinuous in $z' = 0$).

Denoting by $¥_q(t), q \in \Theta_b$, the *first coordination function*:

$$¥_q(t) := -L_{z'}(t,q(t),q'(t)) + \int_0^t L_z(s,q(s),q'(s))ds \quad (3)$$

and by $¥_q^+(t)$ and $¥_q^-(t)$ the expressions obtained when considering the lateral derivatives of L with respect to z'

$$L_{z'}^+(t,z,z') := L_{z'}(t,z,z'_+); \;\; L_{z'}^-(t,z,z') := L_{z'}(t,z,z'_-)$$

$$¥_z^+(t) = L_{z'}^+(t,z(t),z'(t)) - \int_0^t L_z(\tau,z(\tau),z'(\tau))d\tau$$

$$¥_z^-(t) = L_{z'}^-(t,z(t),z'(t)) - \int_0^t L_z(\tau,z(\tau),z'(\tau))d\tau.$$

The problem $\wp_b$ was formulated within the framework of nonsmooth analysis [16-18], using the generalized (or Clarke's) gradient, the following results being proven:

Theorem 1. *If q is a solution of $\wp_b$, then $\exists K \in \mathbb{R}^+$ such that:*

$$\begin{cases} ¥_q^+(t) = ¥_q^-(t) = K & \text{if} \quad q'(t) \neq 0 \\ ¥_q^+(t) \leq K \leq ¥_q^-(t) & \text{if} \quad q'(t) = 0. \end{cases}$$

The following corollary shows a important particular case.

Corollary 1. *If q is a solution of $\wp_b$, and $L_{z'}(t, z, \cdot)$ is continuous, then*

$$\exists K \in \mathbb{R}^+ \ \text{such that:} \ ¥_q(t) = K.$$

In Section 2.7 we shall present the optimization algorithm which is based on Theorem 1 and on the use of a discretized version of the first coordination equation (1).

As we have seen in this introduction, the study of the extremals of the functional F defined over the entire interval $[0, T]$ plays a fundamental role in obtaining the solution. This point is crucial and shall be studied in the following sections on the existence of extremals, the existence of a solution to the problem and several properties related to the shooting method.

2.1. Existence and Uniqueness of Extremals

Let us now consider a series of theorems that establish sufficient conditions of existence of the extremals of the functional F defined over the entire interval $[0, T]$. It should be borne in mind that the extremals satisfying the initial condition $z(0) = 0$ are surely defined in a neighborhood of the origin which need not continue up to the instant T. If this is the case, the problem need not have a solution.

Theorem 2. *If the following conditions hold*

$$\begin{array}{ll} i) & \lim_{z' \to +\infty} [\inf_{\{z,t\}} L_{z'}(t, z, z')] = 0 \\ ii) & \lim_{z' \to -\infty} [\sup_{\{z,t\}} L_{z'}(t, z, z')] = -\infty \\ iii) & 0 < \eta_1 < E(t, Q) < \eta_2 < \infty, \forall Q \in C^2([0, T], \mathbb{R}) \end{array}$$

then, $\forall \lambda \in \mathbb{R}$, there exists a unique extremal $q \in C^2([0, T], \mathbb{R})$ which satisfies:

$$q(0) = 0 \text{ and } q'(0) = \lambda.$$

Proof.

Euler's equation associated with the functional F gives rise to the second-order differential equation

$$L_z - L_{tz'} - z' L_{zz'} - z'' L_{z'z'} = 0$$

Hence

$$z'' = \frac{L_z - L_{tz'} - z' L_{zz'}}{-L_{z'z'}}$$

Since $L_{z'z'} \neq 0$, the equation can be written as $z'' = f(t, z, z')$ with $f \in C^1([0, T] \times \mathbb{R}^2)$. It thus follows that the Cauchy problem

$$\left.\begin{array}{l} z'' = f(t, z, z') \\ z(0) = 0; z'(0) = \lambda \end{array}\right\}$$

has a unique solution in a right neighborhood of zero.

Let (q, I) be the maximal solution (defined for $t \geq 0$) of the above Cauchy problem. It will be sufficient to see that $I = [0, T]$ or, equivalently, that the maximal solution is the global solution. Let us argue by contradiction. Assume that $I = [0, \tau)$. If this is the case, there exists a sequence $\{t_n\}_1^\infty \subset I$ with $\lim_{n\to\infty} t_n = \tau$ satisfying

$$\lim_{n\to\infty} |q'(t_n)| = \infty.$$

According to (2), $q(t)$ should satisfy $\forall t \in I$

$$-L_{z'}(t, q(t), q'(t)) \cdot \exp\left[-\int_0^t \frac{L_z(s, q(s), q'(s))}{L_{z'}(s, q(s), q'(s))} ds\right] = K \neq 0$$

or

$$K = -L_{z'}(t, q(t), q'(t)) \cdot E(t, q)$$

Let us analyze the two possible cases:

A) $\lim_{n\to\infty} q'(t_n) = \infty$.

Consider the sequence S_n identically equal to K

$$S_n = -L_{z'}(t_n, q(t_n), q'(t_n)) \cdot E(t_n, q(t_n))$$

Since L is decreasing with respect to z',

$$\inf_{\{z,t\}} [L_{z'}(t, z, q'(t_n))] \leq L_{z'}(t_n, q(t_n), q'(t_n)) < 0$$

and, by hypothesis i),

$$\lim_{\substack{n\to\infty \\ q'(t_n)\to\infty}} \inf_{\{z,t\}} [L_{z'}(t, z, q'(t_n))] = 0,$$

we have that

$$\lim_{n\to\infty} L_{z'}(t_n, q(t_n), q'(t_n)) = 0.$$

Recalling hypothesis iii),

$$0 < \eta_1 < E(t_n, q(t_n)) < \eta_2 < \infty,$$

we conclude that

$$\lim_{n\to\infty} S_n = 0$$

which contradicts the condition

$$S_n = K \neq 0, \forall n.$$

B) $\lim_{n\to\infty} q'(t_n) = -\infty$.

We arrive at a contradiction by now taking into account that $\forall n$

$$\sup_{\{z,t\}} [L_{z'}(t, z, q'(t_n))] \geq L_{z'}(t_n, q(t_n), q'(t_n))$$

and, by hypothesis ii),

$$\lim_{\substack{n\to\infty \\ q'(t_n)\to-\infty}} \sup_{\{z,t\}} [L_{z'}(t, z, q'(t_n))] = -\infty$$

$$\lim_{n\to\infty} L_{z'}(t_n, q(t_n), q'(t_n)) = -\infty \Rightarrow \lim_{n\to\infty} S_n = \infty$$

which contradicts that fact that S_n are constants. □

Hypotheses i) and ii) of the above theorem refer to the "extreme" asymptotic behavior of $L_{z'}$ with respect to the rate of water discharge, which must be independent of time and the water volume.

More precisely, hypothesis i) means that if the rate of water discharge is high, the fuel cost hardly varies in comparison with the variation in the rate of water discharge. Hypothesis ii) can be interpreted as follows: for high pumped rates, the pumped-storage plant increases fuel consumption disproportionately (consumption shoots up).

Hypothesis iii) means that the rate of water discharge influences the function of the effective hydraulic generation no less than the water volume does. In fact, the hypothesis is true if the estimate

$$\exists \eta > 0 \text{ such that } \left| \frac{\partial H(t, z, z')}{\partial z} \right| < \eta \frac{\partial H(t, z, z')}{\partial z'}, \forall (t, z, z')$$

holds.

In some particular cases, the hypotheses of the above theorem can be relaxed. Let us see two cases of special interest.

Theorem 3. *If the function of effective hydraulic generation has the form*[3]

$$H(t, z, z') = f(z) \cdot z'$$

then the following properties hold:

i) For every extremal q of F, there exists a constant $K_q = P_d(0) - f(q(0)) \cdot q'(0)$, for which

$$P_d(t) - f(q(t)) \cdot q'(t) = K_q, \forall t \in [0, T]$$

i.e. the optimal thermal power is constant.

ii) If $\exists \vartheta > 0$ is such that $\forall z$, $f(z) > \vartheta$, then $\forall \lambda \in \mathbb{R}$ there exists a unique extremal $q \in C^2([0, T], \mathbb{R})$ which satisfies the conditions:

$$q(0) = 0 \text{ and } q'(0) = \lambda.$$

[3] In this model we consider the hydro-plants that generate power proportional to the rate of water discharge and the depth of water. The losses are either neglected or simply do not exist. Among these power plants, it is worth distinguishing those of fixed charge, for which $f(z) = Const.$

Proof.

i) It follows from the second coordination equation that

$$\Psi'\left(P_d(t) - f(q(t)) \cdot q'(t)\right) \cdot f(q(t)) \exp\left(-\int_0^t \frac{f'(q(s)) \cdot q'(s)}{f(q(s))} ds\right) = K$$

$$\Psi'\left(P_d(t) - f(q(t)) \cdot q'(t)\right) \cdot f(q(t)) \exp\left(-\ln f(q(t)) + \ln f(0)\right) = K$$

$$\Psi'\left(P_d(t) - f(q(t)) \cdot q'(t)\right) \cdot f(q(t)) \cdot \frac{f(0)}{f(q(t))} = K$$

$$\Psi'\left(P_d(t) - f(q(t)) \cdot q'(t)\right) = \widetilde{K}$$

$$P_d(t) - f(q(t)) \cdot q'(t) = \left(\Psi'\right)^{-1}\left(\widetilde{K}\right) = K_q$$

ii) Arguing by analogy with the proof of Theorem 2, we see that the maximal right solution is global.

In fact, assuming the contrary we have that the maximal interval would be of the form $[0, \tau)$ and such that $\lim_{t \to \tau^-} |q'(t)| = \infty$. This also means that

$$\lim_{t \to \tau^-} \left|f(q(t)) \cdot q'(t)\right| = \infty,$$

which obviously contradicts the fact that

$$P_d(t) - f(q(t)) \cdot q'(t)$$

is constant. □

Theorem 4. *If the function of effective hydraulic generation has the form*[4]

$$H(t, z, z') = \Phi(f(z) \cdot z'),$$

then the following properties arise*:*

i) If q *is an extremal of* F*, its second coordination equation is*

$$\Psi'\left(P_d(t) - \Phi(f(q(t)) \cdot q'(t))\right) \cdot \Phi'(f(q(t)) \cdot q'(t)) = K$$

ii) If $\exists \vartheta > 0$ *such that* $\forall z,\ f(z) > \vartheta$*, and the auxiliary functions*

$$\Upsilon_t(x) = \Psi(P_d(t) - \Phi(x))$$

possess the properties

$$1) \quad \lim_{x \to \infty} \Upsilon_t'(x) = 0$$
$$2) \quad \lim_{x \to -\infty} |\Upsilon_t'(x)| = \infty$$

then $\forall \lambda \in \mathbb{R}$ *there exists a unique extremal* $q \in C^2([0, T], \mathbb{R})$ *which satisfies the conditions*

$$q(0) = 0 \text{ and } q'(0) = \lambda$$

[4]This model includes the hydro-plants mentioned in the previous footnote, though now we also admit the existence of losses.

Proof.

i) By analogy with the previous theorem.

ii) We only have to observe that

$$\Psi'\left(P_d(t) - \Phi(f(q(t)) \cdot q'(t))\right) \cdot \Phi'(f(q(t)) \cdot q'(t)) = -\Upsilon_t'(f(q(t)) \cdot q'(t))$$

so that

$$-\Upsilon_t'(f(q(t)) \cdot q'(t)) = K$$

and the rest of the proof coincides with that of the previous theorem. □

2.2. Shooting Mappings

The existence of extremals does not guarantee the existence of a solution to the problem. It may occur that the extremals do not satisfy one of the necessary conditions such as, for isntance, the possibility of inclusion in a field of extremals. It may also occur that none of the extremals satisfies the boundary conditions (the restriction on the admissible volume). In the present section, we analyze some additional hypotheses whose fulfillment guarantees that the extremals represent the strong relative minimum and other hypotheses which assure the compatibility with any restriction on the admissible volume.

Throughout the present section we shall assume that $\forall\lambda \in \mathbb{R}$, $\exists| q_\lambda$ an extremal for the functional F which satisfies the conditions

$$q_\lambda(0) = 0 \text{ and } q_\lambda'(0) = \lambda$$

Definition 2. *The mapping*

$$\begin{array}{rccc} \varphi_t: & \mathbb{R} & \longrightarrow & \mathbb{R} \\ & \lambda & \longrightarrow & q_\lambda(t) \end{array}$$

where q_λ stands for the extremal of the functional F and satisfies the conditions

$$q_\lambda(0) = 0 \textit{ and } q_\lambda'(0) = \lambda$$

is called the shooting mapping[5] *with respect to t of the functional F.*

Lemma 1. *If the shooting mappings φ_t of the functional F are strictly increasing, then the extremals q_λ form a field* whose *center* is *at the origin.*

Proof.

It is clear that the center is the origin $(0,0)$ because

$$q_\lambda(0) = 0, \forall\lambda$$

Moreover, $\forall t \in (0,T]$ and $\forall\lambda_1 < \lambda_2$:

$$q_{\lambda_1}(t) = \varphi_t(\lambda_1) < \varphi_t(\lambda_2) = q_{\lambda_2}(t)$$

□

[5]We use this term after the well-known method of shooting for solving differential equations with boundary conditions. By virtue of the Theorem concerning differentiability with respect to the initial conditions, these applications are of the class C^1.

Theorem 5. *If the shooting mappings φ_t of F are strictly increasing, then every extremal q_λ is a unique local solution of the problem $\wp_{q_\lambda(T)}$.*

Proof.

To make an extremal q a strong (relative) minimum, it suffices to claim that it can be included in the field of extremals and satisfies

$$L_{z'z'}(t, q(t), p) > 0, \forall t \in [0, T], \forall p \in \mathbb{R}$$

The possibility of inclusion in the field of extremals is shown in Lemma 1.

The second condition is an immediate by-product of the properties that we initially imposed on the functional.

The uniqueness is obvious once we take into consideration the fact that the shooting mappings are increasing functions. □

Proposition 1. *If the shooting mappings φ_t of F are strictly increasing, then the set*

$$B = \{b | \wp_b \textit{ has a local solution}\}$$

is an open interval (not necessarily bounded).

Proof.

Invoking the previous theorem, we only have to check that the values admissible for the extremals in T constitute an open interval.

In fact, $B = \varphi_T(\mathbb{R})$ is an interval because it is the image of the connected set $\mathbb{R}$ under the continuous mapping φ_T. On the other hand, B is an open set because the shooting mapping is an increasing function. □

2.3. Existence and Uniqueness of Local Solution

Proposition 1 left open the possibility that problem $\wp_b$ does not have a solution for certain values of b. Let us establish a theorem which gives sufficient conditions of existence of a solution under any restriction on the admissible volume.

Definition 3. *The mapping*

$$\begin{array}{cccc} \psi_t : & \mathbb{R} & \longrightarrow & \mathbb{R} \\ & \lambda & \longrightarrow & q'_\lambda(t) \end{array}$$

where q_λ stands for the extremal of the functional F satisfying the conditions

$$q_\lambda(0) = 0 \textit{ and } q'_\lambda(0) = \lambda,$$

is called the mapping of the rate of water discharge in t of the functional F (or simply the rate mapping).

Lemma 2. *If the rate mappings ψ_t are strictly increasing, so are the shooting mappings φ_t.*

Proof.

$$\varphi_t(\lambda) = q_\lambda(t) = \int_0^t q'_\lambda(x)dx = \int_0^t \psi_x(\lambda)dx.$$

Making use of the fact that

$$\lambda_1 < \lambda_2 \Longrightarrow \psi_x(\lambda_1) < \psi_x(\lambda_2), \forall x \in [0,T]$$

$$\varphi_t(\lambda_1) = \int_0^t q'_{\lambda_1}(x)dx = \int_0^t \psi_x(\lambda_1)dx < \int_0^t \psi_x(\lambda_2)dx = \int_0^t q'_{\lambda_2}(x)dx = \varphi_t(\lambda_2)$$

□

Lemma 3. *Let* $\{f_\lambda\}_{\lambda\in\mathbb{R}}$, $(f_\lambda : [0,T] \longrightarrow \mathbb{R})$ *be a family of continuous functions satisfying the conditions*

$$\begin{array}{ll} i) & \forall t \in [0,T],\ \lim_{\lambda\to\infty} f_\lambda(t) = +\infty \\ ii) & \lambda_1 < \lambda_2 \Longrightarrow \forall t \in [0,T],\ f_{\lambda_1}(t) < f_{\lambda_2}(t). \end{array}$$

Then

$$\lim_{\lambda\to\infty}\int_0^T f_\lambda(t)dt = +\infty$$

Proof.

Let us show that $\forall K > 0$ there exists $\widetilde{\lambda}$ such that

$$\lambda > \widetilde{\lambda} \Longrightarrow \int_0^T f_\lambda(t)dt > K$$

Given $M > 0$, let us consider for every $t \in [0,T]$, λ_t such that $f_{\lambda_t}(t) > 2M$. By continuity of f_{λ_t}, there exists $\varepsilon_t > 0$ such that

$$f_{\lambda_t}(\widetilde{t}) > M, \forall \widetilde{t} \in (t-\varepsilon_t\,, t+\varepsilon_t\,) \cap [0,T]$$

Since the family is increasing

$$\lambda > \lambda_t \Longrightarrow f_\lambda(\widetilde{t}) > M, \forall \widetilde{t} \in (t-\varepsilon_t\,, t+\varepsilon_t\,) \cap [0,T]$$

Let us now consider the family of open sets $\{A_t\}_{t\in[0,T]}$, where

$$A_t = (t-\varepsilon_t\,, t+\varepsilon_t\,)$$

It is clear that $\{A_t\}_{t\in[0,T]}$ is a cover of the compact set $[0,T]$, whence the existence of a finite subcover $\{A_{t_i}\}$ with $i = 1,\ldots,n$. Let

$$\widetilde{\lambda} = \max\{\lambda_{t_1},\ldots,\lambda_{t_n}\}$$

For every $\lambda > \widetilde{\lambda}$, and $\widetilde{t} \in [0,T]$ there exists $i \in \{1,\ldots,n\}$ such that

$$\widetilde{t} \in (t_i - \varepsilon_{t_i}\,, t_i + \varepsilon_{t_i}\,)$$

and since

$$\lambda > \widetilde{\lambda} > \lambda_{t_i} \Longrightarrow f_\lambda(\widetilde{t}) > M,$$

we have that

$$\int_0^T f_\lambda(t)dt \geq \int_0^T Mdt = T\cdot M$$

where it is sufficient to choose $M = K/T$. □

Theorem 6. *If* $\forall\lambda \in \mathbb{R}$, *there exists a unique extremal satisfying* $q(0) = 0$ *and* $q'(0) = \lambda$, *the rate mappings* ψ_t *are strictly increasing and the following hypotheses are fulfilled*

$$\begin{array}{ll} i) & 0 < \eta_1 < E(t, Q) < \eta_2 < \infty, \forall Q \in C^2([0, T], \mathbb{R}) \\ ii) & \forall z', \forall t: \inf_z H_{z'}(t, z, z') = I(t, z') > 0 \\ iii) & \forall z', \forall t: \inf_z H(t, z, z') = \widetilde{I}(t, z') > -\infty \\ iv) & \forall z', \forall t: \sup_z H(t, z, z') = S(t, z') < \infty \\ v) & \forall z', \forall t: \sup_z H_{z'}(t, z, z') = \widetilde{S}(t, z') < \infty \\ vi) & \lim_{\lambda\to\infty} L_{z'}(0, 0, \lambda) = 0 \\ vii) & \lim_{\lambda\to-\infty} L_{z'}(0, 0, \lambda) = -\infty \end{array}$$

then $\forall b \in \mathbb{R}$, *problem* $\wp_b$ *has a unique local solution.*

Proof.

Let us consider the extremals q_λ of F which possess the properties:

$$q_\lambda(0) = 0 \text{ and } q'_\lambda(0) = \lambda.$$

The sole thing to prove is that $\varphi_T(\mathbb{R}) = \mathbb{R}$. To this end, taking into account the continuity of φ_T, it suffices to prove that

$$\lim_{\lambda\to\infty} q_\lambda(T) = +\infty \text{ and } \lim_{\lambda\to-\infty} q_\lambda(T) = -\infty.$$

Let K_λ be the coordination constants of the extremals q_λ, i.e.,

$$K_\lambda = H_{z'}(0, 0, \lambda)\Psi'(P_d(0) - H(0, 0, \lambda)) = -L_{z'}(0, 0, \lambda)$$

It is obvious that

$$\lim_{\lambda\to\infty} K_\lambda = -\lim_{\lambda\to\infty} L_{z'}(0, 0, \lambda) = 0$$

and

$$\lim_{\lambda\to-\infty} K_\lambda = -\lim_{\lambda\to-\infty} L_{z'}(0, 0, \lambda) = \infty$$

Let us proceed by arguing by contradiction. Let

$$\exists t_0 \in [0, T] \text{ such that } \forall\lambda > 0,\ q'_\lambda(t_0) < C$$

We have that $\forall\lambda > 0$

$$K_\lambda = H_{z'}(t_0, q_\lambda(t_0), q'_\lambda(t_0)) \cdot \Psi'(P_d(t_0) - H(t_0, q_\lambda(t_0), q'_\lambda(t_0))) \cdot E(t_0, q_\lambda) > 0$$

and, since $H_{z'z'} \leq 0$ and Ψ' is increasing,

$$K_\lambda > H_{z'}(t_0, q_\lambda(t_0), C) \cdot \Psi'(P_d(t_0) - H(t_0, q_\lambda(t_0), C)) \cdot E(t_0, q_\lambda)$$
$$H_{z'}(t_0, q_\lambda(t_0), C) \geq \inf_x H_{z'}(t_0, x, C) = I(t_0, C) > 0$$
$$H(t_0, q_\lambda(t_0), C) \leq \sup_x H(t_0, x, C) = S(t_0, C) < \infty$$
$$\Psi'(P_d(t_0) - H(t_0, q_\lambda(t_0), C)) > \Psi'(P_d(t_0) - S(t_0, C)) > 0$$

By virtue of the hypotheses

$$0 < \eta_1 < E(t_0, q_\lambda) < \eta_2 < \infty,$$

whence

$$K_\lambda > \eta_1 \cdot I(t_0, C) \cdot \Psi'(P_d(t_0) - S(t_0, C)) > 0$$

and thus

$$\lim_{\lambda \to \infty} K_\lambda \geq \eta_1 \cdot I(t_0, C) \cdot \Psi'(P_d(t_0) - S(t_0, C)) > 0,$$

which is a contradiction. Thus, the assumed estimate

$$q'_\lambda(t_0) < C, \forall \lambda > 0$$

is not truc for any $t_0 \in [0, T]$ and

$$\forall t \in [0, T] \lim_{\lambda \to \infty} q'_\lambda(t) = +\infty.$$

Making use of Lemma 3, we have that

$$\lim_{\lambda \to \infty} q_\lambda(T) = +\infty.$$

In the same way, we prove that

$$\lim_{\lambda \to -\infty} q_\lambda(T) = -\infty.$$

□

The hypotheses from ii) to iv) once again impose certain restrictions concerning the influence of the volume on the function of the effective hydraulic generation. More precisely, for every (t, z') the functions of the volume

$$h_{t,z'}(z) = H(t, z, z'), f_{t,z'}(z) = H_{z'}(t, z, z') \text{ and } \frac{1}{f_{t,z'}(z)}$$

are bounded.

The existence and uniqueness of the relative minimum do not guarantee that it is a solution of the problem. Since the set of admissible functions is not compact, the solution need not exist. This is still an open problem, which we shall attempt to resolve. Our conjecture is that the problem has an affirmative answer.

Conjecture. Under the hypotheses of Theorem 6

$$\forall b \in \mathbb{R}, \text{ problem } \wp_b \text{ has a unique solution.}$$

2.4. A Particular Case

Now we consider the particular case when

$$H(t, z, z') = f(z) \cdot z'$$

This particular case embraces the hydro pumped-storage plants of all types (without losses) whose generated power – or consumption during pumping – is proportional to the rate of water discharge and the height of the deposit. The latter is assumed to depend only on the turbined volume. The simplicity of the model allows us to apply the technique which, without any doubt, is the most natural for the study of the behavior of the shooting mappings: derivation with respect to the initial conditions.

Theorem 7. *If* $H(t, z, z') = f(z) \cdot z'$ $(f(z) > 0)$, *then:*

i) *the shooting mappings are strictly increasing,*

ii) *in addition, if there exist* $\varepsilon_1, \varepsilon_2 \in \mathbb{R}$ *such that*

$$0 < \varepsilon_1 < f(z) < \varepsilon_2, \forall z \in \mathbb{R}$$

then $\forall b \in \mathbb{R}$, $\wp_b$ *has a unique local solution.*

Proof.

i) It will be sufficient to see that

$$\varphi_t'(\lambda) > 0, \ \forall \lambda \in \mathbb{R}, \forall t \in (0, T]$$

Let us recall Theorem 3, which states that, in the present case, the extremals satisfy the equation

$$P_d(t) - f(z(t)) \cdot z'(t) = P_d(0) - f(z(0)) \cdot z'(0), \forall t \in [0, T]$$

If $z(t) = \varphi_t(\lambda)$ is the extremal which satisfies $z(0) = 0$ and $z'(0) = \lambda$, it also satisfies the differential equation

$$P_d(t) - f(z) \cdot z' = P_d(0) - f(0) \cdot \lambda$$

which is an equation in separable variables and can be solved via integration:

$$\int P_d(t)dt - \int f(z)dz = t \cdot [P_d(0) - f(0) \cdot \lambda] + c.$$

Derivating with respect to λ (the initial condition), we have

$$-f(\varphi_t(\lambda))\frac{d\varphi_t(\lambda)}{d\lambda} = -t \cdot f(0).$$

Thus,

$$\varphi_t'(\lambda) = \frac{-t \cdot f(0)}{-f(\varphi_t(\lambda))} > 0.$$

ii) The hypothesis of Theorem 3 holds

$$0 < \varepsilon_1 < f(z)$$

yielding that for any $\lambda \in \mathbb{R}$ there exists a unique extremal q which satisfies $q(0) = 0$ and $q'(0) = \lambda$.

We now only have to check that $\varphi_T(\mathbb{R}) = \mathbb{R}$. To this end, it is sufficient to show that

$$\lim_{\lambda\to\infty} \varphi_T(\lambda) = +\infty \text{ and } \lim_{\lambda\to-\infty} \varphi_T(\lambda) = -\infty.$$

Note that

$$\varphi_T'(\lambda) = \frac{T\cdot f(0)}{f(\varphi_T(\lambda))} > \frac{T\cdot\varepsilon_1}{\varepsilon_2} > 0,$$

which readily follows from the fact that φ_T is bounded neither from below nor from above. □

2.5. Solutions for Convex Functionals

Convex functionals form the ideal framework for addressing optimization (minimization) problems. The key to the importance of these functionals is the equivalence between the concepts of the critical point and the relative and absolute minimums. This equivalence guarantees that if an admissible extremal exists, it is the solution of the problem.

Theorem 8. *If the functional F is convex in Θ_b and $q \in \Theta_b$ is an extremal of F, then q is a solution of problem $\wp_b$. (The solution is unique if the convexity is strict.).*

Proof.

We only have to bear in mind that F is differentiable in the sense of Gâteaux on the affine sub-manifold $\Theta_b \cap C^1[0,T]$ and that the differential vanishes at the extremals of the functional. Therefore, q is a critical point of F, whose convexity guarantees that it is the absolute minimum of F on the convex set $\Theta_b \cap C^1[0,T]$ (this is still insufficient because the problem is posed in $KC^1[0,T]$).

Next, invoking the Lemma on angle rounding [25],

$$F(q) = \inf_{\substack{z\in\Theta_b \\ z\in C^1[0,T]}} F(z) = \inf_{\substack{z\in\Theta_b \\ z\in KC^1[0,T]}} F(z)$$

q is also the absolute minimum in Θ_b and is, therefore, a solution of problem $\wp_b$ (the unique solution if the convexity is strict). □

The convexity of a functional is difficult to check. Nonetheless, the convexity of F is assured by the assumptions accepted in the introduction to this section related with the concavity of H.

Proposition 2. *For the convexity in Θ_b of the functional $F : \Theta_b \longrightarrow \mathbb{R}$ defined by*

$$F(z) = \int_0^T \Psi\left(P_d(t) - H(t, z(t), z'(t))\right) dt$$

it is sufficient for Ψ to be increasing and convex and H concave with respect to (z, z').

Proof.

Let $\alpha \in [0,1]$ and $g_1, g_2 \in \Theta_b$

$$F(\alpha g_1 + (1-\alpha)g_2) =$$

$$= \int_0^T \Psi\left(P_d(t) - H\left(t, \alpha g_1(t) + (1-\alpha)g_2(t), \alpha g_1'(t) + (1-\alpha)g_2'(t)\right)\right) dt$$

Since H is concave with respect to (z, z'), for every t we have

$$H\left(t, \alpha g_1(t) + (1-\alpha)g_2(t), \alpha g_1'(t) + (1-\alpha)g_2'(t)\right) \geq \\ \geq \alpha H\left(t, g_1(t), g_1'(t)\right) + (1-\alpha)H\left(t, g_2(t), g_2'(t)\right)$$

from which it follows, since Ψ is increasing, that

$$F\left(\alpha g_1 + (1-\alpha)g_2\right) \leq \\ \leq \int_0^T \Psi\left(P_d(t) - \alpha H(t, g_1(t), g_1'(t)) - (1-\alpha)H(t, g_2(t), g_2'(t))\right) dt = \\ = \int_0^T \Psi\left(\alpha P_d(t) - \alpha H(t, g_1(t), g_1'(t)) + (1-\alpha)P_d(t) - (1-\alpha)H(t, g_2(t), g_2'(t))\right) dt \\ = \int_0^T \Psi\left(\alpha\left(P_d(t) - H(t, g_1(t), g_1'(t))\right) + (1-\alpha)\left(P_d(t) - H(t, g_2(t), g_2'(t))\right)\right) dt.$$

Now using the convexity of Ψ

$$F\left(\alpha g_1 + (1-\alpha)g_2\right) \leq \\ \leq \int_0^T \alpha\Psi\left(P_d(t) - H(t, g_1(t), g_1'(t))\right) dt + \int_0^T (1-\alpha)\Psi\left(P_d(t) - H(t, g_2(t), g_2'(t))\right) dt = \\ = \alpha F(g_1) + (1-\alpha)F(g_2)$$

and, finally,

$$F\left(\alpha g_1 + (1-\alpha)g_2\right) \leq \alpha F(g_1) + (1-\alpha)F(g_2), \forall \alpha \in [0,1].$$

□

If $H(t, z, z')$ is independent of z, it is concave with respect to (z, z') (recall the assumption about the concavity with respect to z') and the functional is therefore convex.

In this section, the fact that the shooting mappings are strictly increasing functions was found to be unnecessary to determine whether an extremal delivers the minimum to a functional. Nonetheless, let us see that this property can be easily derived in the case where the functional is strictly convex.

Proposition 3. *If F is strictly convex, then the shooting mappings are strictly increasing functions.*

Proof.

Let us argue by contradiction. Let us assume that $\lambda_1 < \lambda_2$ and $\tau > 0$ with $\varphi_\tau(\lambda_1) \geq \varphi_\tau(\lambda_2)$. Let q_{λ_1} and q_{λ_2} be the extremals of F satisfying

$$\begin{cases} q_{\lambda_1}(0) = q_{\lambda_2}(0) = 0 \\ q_{\lambda_1}'(0) = \lambda_1 < \lambda_2 = q_{\lambda_2}'(0) \end{cases}$$

We have that

$$q_{\lambda_1}(\tau) \geq q_{\lambda_2}(\tau)$$

By virtue of the continuity of q'_{λ_i}, $\exists t_0 \in (0, \tau)$ such that $\forall t \in [0, t_0]$

$$q'_{\lambda_1}(t) < q'_{\lambda_2}(t)$$

and, consequently, $\forall t \in [0, t_0]$, it also follows that

$$q_{\lambda_1}(t) < q_{\lambda_2}(t)$$

Moreover, since

$$q_{\lambda_1}(\tau) \geq q_{\lambda_2}(\tau)$$

it follows from the continuity of $q_{\lambda_i}(t)$ that there exists $t_1 \in (t_0, \tau]$ such that

$$q_{\lambda_1}(t_1) = q_{\lambda_2}(t_1)$$

which contradicts the uniqueness result of Theorem 8, applied to the functional F on the interval $[0, t_1]$. □

2.6. Solutions for Non-Convex Functionals

Let us now present a result which makes it evident that the convexity of a functional is not necessary to guarantee that the extremals are solutions of the problem. As we have already seen, this is closely related to the fact that the shooting mappings are strictly increasing.

Theorem 9. *Let the following conditions hold:*
i) $L_{z'z}(t, z, z') \leq 0$ *(*$L_{z'}$ *is decreasing with respect to* z*);*
ii) if $q_1'(t) \leq q_2'(t)$, $\forall t \in [0, \tau]$ *it follows that* $E(\tau, q_1) \leq E(\tau, q_2)$.
Then the rate mappings ψ_t *(of* F*) are strictly increasing in* t.

Proof.

Let us argue by contradiction and prove that if q_1 and q_2 are extremals satisfying

$$\begin{aligned} q_1(0) &= q_2(0) = 0 \\ q_1'(0) &= \lambda_1 < \lambda_2 = q_2'(0) \end{aligned}$$

then $q_1'(t) < q_2'(t), \forall t \in [0, T]$ whence the assertion of the theorem.

Denoting by K_i the coordination constant of the extremal q_i, we have

$$\begin{aligned} K_1 &= H_{z'}(0,0,\lambda_1) \cdot \Psi'(P_d(0) - H((0,0,\lambda_1)) = -L_{z'}(0,0,\lambda_1) \\ K_2 &= H_{z'}(0,0,\lambda_2) \cdot \Psi'(P_d(0) - H((0,0,\lambda_2)) = -L_{z'}(0,0,\lambda_2) \end{aligned}$$

whence we deduce, taking into account that $L_{z'z'} > 0$, that $K_1 > K_2$.

Let us now assume that $\exists t_0 > 0$ so that

$$q_1'(t_0) \geq q_2'(t_0)$$

It follows from the continuity of q_i' that there exists some $\zeta \in (0, t_0]$ such that

$$\chi = q_1'(\zeta) = q_2'(\zeta) \text{ and } q_1'(t) < q_2'(t), \forall t \in [0, \zeta)$$

Taking the second coordination equation at the instant ζ

$$K_1 = -L_{z'}(\zeta, q_1(\zeta), \chi) \cdot E(\tau, q_1)$$
$$K_2 = -L_{z'}(\zeta, q_2(\zeta), \chi) \cdot E(\tau, q_2)$$

Let us now observe that $q_1(\zeta) < q_2(\zeta)$, whence we derive, taking into account that $L_{z'z} \leq 0$, which means the strict increment of $-L_{z'}$ with respect to z, that

$$-L_{z'}(\zeta, q_1(\zeta), \chi) \leq -L_{z'}(\zeta, q_2(\zeta), \chi).$$

Bearing in mind ii) we have that $K_1 \leq K_2$, which contradicts the fact that

$$K_1 > K_2.$$

□

2.7. Optimization Algorithm

Determining the unique extremal that satisfies the restriction on the volume $z(T) = b$ is realized by obtaining the unique zero of the function

$$\varphi_T(\lambda) - b$$

This is the framework of the shooting method, which consists in varying the initial condition (the initial rate of water discharge) until the moment when the second boundary condition is fulfilled. On the other hand, the strict increment in T of the shooting mappings obviously simplifies the search for the solution. As always, if we address the problem of solving Euler's equation for the functional under the initial and boundary conditions $z(0) = 0$ and $z'(0) = \lambda$, the difficulty consists in determining $\varphi_T(\lambda)$.

There is a long list of numerical methods appropriate for solving this problem approximately. We use a similar numerical method to those used to solve differential equations in combination with an appropriate adaptation of the classical shooting method.

Step 1) *Approximate construction of* z_K (*the adapted Euler method*).

The problem will consist in finding for each K the function z_K that satisfies $z_K(0) = 0$ and the conditions of Theorem 1. In general, the construction of z'_K cannot be carried out all at once over the entire interval $[0, T]$. The construction must necessarily be carried out by constructing and successively concatenating the extremal arcs ($z'(t) \neq 0$) and arcs where the plant is neither generating nor pumping ($z'(t) = 0$) until completing the interval $[0, T]$. This is relatively simple to implement.

From the computational point of view, the construction of z_K can be performed using a discretized version of the first coordination equation (1):

$$\int_0^t L_z(t, z(s), z'(s))ds - L_{z'}(t, z(t), z'(t)) = K = -L_{z'}(0, 0, \lambda)$$

The approximate construction of each z_K, which we shall call $\widetilde{z}_K$, is carried out by means of polygonals (Euler's method). We consider the triple recurring sequence (X_n, Y_n, I_n)

with

$$\begin{aligned} n &= 0, \cdots, N-1, \\ h &= T/N, \\ t_n &= n \cdot h, \end{aligned}$$

which represents the following approximations:

$$\begin{aligned} z_K(t_n) &\approx \widetilde{z}_K(t_n) := X_n, \\ z'_K(t_n) &\approx \widetilde{z}'_K(t_n) := Y_n, \\ z_K(t) \approx \widetilde{z}_K(t) &:= X_{n-1} + (t - t_{n-1}) \cdot Y_{n-1};\ t \in [t_{n-1}, t_n], \\ \int_0^{t_n} L_z(s, z_K(s), z'_K(s))ds \approx I_n &:= \int_0^{t_n} L_z(s, \widetilde{z}_K(s), \widetilde{z}'_K(s))ds. \end{aligned}$$

and which obeys the following relation of recurrence:

$$\begin{aligned} X_0 &= 0;\ I_0 = 0 \\ Y_n &= \text{ solution of: } I_n - L_{z'}(t_n, X_n, \chi) = K \\ X_{n+1} &= X_n + h \cdot Y_n \\ I_{n+1} &= I_n + \int_{t_n}^{t_{n+1}} L_z(s, X_n + (s - t_n) \cdot Y_n, Y_n)ds. \end{aligned}$$

Step 2) *Construction of a sequence* $\{K_j\}_{j\in\mathbb{N}}$ *such that* $z_{K_j}(T)$ *converges to* b *(the adapted shooting method).*

Varying the coordination constant K, we would search for the extremal that fulfils the second boundary condition $z_K(T) = b$. The procedure is similar to the shooting method used to resolve a two-point boundary value problem (TPBVP). We implemented a SSM, obtaining good results. Effectively, we may consider the function $\varphi(K) := z_K(T)$ and calculate the root of

$$\varphi(K) - b = 0,$$

which may be realized approximately using elemental procedures. The secant method was used in the present study; the algorithm shows a rapid convergence to the optimal solution for a wide range of $K_{\min}$ and $K_{\max}$.

In a previous paper [26], we presented a qualitative aspect of the solution of the problem $\wp_b$, establishing the result called *smooth transition*. We proved that, under certain conditions, the discontinuity of the derivative of the Langrangian does not translate as discontinuity in the derivative of the solution. In fact, it is verified that the derivative of the extremal where the minimum is reached presents an interval of constancy, the constant being the value for which $L_{z'}(t, z, \cdot)$ presents discontinuity. The character C^1 of the solution is thus guaranteed[6]. We shall present examples of this result in Section 4.

The simplicity and easy realization of the algorithm is worth noting; it is also more versatile than others that are already considered classical. Its true importance will become

[6] Note that this result has a very clear interpretation in terms of pumped-storage plants: under optimum operating conditions, these plants never switch brusquely from generating power to pumping water or vice versa, but rather carry out a smooth transition, remaining inactive during a certain period of time.

evident in the study of problems with restrictions, as we shall see in the next section. Let us highlight some of the most outstanding features of the algorithm:

• The absence of z'', despite of the fact that we are dealing with second order equations.

• The presence of the derivative of the demanded power is avoided, which allows us to satisfactorily solve problems in which the demanded power is not represented by a differentiable function.

• Pumped storage hydro-plants, which are usually modeled my means of a function of hydraulic generation not differentiable at zero, can be included in our study without this circumstance producing any serious effect.

3. Problem with Restrictions

In the previous section, the domains of definition of the functions involved were very wide. We allowed the function of effective hydraulic generation $H(t, z, z')$ to be defined for negative values of z' (the rate of water discharge), which occurs in practice for pumped-storage plants. We also allowed the function of thermal cost Ψ to admit negative values of the argument (the thermal power). We proceed in this section with a single hydro-plant, assuming that the same properties that were accepted in the previous section now hold for the functions $H(t, z, z')$ and Ψ in the interior of their domains of definition.

However, we impose certain restrictions on these domains of definition or, what is the same, on the admissible functions:

- We only admit nonnegative thermal power, i.e.:

$$\Psi : \mathbb{R}^+ \longrightarrow \mathbb{R}^+$$

- We shall solely admit nonnegative volumes and rates of water discharge, i.e.:

$$H : \Omega_H \subset [0, T] \times \mathbb{R}^+ \times \mathbb{R}^+ \longrightarrow \mathbb{R}^+$$

We preserve the two crucial hypotheses: the strict positivity of $H_{z'}$ and the strict increment of $L_{z'}$ in z'.

Definition 4. *We shall denote by Π_b the problem of minimization of the functional*

$$F(z) = \int_0^T L(t, z(t), z'(t))dt$$

with L of the form

$$L(t, z(t), z'(t)) = \Psi(P_d(t) - H(t, z(t), z'(t)))$$

over the set

$$\hat{\Theta}_b = \{z \in KC^1[0, T] \ / \ z(0) = 0, z(T) = b, z'(t) \geq 0 \ \wedge \ H(t, z(t), z'(t)) \leq P_d(t)\}$$

Moreover, let us impose the following assumptions:

- For every instant and volume, the hydro-plant can fully generate all the power demanded, i.e.:

$$\forall (t, z) \in [0, T] \times \mathbb{R}^+, \ \exists z' \text{ such that } H(t, z, z') = P_d(t) \tag{4}$$

- For zero rates of water discharge, the hydraulic power must always be zero and less than the demanded power:

$$\forall(t,z) \in [0,T] \times [0,b], 0 = H(t,z,0) < P_d(t) \tag{5}$$

We shall now study the existence of a solution and analyze the influence of the interior and boundary solutions.

3.1. Existence of Solution

The assumption about the uniform boundedness of the admissible functions (the rates of water discharge) yields a significant simplification of the existence problem. Under these conditions, the non-emptiness of the set of admissible functions is sufficient to ensure the existence of a solution [25].

Theorem 10. *If* $\forall \tau \in [0,T]$, $\exists M_\tau$ *such that*

$$H(t,q(t),q'(t)) \leq P_d(t) \text{ in } [0,\tau\} \Longrightarrow q'(t) < M_\tau \text{ in } [0,\tau\}$$

then:

i) The differential equation $H(t,z(t),z'(t)) - P_d(t) = 0$, *with* $z(0) = 0$, *has a unique solution in* $[0,T]$.

ii) If we denote by $\omega(t)$ *the solution of the above equation and set*[7] $\ell = \omega(T)$, *then*

$$\forall b \in [0,\ell] \text{ the problem } \Pi_b \text{ has a solution.}$$

Proof.

i) We have to deal with a differential equation given in the implicit form:

$$0 = \Phi(t,z,z') = H(t,z,z') - P_d(t)$$

and, taking into account hypothesis (4) and the fact that $H_{z'} > 0$,

$$\forall(t,z)\ \exists| z' \text{ such that } \Phi(t,z,z') = 0$$

so that $z' = f(t,z)$ in $[0,T] \times \mathbb{R}$. Moreover,

$$\frac{\partial \Phi}{\partial z'} = \frac{\partial H(t,z,z')}{\partial z'} \neq 0$$

which allows us to apply the Implicit Function Theorem and to assure, moreover, that $f : [0,T] \times \mathbb{R}^+ \rightarrow \mathbb{R}$ is of the class C^1.

Thus, we have to deal in fact with the initial-value problem for a differential equation which has a unique solution in a neighborhood of 0. Let $\omega(t)$ be the solution and $[0,\tau)$ be its maximal domain. We thus have that

$$\lim_{t \to \tau^-} \omega(t) = \infty,$$

[7] ℓ represents the volume of water that would be consumed by the hydro-plant to produce all the demanded power without the intervention of the thermal power plant.

which, as we will see, is a contradiction, whence the domain is the interval $[0, T]$.

In fact,

$$H(t, \omega(t), \omega'(t)) = P_d(t), \forall t \in [0, \tau)$$

whence $\omega'(t) \leq M_\tau \ \forall t \in [0, \tau)$. It thus follows that

$$\omega(t) \leq t M_\tau, \ \forall t \in [0, \tau)$$

which contradicts the assumption $\lim_{t \to \tau^-} \omega(t) = \infty$.

ii) By hypothesis

$$\exists M_T \in \mathbb{R} \text{ such that } \forall q \in \hat{\Theta}_b \text{ there holds } 0 \leq q'(t) < M_T, \ \forall t \in [0, T],$$

which allows us to assert that the problem has a solution provided that the set of admissible functions is non-empty. Let us check that $\forall b \in [0, \ell]$, so we have that $\hat{\Theta}_b \neq \emptyset$.

Invoking the continuity of $\omega(t)$ in $[0, T]$, we have that there exists $\delta \in [0, T]$ such that $\omega(\delta) = b$. Let us now consider

$$\omega_b(t) = \begin{cases} \omega(t) & \text{if} \quad t \in [0, \delta] \\ b & \text{if} \quad t \in [\delta, T] \end{cases}$$

Obviously, $\omega_b(t) \in \hat{\Theta}_b$. □

3.2. Interior Solutions

In previous papers [27,28], we saw that the absence of corner points is guaranteed for the minimizing function. This result is known [29] if the admissible functions are prohibited from entering the domain bounded by the curve: the extremal is tangential to the boundary of the domain. In our case, the restrictions on the admissible functions are more complicated, since they also affect their derivatives. The transformation of some arcs into others is smooth in the sense that the derivative of the minimizing function (the rate of water discharge) is continuous at the extreme points of these arcs. In the present section, we shall study the conditions which guarantee non-existence of the boundary solutions.

Theorem 11. *If the following hypotheses are true*

$$\begin{array}{ll} i) & \Psi'(0) = 0. \\ ii) & \lim_{z' \to a} \left[\inf_{\{z,t\}} H_{z'}(t, z, z')\right] = +\infty \Longleftrightarrow a = 0. \\ iii) & 0 < \eta_1 < E(t, Q) < \eta_2 < \infty, \forall Q \in C^2([0, T], \mathbb{R}). \end{array}$$

then $\forall(\mu, Z, \lambda) \in [0, T] \times [0, b] \times (0, \infty)$ *such that* $H(\mu, Z, \lambda) < P_d(\mu)$, *there exists a unique extremal of* F

$$q(t) \in C^2([0, T], \mathbb{R}), \textit{with } 0 < q'(t) \ \wedge \ H(t, q(t), q'(t)) < P_d(t)$$

which satisfies the conditions:

$$q(\mu) = Z \textit{ and } q'(\mu) = \lambda.$$

Proof.

Like in Theorem 2, we have the equation: $z'' = f(t, z, z')$ where $f \in C^1(\Omega)$ with

$$\Omega = \{(t, z, z')/\ 0 < z'(t)\ \wedge\ H(t, z, z') < P_d(t) \text{ and } t \in [0, T]\},$$

so that the Cauchy problem:

$$\left.\begin{array}{l} z'' = f(t, z, z') \\ z(\mu) = Z; z'(\mu) = \lambda \end{array}\right\}$$

has a unique solution in a neighborhood of μ.

Let (q, I) be the maximal solution of the Cauchy problem. It suffices to see that $I = [0, T]$ or, what is the same, that the maximal solution is global.

We proceed by arguing by contradiction and assuming that $(q, [\mu, \tau))$ is the maximal right solution (the same argument applies for the left solution). If this is the case, there exists a sequence $\{t_n\}_1^\infty \subset [\mu, \tau)$ with $\lim\limits_{n\to\infty} t_n = \tau$ for which

$$\lim_{n\to\infty} (t_n, q(t_n), q'(t_n)) \in \partial\Omega.$$

In other words, one of the following inequalities holds:

$$\begin{array}{ll} 1) & \lim\limits_{n\to\infty} H(t_n, q(t_n), q'(t_n)) - P_d(t_n) = 0 \\ 2) & \lim\limits_{n\to\infty} q'(t_n) = 0. \end{array}$$

Let us now analyze the two possibilities corresponding to these inequalities:

1) $\lim\limits_{t_n\to\tau} H(t_n, q(t_n), q'(t_n)) = P_d(\tau)$.

By virtue of (2), we have that $\forall n$,

$$-L_{z'}(t_n, q(t_n), q'(t_n)) \cdot \exp\left[-\int_\mu^{t_n} \frac{H_z(s, q(s), q'(s))}{H_{z'}(s, q(s), q'(s))} ds\right] = K.$$

Let us take into consideration the constant sequence

$$K_n = -L_{z'}(t_n, q(t_n), q'(t_n)) \cdot \exp - \left[\int_\mu^{t_n} \frac{H_z(s, q(s), q'(s))}{H_{z'}(s, q(s), q'(s))} ds\right]$$

$$-L_{z'}(t_n, q(t_n), q'(t_n)) = H_{z'}(t_n, q(t_n), q'(t_n)) \cdot \Psi'(P_d(t_n) - H(t, q(t_n), q'(t_n)))$$

with

$$K_n = K = H_{z'}(\mu, Z, \lambda) \cdot \Psi'(P_d(\mu) - H(\mu, Z, \lambda) > 0.$$

Bearing in mind (hypothesis i) that:

$$\lim_{t_n\to\tau} \Psi'(P_d(t_n) - H(t, q(t_n), q'(t_n))) = \Psi'(0) = 0$$

that

$$\exp\left(-\int_\mu^{t_n} \frac{H_z(s, q(s), q'(s))}{H_{z'}(s, q(s), q'(s))} ds\right)$$

is bounded (immediately follows from hypothesis iii)), that

$$\lim_{t_n\to\tau} H(t_n, q(t_n), q'(t_n)) = P_d(\tau) \Longrightarrow \lim_{t_n\to\tau} q'(t_n) > 0$$

a by-product of (5), and that

$$\lim_{t_n\to\tau} H_{z'}(t_n, q(t_n), q'(t_n)) < \infty$$

(bounded by hypothesis ii), we therefore have that

$$\lim_{n\to\infty} K_n = 0$$

which contradicts the fact that K_n is constant.

2) If $\lim_{t_n\to\tau} q'(t_n) = 0$,
we have, by hypothesis ii), that

$$\lim_{t_n\to\tau} H_{z'}(t_n, q(t_n), q'(t_n)) = \infty$$

Moreover, it follows from (5) that

$$\lim_{t_n\to\tau} \left[P_d(t_n) - H(t_n, q(t_n), q'(t_n))\right] = \varsigma > 0$$

Since Ψ' is increasing and $\Psi'(0) = 0$, we now have:

$$\lim_{t_n\to\tau} \Psi'(P_d(t_n) - H(t, q(t_n), q'(t_n))) = \Psi'(\varsigma) > 0$$

whence

$$\lim_{n\to\infty} K_n = +\infty$$

which again is a contradiction to the fact that K_n is constant. □

Corollary 2. *If under the hypotheses of the previous theorem q is a solution of problem Π_b, then the following properties hold*[8]*:*

$$\begin{array}{ll} i) & q'(t) = 0, \ \forall t \in [0, T]. \\ ii) & H(t, q(t), q'(t)) = P_d(t), \ \forall t \in [0, T]. \\ iii) & 0 < H(t, q(t), q'(t)) < P_d(t), \ \forall t \in [0, T]. \end{array}$$

Proof.

Previous papers [27,28] showed that the solution q is of the class C^1. By (5), it is also clear that no consequent boundary arcs of different types can exist.

Let us assume that the boundary and interior arcs coexist in the solution. Let $\tau \in (t_1, t_2)$ be such that

$$0 < H(t, q(t), q'(t)) < P_d(t), \ \forall t \in (t_1, t_2)$$

with the boundary point t_2 (for example).

[8]The solution is either entirely the boundary solution, or the interior solution.

Let us now consider the unique extremal $p \in C^2([0,T],\mathbb{R})$ for which

$$0 < H(t,p(t),p'(t)) < P_d(t),\ \forall t \in [0,T]$$

and which satisfies the conditions $p'(\tau) = q'(\tau)$ and $p(\tau) = q(\tau)$. Obviously, $p \equiv q$ on the interval (t_1,t_2).

If we recall that

$$q'(t_2) = 0 \text{ or } H(t_2,q(t_2),q'(t_2)) - P_d(t_2) = 0$$

and that q' is continuous, we shall have that every sequence $\{t_n\} \subset (t_1,t_2)$ satisfying $\lim_{n\to\infty} t_n = t_2$ likewise satisfies

$$\lim_{n\to\infty} q'(t_n) = 0 \text{ or } \lim_{n\to\infty} H(t_n,q(t_n),q'(t_n)) - P_d(t_n) = 0.$$

Hence

$$\lim_{n\to\infty} p'(t_n) = 0 \text{ or } \lim_{n\to\infty} H(t_n,p(t_n),p'(t_n)) - P_d(t_n) = 0.$$

It thus follows that

$$p'(t_2) = 0 \text{ or } H(t_2,p(t_2),p'(t_2)) = P_d(t_2)$$

which contradicts the assumption that p is the interior extremal. □

3.3. Boundary Solutions

When an extremal $q(t)$ admits bilateral variations over the entire interval $[0,T]$, $\forall t \in [0,T]$ the second coordination equation holds

$$-L_{z'}(t,z(t),z'(t)) \cdot \exp\left[-\int_0^t \frac{H_z(s,z(s),z'(s))}{H_{z'}(s,z(s),z'(s))}ds\right] = -L_{z'}(0,z(0),z'(0)) = K,$$

where the coordination constant

$$K = \Psi'\left(P_d(0) - H(0,0,q'(0))\right) \cdot H_{z'}(0,0,q'(0)).$$

If we did not have the restrictions

$$z'(t) \geq 0 \ \wedge\ H\left(t,z(t),z'(t)\right) \leq P_d(t),$$

we could use the shooting method to solve the problem. Varying the initial condition of the derivative $z'(0)$ (initial flow rate), we would search for the extremal that fulfils the second boundary condition $z(T) = b$ (final volume). However, we cannot use this method in our case, since, on account of the restrictions, the extremals may not admit bilateral variations, i.e. they may present boundary arcs.

If the minimizing function contains boundary arcs, the following questions arise:

- Do all the interior arcs[9] have the same coordination constant?

[9] Arcs C_1 and C_3 in Figure 1.

- At what moments is the boundary to be penetrated and abandoned?

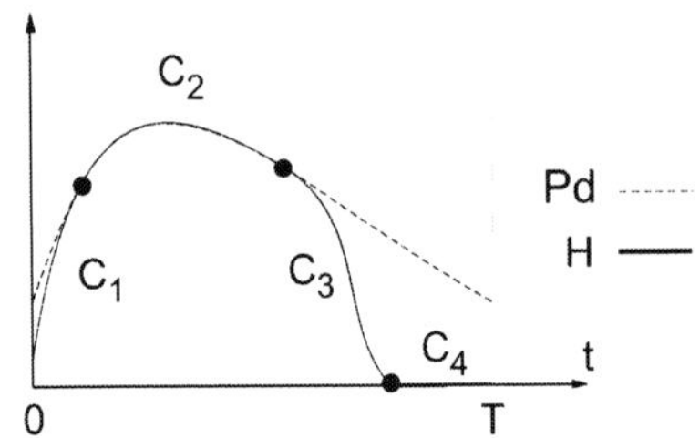

Figure 1. Boundary and interior arcs.

In a previous paper [13], we formulated the problem within the framework of Optimal Control Theory, obtaining satisfactory answers to these questions without any additional condition. The problem Π_b was stated in the following terms:

$$\min_{z\in\hat{\Theta}_b} F(z) = \min_{z\in\hat{\Theta}_b} \int_0^T \Psi\left[P_d(t) - H(t,z(t),z'(t))\right] dt$$

$$= \min_{z\in\hat{\Theta}_b} \int_0^T L(t,z(t),z'(t))dt$$

$$\hat{\Theta}_b = \{z \in KC^1[0,T] \mid z(0)=0, z(T)=b, 0 \le H(t,z(t),z'(t)) \le P_d(t), \forall t \in [0,T]\}.$$

The function

$$\mathbb{Y}_q(t) := -L_{z'}(t,q(t),q'(t)) \cdot \exp\left[-\int_0^t \frac{H_z(s,q(s),q'(s))}{H_{z'}(s,q(s),q'(s))} ds\right] \tag{6}$$

was called *the second coordination function of* $q \in \hat{\Theta}_b$.

Using Pontryagin's Minimum Principle (PMP) to solve the constrained problem, we obtained the following theorem.

Theorem 12. *If* q *is a solution of* (Π_b)*, then* $\exists K \in \mathbb{R}^+$ *such that:*

$$\mathbb{Y}_q(t) \textit{is} \begin{cases} \le K & \textit{if } H(t,q(t),q'(t)) = 0 \\ = K & \textit{if } 0 < H(t,q(t),q'(t)) < P_d(t) \\ \ge K & \textit{if } H(t,q(t),q'(t)) = P_d(t). \end{cases}$$

Theorem 12 in combination with the shooting method plays the same role as Theorem 1 in the previous section in obtaining the optimal solution. On the basis of this theorem, we are now ready to develop the optimization algorithm.

3.4. Optimization Algorithm

To obtain the optimum operating conditions of the hydro-plant, we shall use the second coordination equation (2)

$$-L_{z'}(t,z(t),z'(t)) \cdot \exp\left[-\int_0^t \frac{H_z(s,z(s),z'(s))}{H_{z'}(s,z(s),z'(s))} ds\right] = -L_{z'}(0,z(0),z'(0)) = K.$$

The peculiar form of the solution, expressed in Theorem 12, allows us to undertake its approximate calculation using similar numerical methods to those used to solve differential equations in combination with an appropriate adaptation of the classical shooting method. More precisely, we shall undertake two processes of approximation:

- Approximate construction of z_K (the adapted Euler method).
- Construction of a sequence $\{K_j\}_{j\in\mathbb{N}}$ such that $z_{K_j}(T)$ converges to b (the adapted shooting method).

Step 1) Approximate construction of z_K (the adapted Euler method).

The problem will consist in finding for each K the function z_K that satisfies $z_K(0) = 0$, and the conditions of Theorem 12. From the computational point of view, the construction of z_K can be performed using a discretized version of equation (2). In general, the construction of z'_K cannot be carried out all at once over the entire interval $[0, T]$. The construction must necessarily be carried out by constructing and successively concatenating the extremal arcs and boundary arcs until completing the interval $[0, T]$. If the values obtained for z and z' do not obey the constraints, we force the solution z_K to belong to the boundary until the moment when the conditions of leaving the domain (established in Theorem 12) are fulfilled.

The approximate construction of each z_K, which we shall call $\widetilde{z}_K$, is carried out by means of polygonals (Euler's method). We denote

$$\mathbb{Y}_{\widetilde{z}_K}(t_n) = -L_{z'}(t_n, X_n, Y_n) \cdot \exp\left[-I_n\right]$$

and we consider the triple recurring sequence (X_n, Y_n, I_n) with

$$\begin{aligned} n &= 0, \cdots, N-1, \\ h &= \frac{T}{N}, \\ t_n &= n \cdot h, \end{aligned}$$

which represents the following approximations:

$$\begin{aligned} z_K(t_n) &\approx \widetilde{z}_K(t_n) := X_n \\ z'_K(t_n) &\approx \widetilde{z}'_K(t_n) := Y_n \\ z_K(t) \approx \widetilde{z}_K(t) &:= X_{n-1} + (t - t_{n-1}) \cdot Y_{n-1};\ t \in [t_{n-1}, t_n] \\ \int_0^{t_n} \frac{H_z(s, z_K(s), z'_K(s))}{H_{z'}(s, z_K(s), z'_K(s))} ds \approx I_n &:= \int_0^{t_n} \frac{H_z(s, \widetilde{z}_K(s), \widetilde{z}'_K(s))}{H_{z'}(s, \widetilde{z}_K(s), \widetilde{z}'_K(s))} ds \end{aligned}$$

and which obeys the following relation of recurrence:

$$X_0 = 0;\ I_0 = 0$$

- If $H(t_{n-1}, X_{n-1}, Y_{n-1}) = H_{\min} \rightarrow$

$$Y_n = \text{ solution of: } H(t_n, X_n, \chi) = H_{\min}$$

$$\text{If} \begin{cases} \mathbb{Y}_{\widetilde{z}_K}(t_n) < K & \rightarrow \quad Y_n = \text{ solution of: } H(t_n, X_n, \chi) = H_{\min} \\ \mathbb{Y}_{\widetilde{z}_K}(t_n) \geq K & \rightarrow \quad Y_n = \text{ solution of: } -L_{z'}(t_n, X_n, \chi) \cdot \exp\left[-I_n\right] = K \end{cases}$$

- If $H_{\min} < H(t_{n-1}, X_{n-1}, Y_{n-1}) < H_{\max} \rightarrow$

$$Y_n = \text{ solution of: } -L_{z'}(t_n, X_n, \chi) \cdot \exp\left[-I_n\right] = K$$

$$\text{If} \begin{cases} H(t_n, X_n, Y_n) < H_{\min} & \rightarrow & Y_n = \text{ solution of: } \\ & & H(t_n, X_n, \chi) = H_{\min} \\ H_{\min} \leq H(t_n, X_n, Y_n) \leq H_{\max} & \rightarrow & Y_n = \text{ solution of: } \\ & & -L_{z'}(t_n, X_n, \chi) \cdot \exp\left[-I_n\right] = K \\ H(t_n, X_n, Y_n) > H_{\max} & \rightarrow & Y_n = \text{ solution of: } \\ & & H(t_n, X_n, \chi) = H_{\max} \end{cases}$$

- If $H(t_{n-1}, X_{n-1}, Y_{n-1}) = H_{\max} \rightarrow$

$$Y_n = \text{ solution of: } H(t_n, X_n, \chi) = H_{\max}$$

$$\text{If} \begin{cases} \mathbb{Y}_{\tilde{z}_K}(t_n) > K & \rightarrow & Y_n = \text{ solution of: } H(t_n, X_n, \chi) = H_{\max} \\ \mathbb{Y}_{\tilde{z}_K}(t_n) \leq K & \rightarrow & Y_n = \text{ solution of: } -L_{z'}(t_n, X_n, \chi) \cdot \exp\left[-I_n\right] = K \end{cases}$$

$$X_{n+1} = X_n + h \cdot Y_n$$

$$I_{n+1} = I_n + \int_{t_n}^{t_{n+1}} \frac{H_z(s, X_n + (s - t_n) \cdot Y_n, Y_n)}{H_{z'}(s, X_n + (s - t_n) \cdot Y_n, Y_n)} ds.$$

Step 2) Construction of a sequence $\{K_j\}_{j \in \mathbb{N}}$ such that $z_{K_j}(T)$ converges to b (the adapted shooting method).

Varying the coordination constant K, we would search for the extremal that fulfils the second boundary condition $z_K(T) = b$. The procedure is similar to the shooting method used to resolve a two-point boundary value problem (TPBVP). We implemented a SSM, obtaining good results. Effectively, we may consider the function $\varphi(K) := z_K(T)$ and calculate the root of

$$\varphi(K) - b = 0, \tag{7}$$

which may be realized approximately using elemental procedures.

In this study, the secant method was used to calculate the approximate value of K for which (7) is verified. The algorithm shows a rapid convergence to the optimal solution if we choose the next $K_{\min}$ and $K_{\max}$.

We set $H\left(t, z(t), z'(t)\right) = H_{\max}, \forall t \in [0, T]$. We calculate $Y_z(t), \forall t \in [0, T]$ and we choose

$$K_{\min} = \min_t Y_z(t).$$

We set $H\left(t, z(t), z'(t)\right) = H_{\min}, \forall t \in [0, T]$. We calculate $Y_z(t), \forall t \in [0, T]$ and we choose

$$K_{\max} = \max_t Y_z(t).$$

Note that for $\forall K$ admissible (with the hypothesis $H_{z'z'}(t, z, \cdot) < 0$), we have

$$K_{\min} < K < K_{\max}.$$

The following proposition guarantees that, at the nodes $\{t_n\}_{n=0}^{N-1}$, the approximation $\tilde{z}_K$ satisfies the condition established in Theorem 12.

Proposition 4. $\tilde{z}_K$ *satisfies in* $\{t_n\}_{n=0}^{N-1}$ *the following:*

$$\mathbb{Y}_{\tilde{z}_K}(t_n) is \begin{cases} \leq K & \text{if } H(t_n, X_n, Y_n) = H_{\min} \\ = K & \text{if } H_{\min} < H(t_n, X_n, Y_n) < H_{\max} \\ \geq K & \text{if } H(t_n, X_n, Y_n) = H_{\max} \end{cases}$$

Proof. Let us bear in mind that

$$\mathbb{Y}_{z_K}(t_n) \approx \mathbb{Y}_{\tilde{z}_K}(t_n) = L_{z'}(t_n, X_n, Y_n) \cdot \exp\left[-I_n\right].$$

If

$$H_{\min} < H(t_n, X_n, Y_n) < H_{\max} \Longrightarrow \mathbb{Y}_{\tilde{z}_K}(t_n) = -L_{z'}(t_n, X_n, Y_n) \cdot \exp\left[-I_n\right] = K.$$

Considering now that $H_{z'}(t_n, X_n, \cdot)$ is decreasing, we have that:

If

$$H(t_n, X_n, Y_n) = H_{\min} \Longrightarrow \exists \xi \mid H_{\min} \leq H(t_n, X_n, \xi)$$

such that

$$-L_{z'}(t_n, X_n, \xi) \cdot \exp\left[-I_n\right] = K \Longrightarrow \mathbb{Y}_{\tilde{z}_K}(t_n) \leq K.$$

If

$$H(t_n, X_n, Y_n) = H_{\max} \Longrightarrow \exists \xi \mid H(t_n, X_n, \xi) \leq H_{\max}$$

such that

$$-L_{z'}(t_n, X_n, \xi) \cdot \exp\left[-I_n\right] = K \Longrightarrow \mathbb{Y}_{\tilde{z}_K}(t_n) \geq K.$$

□

4. Examples

In this section we present various examples and implemented the proposed algorithms using Mathematica ©. We consider a wide variety of situations which are resolved in a very satisfactory way in all the cases.

4.1. Example 1: A Problem without Restrictions

Let us now see a problem whose solution may be constructed in a simple way taking into consideration the optimization algorithm presented in Section 2.7. A program that solves the optimization problem was written using the Mathematica © package and was then applied to one example of a hydrothermal system made up of the thermal equivalent and a hydraulic pumped-storage plant.

For the fuel cost model of the equivalent thermal plant, we use the quadratic model:

$$\Psi(P(t)) = \alpha_{eq} + \beta_{eq} P(t) + \gamma_{eq} P(t)^2$$

The data are summarized in Table 1, the units for the coefficients being: α_i in $(\$/h)$, β_i in $(\$/h.Mw)$, and γ_i in $(\$/h.Mw^2)$.

Table 1. Coefficients of the thermal plant.

Plant	α_{eq}	β_{eq}	γ_{eq}
Equivalent	9195.48	19.6813	0.00143078

In this first example, we use a *fixed head* model without transmission losses for the hydro-plant. Thus, the function of effective hydraulic generation H is a lineal function of the rate of water discharge. In practice, H differs depending on the positivity or negativity (pumping) of the rate of water discharge, and hence the function H is defined piecewise

$$H(t,z') := \begin{cases} A(t)\cdot z' & \text{if } z' \geq 0 \\ M\cdot z' & \text{if } z' < 0 \end{cases}$$

where

$$A(t) := \frac{B_y}{G}(S_0 + t\cdot i); \quad M := (1.1)\cdot \frac{B_y S_0}{G}.$$

The units for the coefficients of the hydro-plant are: the factor of water-conversion of the pumped-storage plant M in $(h.Mw/m^3)$, the efficiency G in $(m^4/h.Mw)$, the restriction on the volume b in (m^3), the natural inflow i in (m^3/h), the initial volume S_0 in (m^3), and the coefficient B_y in (m^{-2}) (a parameter that depends on the geometry of the reservoir). The data for the hydro-plant are summarized in Table 2.

Table 2. Coefficients of the hydro-plant.

G	S_0	B_y	b
526315	$200\ 10^9$	$149.5\ 10^{-12}$	$7.78659\ 10^6$

We analyze two cases. In the first case, the natural inflow: $i = 0$ and in the second case, the natural inflow: $i = 3.1313\cdot 10^8$.

An optimization interval of $T = 24\ h.$ was considered, with a discretization of $24\cdot 8$ subintervals. For the sake of simplicity, the values of the power demand $P_d\ (Mw)$ were fitted to the following curve:

$$P_d(t) := 1000 + 3(1 + 2Sin[4\pi t/24] - t(24-t)2Cos[4\pi t/24])$$

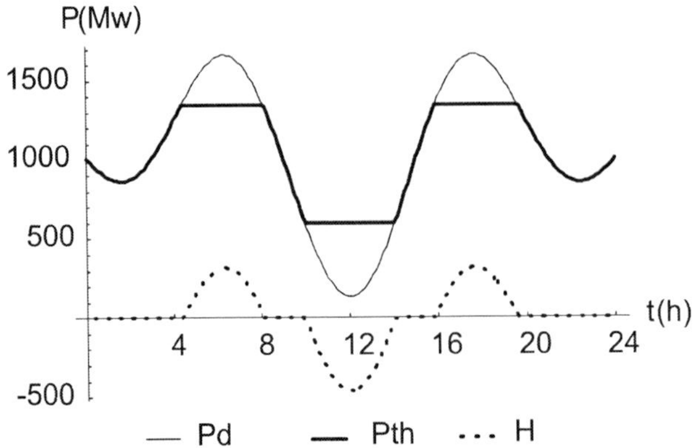

Figure 2. Optimal solution to the first case.

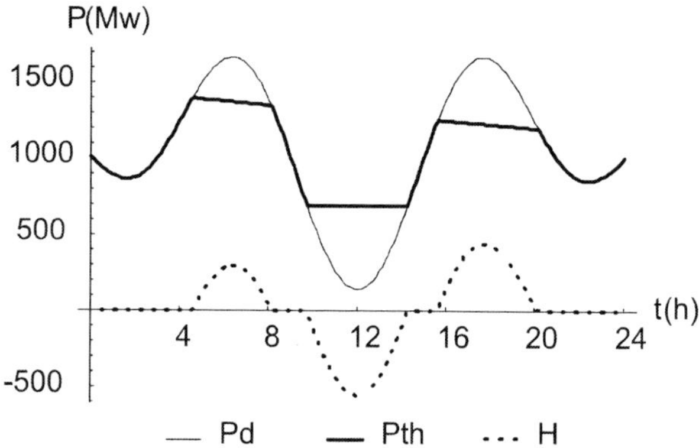

Figure 3. Optimal solution to the second case.

Figure 2 presents the optimum solution to the first case. As established in Theorem 3, we can see that the optimum thermal power P_{th} remains constant at all the instants in which the optimum hydraulic power H is positive and at the instants in which pumping takes place.

Figure 3 presents the optimum solution to the second case. We can see that the optimum thermal power remains constant at all the instants in which pumping takes place, but is decreasing in the zone where the optimum hydraulic power is positive, as a result of the presence of natural inflow $i \cdot t$.

The cost of the optimal solution to the first case is $\$ 751724$ and the cost of the second is $\$ 750903$. As is natural, fuel consumption is lower in the second case, since the performance of the hydro-plant increases due to the existence of natural inflow $i{\cdot}t$. It can also be observed that more intense pumping takes place in the second example. This is reasonable if we bear in mind that the water is transformed into energy more efficiently in this model.

The algorithm shows a rapid convergence to the optimal solution. The secant method was used to calculate the approximate value of K for which

$$q_K(T) - b = 0.$$

13 iterations were found to be sufficient in the first case and 16 iterations in the second to obtain the prescribed error:

$$|q_K(T) - b| < 10^{-2}\ (m^3).$$

The convergence of the algorithm to the prescribed final volume b is shown in Figure 4 for the first case.

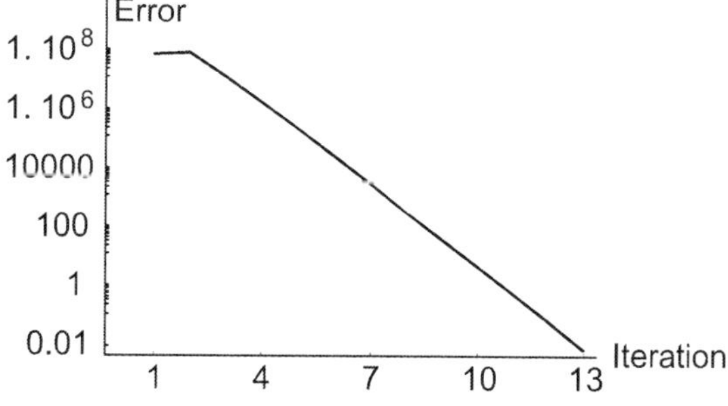

Figure 4. Convergence of the algorithm.

The CPU time used was 6.9 sec. and 8.0 sec. respectively, on a personal computer (Pentium IV, 3.40 GHz, 1 GB RAM). The coordination constant was $K = 1336.811821\ 10^{-6}$ and $K = 1353.870282\ 10^{-6}$ respectively.

4.2. Example 2: A Problem with Restrictions

In the second example, we consider certain restrictions on the domains of definition: the problem Π_b.

For the fuel cost model of the equivalent thermal plant $\Psi : \mathbb{R}^+ \longrightarrow \mathbb{R}^+$, we use the same quadratic model as in Example 1.

However, we use a more complex hydro-model in this second example. We use a *variable head* model, the hydro-plant's active power generation P_h being given by

$$P_h(t) = A(t)z'(t) - Bz'(t)z(t) - Cz'^2(t)$$

where the coefficients A, B and C are

$$A(t) = \frac{B_y}{G}(S_0 + t \cdot i), \;\; B = \frac{B_y}{G}, \;\; C = \frac{B_T}{G}$$

In variable head models, the term $-Bz'(t)z(t)$ represents the negative influence of the consumed volume and reflects the fact that consuming water lowers the effective height and hence the performance of the plant. We consider the transmission losses for the hydro-plant to be expressed by Kirchmayer's model, with the following loss equation: $b_l \cdot (P_h(t))^2$. Thus,

$$H(t) = P_h(t) - b_l \cdot (P_h(t))^2$$

The units for the coefficients of the hydro-plant are: the efficiency G in $(m^4/h.Mw)$, the restriction on the volume b in (m^3), the loss coefficient b_l in $(1/Mw)$, the natural inflow i in (m^3/h), the initial volume S_0 in (m^3), and the coefficients B_y in (m^{-2}) and B_T in $(h \cdot m^{-2})$, parameters which depend on the geometry of the reservoir. The data on the hydro-plant are summarized in Table 3.

Table 3. Coefficients of the hydro-plant.

G	i	S_0	B_T	B_y	b_l
526315	10190000	200 10^9	581.740 10^{-10}	149.5 10^{-12}	0.0002

The values of the power demand were fitted to the same curve as in Example 1:

$$P_d(t) = 1000 + 3(1 + 2Sin[4\pi t/24] - t(24 - t)2Cos[4\pi t/24])$$

First, an optimization interval of 24 h. was considered, with a discretization of $24 \cdot 8$

subintervals, and a final volume $b = 30 \cdot 10^6\ m^3$.

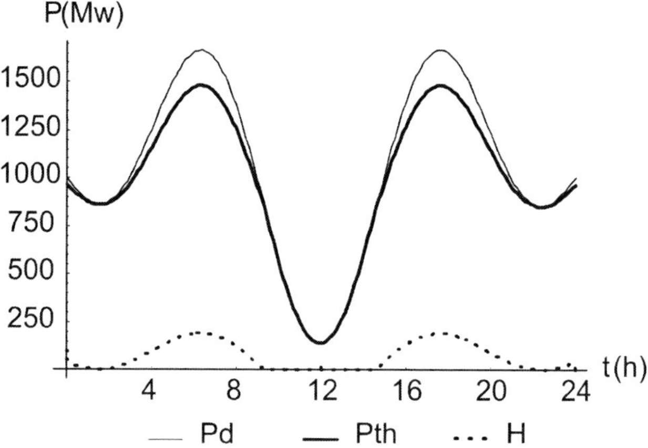

Figure 5. Optimal solution with $b = 30 \cdot 10^6 m^3$.

Figure 5 presents the plots of power demand (P_d), optimal thermal power (P_{th}) and optimal effective hydraulic power (H). We can see that from 9 h. until 15 h., corresponding to the hours of lowest power demand (i.e. with the most pronounced trough), the hydro-plant stops functioning and the thermal plant assumes all the power demand. This is done to reserve water for when power demands are very high, which corresponds to the peaks that can be seen in the figure. In this case, the cost is \$724074.

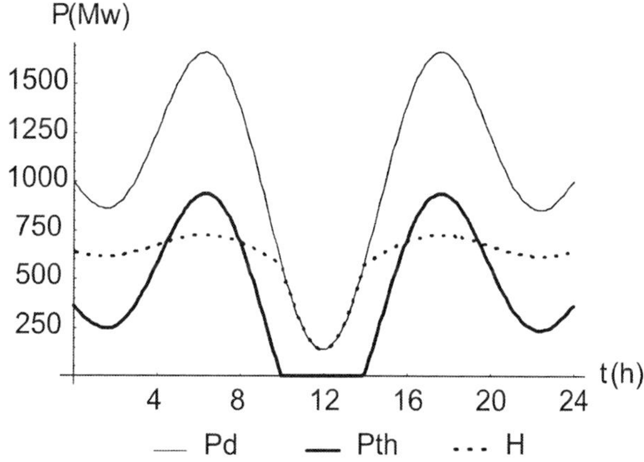

Figure 6. Optimal solution with $b = 300 \cdot 10^6 m^3$.

However, if we take a larger final volume, $b = 300 \cdot 10^6\ m^3$, the solution is that depicted in Fig. 6. Here we see that as there is sufficient water, the hydro-plant does not stop functioning at any time whatsoever, though the thermal plant shuts off in the most pronounced trough, i.e. from 11 h. until 13 h. In this case, the fuel cost is \$443077, which logically is considerably lower.

8 iterations were found to be sufficient in the first case and 5 iterations in the second to obtain the prescribed error:

$$|q_K(T) - b| < 10^{-2}\ (m^3).$$

The convergence of the algorithm to the prescribed final volume b is shown in Figure 7 for the second case.

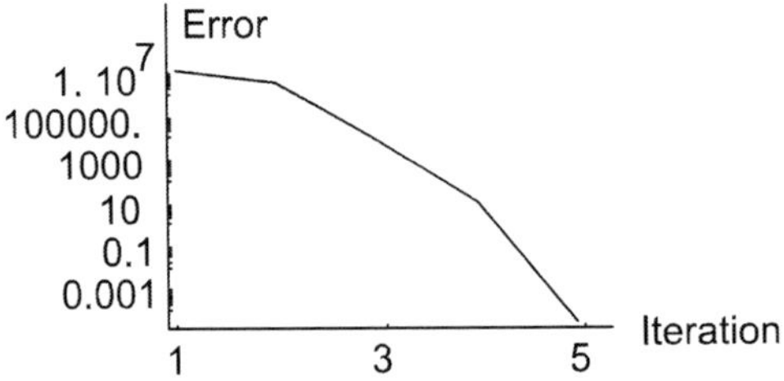

Figure 7. Convergence of the algorithm.

The CPU time used was 3.8 sec . and 7.0 sec . respectively, on a personal computer (Pentium IV, 3.40 GHz, 1 GB RAM). The coordination constant was $K = 1255.984892\ 10^{-6}$ and $K = 824.126288\ 10^{-6}$ respectively.

4.3. Example 3: Fields of Extremals

We now present the fields of extremals for the above examples. In Example 1, the model of effective hydraulic generation H is

$$H(t, z') := \begin{cases} A(t) \cdot z' & \text{if } z' \geq 0 \\ M \cdot z' & \text{if } z' < 0 \end{cases}$$

Solving the first coordination equation for the functional under the initial conditions $z(0) = 0$ and $z'(0) = \lambda$, and varying λ, we have:

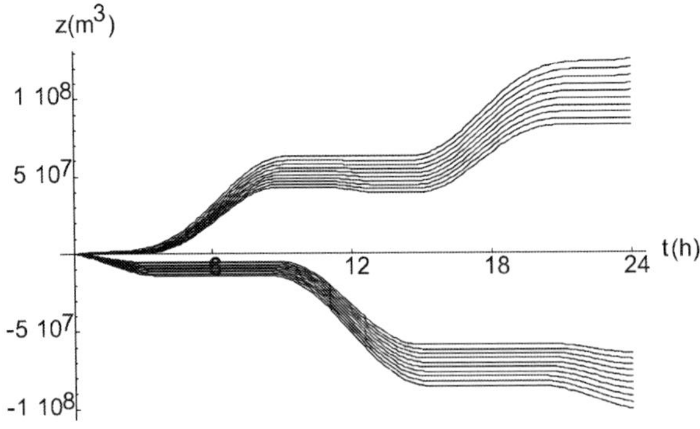

Figure 8. Field of extremals of Example 1.

In Example 2, the model of effective hydraulic generation H is

$$H(t, z, z') = P_h(t) - b_l \cdot (P_h(t))^2$$

with

$$P_h(t) = A(t)z'(t) - Bz'(t)z(t) - Cz'^2(t).$$

Solving the second coordination equation, the field of extremals is in this case:

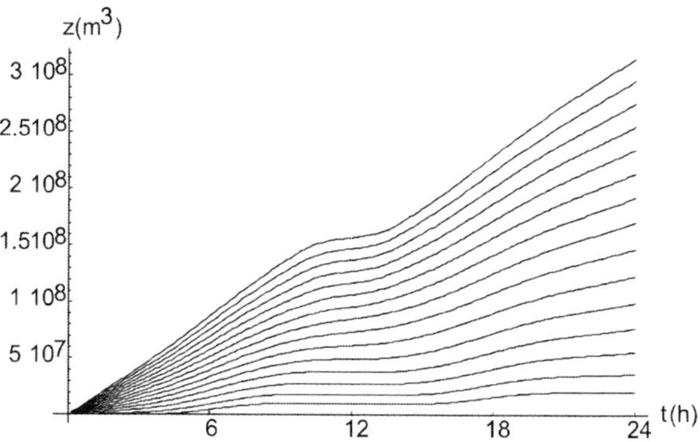

Figure 9. Field of extremals of Example 2.

Let us now consider two models which serve to illustrate some interesting ideas.

(a) First we choose the model of effective hydraulic generation:

$$H(t, z, z') = A(t)z'(t) + Bz(t)^2,$$

with B_y =149.5 10^{-10} and the rest of coefficients analogous to Example 1.

This model provides an example in which the shooting mappings are not strictly increasing and therefore the extremals do not form a central field.

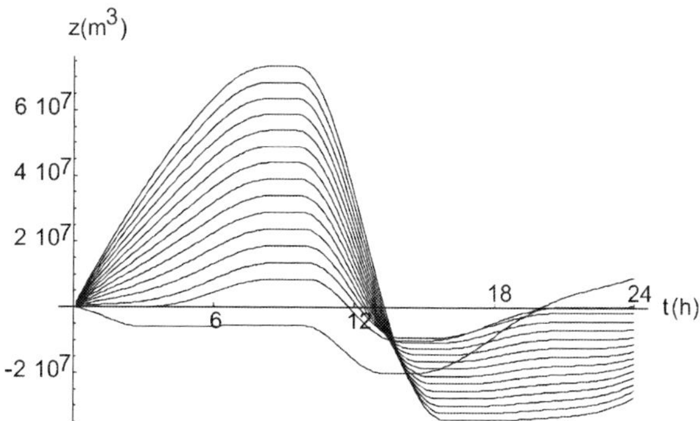

Figure 10. The extremals do not form a central field.

(b) Finally, we choose the model of effective hydraulic generation:

$$H(t, z, z') = A(t)z'(t) \cdot \exp(Cz(t))$$

with C =1 10^{-8} and the rest of coefficients analogous to Example 1.

This model gives us an example of the fact that if the hypothesis of point ii) in Theorem 7 is not satisfied, then for certain initial rates of water discharge the extremals need not exist, or the solution does not exist for some available volumes.

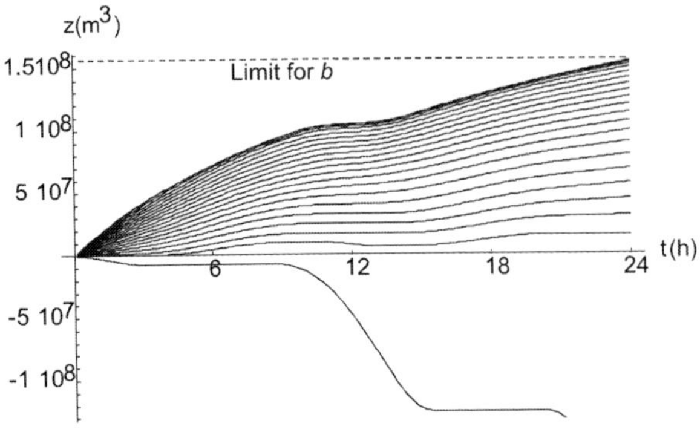

Figure 11. The solution does not exist for some b.

These examples show the importance of the conditions analyzed in this study.

5. Conclusion

The contributions of this study vary in character, being methodological, theoretical and algorithmic.

Methodological. The problem of optimization of the fuel cost in a hydrothermal system was formulated within the framework of optimal control, and we have obtained a simpler and, at the same time, more general formulation of this classical problem.

Theoretical. We have established the conditions that guarantee the existence and uniqueness of solutions in problems with a single hydro-plant, with and without restrictions. In problems without restrictions, Theorem 1 provides us with qualitative information about the solution that is sufficient for its approximate calculation; as does Theorem 10 in problems with restrictions.

Algorithmic. On the basis of the obtained theoretical results, we have developed the algorithms for solving all the formulated problems. These algorithms, implemented using Mathematica ©, allowed us to solve the concrete problems presented in Section 4.

References

[1] Wood, A. J.; Wollenberg, B. F. *Power generation, operation, and control*; John Willey & Sons, New York, 1996.

[2] Ferrero, R. W.; Rivera, J. F.; Shahidehpour, S. M. A dynamic programming two-stage algorithm for long-term hydrothermal scheduling of multireservoir systems. *IEEE Trans. on Power Systems* 1998, 13(4), 1534-1540.

[3] Nilson, O.; Sjelvgren, D. Mixed integer programming applied to short-term planning of a hydro-thermal system. *Proc. of the 1995 IEEE PICA 1995*, 158-163.

[4] Torre, S.; Arroyo, J. M.; Conejo, A. J.; Contreras, J. Price maker self-scheduling in a pool-based electricity market: a mixed-integer LP approach. *IEEE T. Power Syst.* 2002, 17(4), 1037-1042.

[5] Salam, M. S.; Nor, K. M.; Hamdan, A. R. Hydrothermal scheduling based lagrangian relaxation approach to hydrothermal coordination. *IEEE T. Power Syst.* 1998, 13(1), 226-235.

[6] Redondo, N. J.; Conejo, A. J. Short-term hydro-thermal coordination by Lagrangian relaxation: solution of the dual problem. *IEEE T. Power Syst.* 1999, 14(1), 89-95.

[7] Orero, S. O.; Irving, M. R. A genetic algorithm modeling framework a solution technique for short term scheduling of hydro-thermal system. *IEEE T. Power Syst.* 1998, 13, 501–518.

[8] Zoumas, C. E.; Bakirtzis, A. G.; Theocharis, J. B.; Petridis, V. A genetic algorithm solution approach to the hydro-thermal coordination problem. *IEEE T. Power Syst.* 2004, 19(3), 1356–1364.

[9] Sinha, N.; Chakrabarti, R.; Chattopadhyay, P. K. Fast evolutionary programming techniques for short term hydro-thermal scheduling, *IEEE T. Power Syst.* 2003, 18(1), 214–220.

[10] Wong, K. P.; Wong, Y. W. Short term hydro-thermal scheduling-part-I: simulated annealing approach. *Proc. Inst. Electron. Eng.* 1994, 141(5), 497–501.

[11] Werner, T. G.; Verstege, J. F. An evolutionary strategy for short term operation planning of hydro-thermal power systems. *IEEE T. Power Syst.* 1999, 14(4), 1362–1368.

[12] Bayón, L.; Grau, J. M.; Suárez, P. M. A New Algorithm for the Optimization of a Simple Hydrothermal Problem. *Communications in Nonlinear Science and Numerical Simulation* 2004, 9(2), 197-207.

[13] Bayón, L.; Grau, J. M.; Ruiz, M. M.; Suárez, P. M. New Developments in the Application of Pontryagin's Principle for the Hydrothermal Optimization. *IMA J. Math. Control I.* 2005, 22(4), 377-393.

[14] Bayón, L.; Grau, J. M.; Ruiz, M. M.; Suárez, P. M. Optimization of SO_2 and NO_x emissions in Thermal Plants. *J. Math. Chem.* 2006, 40(1), 29-41.

[15] Bayón, L.; Grau, J. M.; Ruiz, M. M.; Suárez, P. M. A Bolza Problem in Hydrothermal Optimization. *Appl. Math. Comput.* 2007, 184(1), 12-22.

[16] Clarke, F. H. *Optimization and nonsmooth analysis*; John Wiley & Sons, New York, 1983.

[17] Loewen, P. D. Optimal control via nonsmooth analysis; *CRM Proceedings & Lecture Notes,* **2**; American Mathematical Society, Providence, RI, 1993.

[18] Clarke, F. H.; Ledyaev, Yu. S.; Stern, R. J.; Wolenski, P. R. Nonsmooth analysis and control theory; *Graduate Texts in Mathematics,* **178**; Springer-Verlag, New York, 1998.

[19] Vinter, R. *Optimal control;* Systems & Control: Foundations & Applications. Birkhäuser Boston, Inc., Boston, MA, 2000.

[20] Ascher, U. M.; Mattheij, R.; Russell, R. D. Numerical Solution of Boundary Value Problems for Ordinary Differential Equations. *Classics in Applied Mathematics,* **13**, SIAM, Philadelphia, 1995.

[21] Stoer, J.; Bulirsch, R. Introduction to Numerical Analysis. *Texts in Applied Mathematics*, **12**, Springer-Verlag, New York, 1993.

[22] Bayón, L.; Grau, J. M.; Suárez, P. M. A new formulation of the equivalent thermal in Optimization of hydrothermal systems. *Math. Probl. Eng.* 2002, 8(3), 181-196.

[23] Bayón, L.; Grau, J. M.; Ruiz, M. M.; Suárez, P. M. New developments on equivalent thermal in hydrothermal optimization: an algorithm of approximation. *J. Comput. Appl. Math.* 2005, 175(1), 63-75.

[24] Bayón, L.; Grau, J. M.; Ruiz, M. M.; Suárez, P. M. Nonsmooth Optimization of Hydrothermal Problems. *J. Comput. Appl. Math.* 2006, 192(1), 11-19.

[25] Galeev, E.; Tijomirov, V. *Breve curso de la teoría de problemas extremales*; Ed. Mir, Moscú, 1991.

[26] Bayón, L.; Grau, J. M.; Ruiz, M. M.; Suárez, P. M., A Constrained and Nonsmooth Hydrothermal Problem. *Lecture Series on Computer and Computational Sciences*, Eds.: Simos, T.E., Maroulis, G.; VSP International Science Publishers, The Netherlands, 2005, 4A, 60-64,.

[27] Bayón, L.; Grau, J. M.; Suárez, P. M. A Necessary Condition for Broken Extremals in Problems Involving Inequality Constraints. *Archives of Inequalities and Applications* 2003, 1(1), 75-84.

[28] Bayón, L.; Grau, J. M.; Ruiz, M. M.; Suárez, P. M. The First Weierstrass-Erdmann Condition in Variational Problems involving Differential Inclusions. *Math. Inequal. Appl.* 2004, 7(3), 457-469.

[29] Elsgoltz, L. *Ecuaciones diferenciales y cálculo variacional;* Ed. Mir, Moscú, 1994.

In: Numerical Simulation Research Progress
Editors: Simone P. Colombo et al, pp. 249-256
ISBN 978-1-60456-783-0

Chapter 6

Exact N-Soliton Solutions of the Sharma-Tasso-Olver-Kadomtsev-Petviashvili (STO-KP) Equation

Abdul-Majid Wazwaz
Department of Mathematics, Saint Xavier University
Chicago, IL 60655

Abstract

In this work we derive a new completely integrable dispersive equation. The equation is obtained by extending the Sharma-Tasso-Olver (STO) equation using the extension sense of the Kadomtsev-Petviashvili (KP) equation. The newly derived Sharma-Tasso-Olver-Kadomtsev-Petviashvili (STO-KP) equation is studied by using the tanh-coth method to obtain kink solutions and periodic solutions. The powerful Hirota bilinear method is used to determine exact N-soliton solutions for this new integrable equation. The work highlights the power of the used methods and the structures of the obtained multiple-front solutions.

1. Introduction

The KdV equation is given by

$$u_t + 6uu_x + u_{xxx} = 0, \tag{1}$$

that includes the nonlinear term uu_x and the linear dispersive term u_{xxx}. On the other hand, the KP equation is given by

$$(u_t + 6uu_x + u_{xxx})_x + u_{yy} = 0, \tag{2}$$

that extends the KdV equation. The solution of Eq. (1) is of the form

$$u(x,t) = 2\frac{\partial^2 \ln f(x,t)}{\partial x^2}. \tag{3}$$

It is well known that the KdV and the KP equations are completely integrable equations, with infinitely many conservation laws, and each equation gives multiple soliton solutions.

The Sharma-Tasso-Olver (STO) equation [1–3] is a nonlinear dispersive equation given by

$$u_t + \alpha(u^3)_x + \frac{3}{2}\alpha(u^2)_{xx} + \alpha u_{xxx} = 0. \tag{4}$$

The STO equation appears in many scientific applications. This equation was handled by many methods such as hyperbolic functions in [1], the symmetry reduction procedure in [2], and by the Backlund transformations method in [3]. The STO equation (4) is completely integrable with infinitely many conservation laws and multiple-soliton solutions of any order.

Following the extension sense of the KP equation (2) we can develop the Sharma-Tasso-Olver-Kadomtsev-Petviashvili (STO-KP) equation in the form

$$\left[u_t + \alpha(u^3)_x + \frac{3}{2}\alpha(u^2)_{xx} + \alpha u_{xxx}\right]_x + u_{yy} = 0. \tag{5}$$

The STO-KP equation is a completely integrable evolution equation that has an infinite number of conservation laws of energy and hence gives multiple-soliton solutions.

Many types of travelling waves are of particular interest in solitary wave theory. Some of these types are *solitons*, which are localized travelling waves, *periodic solutions*, *kink waves*, which rise or descend from one asymptotic state to another, *peakons* that are solitons with peak at the corner, *cuspons* that are solitons with cusps at their crests, and *compactons* that are solitons with compact spatial support such that each compacton is a soliton confined to a finite core or a soliton without exponential wings.

The objectives of this work are twofold. First, we seek to establish single solitons and multiple-soliton solutions for the new completely integrable equation (5) to complete our work in [1]. This will be justified by using the tanh-coth method [4–13]. Second, we aim to implement the Hirota's direct method [14–18] combined with Hereman's simplified version of the direct method [19–20] to obtain multiple-kink solutions.

2. The Methods

For single front solutions, the tanh-coth method [4–13] is usually used. However, the Hirota's direct method is also used for single and multiple-front solutions. In what follows, the methods will be reviewed briefly, where details can be found in [4–20].

2.1. The tanh-coth Method

A wave variable $\xi = x + y - ct$ converts a PDE to an ODE

$$Q(u, u', u'', u''' \cdots) = 0. \tag{6}$$

The last equation is integrated as long as all terms contain derivatives where integration constants are considered zeros.

The standard tanh method [4–7] introduces a new independent variable

$$Y = \tanh(\mu\xi), \xi = x + y - ct, \tag{7}$$

that leads to the change of derivatives:

$$\begin{array}{rcl} \frac{d}{d\xi} & = & \mu(1-Y^2)\frac{d}{dY}, \\ \frac{d^2}{d\xi^2} & = & -2\mu^2 Y(1-Y^2)\frac{d}{dY} + \mu^2(1-Y^2)^2\frac{d^2}{dY^2}. \end{array} \tag{8}$$

The tanh-coth method [12–13] admits the use of a finite expansion of tanh and coth functions

$$u(\mu\xi) = S(Y) = \sum_{k=0}^{M} a_k Y^k + \sum_{k=1}^{M} b_k Y^{-k}, \tag{9}$$

where M is a positive integer that will be determined. To determine the parameter M, we usually balance the linear terms of highest order in the resulting equation with the highest order nonlinear terms.

2.2. The Hirota's Bilinear Method

To formally derive soliton solutions of any integrable equation, we will mainly use the Hirota's direct method [14–18] combined with the simplified version of Hereman et. al [19–20] where it was shown that soliton solutions are just polynomials of exponentials.

We first substitute

$$u(x,t) = e^{kx-ct}, \tag{10}$$

into the linear terms of the equation under discussion to determine the dispersion relation between k and c. We then substitute the single soliton solution

$$u(x,t) = R\frac{\partial \ln f(x,t)}{\partial x} = R\frac{f_x(x,t)}{f(x,t)}, \tag{11}$$

into the equation under discussion, where the auxiliary function $f(x,t)$ is given by

$$f(x,t) = 1 + f_1(x,t) = 1 + e^{\theta_1}, \tag{12}$$

where

$$\theta_i = k_i x - c_i t, i = 1, 2, \cdots, N, \tag{13}$$

and solving the resulting equation to determine the numerical value for R. Notice that the N-soliton solutions can be obtained by using the following forms of $f(x,t)$ into (5)

(i) For dispersion relation, we use

$$u(x,t) = e^{\theta_i}, \theta_i = k_i x - c_i t. \tag{14}$$

(ii) For single soliton, we use

$$f = 1 + e^{\theta_1}. \tag{15}$$

(iii) For two-soliton solutions, we use

$$f = 1 + e^{\theta_1} + e^{\theta_2} + a_{12}e^{\theta_1+\theta_2}. \tag{16}$$

(iv) For three-soliton solutions, we use

$$f = 1 + e^{\theta_1} + e^{\theta_2} + e^{\theta_3} + a_{12}e^{\theta_1+\theta_2} + a_{23}e^{\theta_2+\theta_3} + a_{13}e^{\theta_1+\theta_3} + b_{123}e^{\theta_1+\theta_2+\theta_3}. \quad (17)$$

Notice that we use (14) to determine the dispersion relation, (16) to determine the factor a_{12} to generalize the result for the other factors $a_{ij}, 1 \le i < j \le N$, and finally we use (17) to determine b_{123}, which should result in $b_{123} = a_{12}a_{23}a_{13}$ for three-soliton solutions to exist. The parameter b_{123} should be in terms of the free parameters a_{ij} only, because three-solitons solutions and higher level soliton-solutions should not contain free parameters other than a_{ij}. The existence of three-soliton solutions confirms the fact that N-soliton solutions exist for any order. However, if three-soliton solutions do not exist, then the examined equation is not integrable. In the following, we will apply the aforementioned methods to the STO-KP equation.

3. Using the tanh-coth Method

The STO-KP equation can be reduced to

$$(1-c)u + \alpha u^3 + 3\alpha u u' + \alpha u'' = 0, \quad (18)$$

by using the wave variable $\xi = x + y - ct$ and integrating twice. Balancing u'' with u^3 in (18) we find

$$M + 2 = 3M, \quad (19)$$

so that

$$M = 1. \quad (20)$$

The tanh-coth method admits the use of the finite expansion

$$u(\xi) = a_0 + a_1 Y + \frac{b_1}{Y}, \quad (21)$$

into (18), collecting the coefficients of Y^j and setting these coefficients equal to zero to find the following sets for a_0, a_1, b_1 and μ:

$$\begin{array}{ll}
a_0 = 0,\ a_1 = \sqrt{\frac{c-1}{\alpha}}, & b_1 = 0\ \mu = \sqrt{\frac{c-1}{\alpha}}, \\
a_0 = 0,\ a_1 = \sqrt{\frac{c-1}{\alpha}}, & b_1 = 0\ \mu = \frac{1}{2}\sqrt{\frac{c-1}{\alpha}}, \\
a_0 = \sqrt{\frac{c-\alpha\lambda^2-1}{3\alpha}},\ a_1 = \lambda, & b_1 = 0\ \mu = \lambda, \\
a_0 = 0,\ a_1 = 0, & b_1 = \sqrt{\frac{c-1}{\alpha}}\ \mu = \sqrt{\frac{c-1}{\alpha}}, \\
a_0 = 0,\ a_1 = 0, & b_1 = \sqrt{\frac{c-1}{\alpha}}\ \mu = \frac{1}{2}\sqrt{\frac{c-1}{\alpha}}, \\
a_0 = \sqrt{\frac{c-\alpha\lambda^2-1}{3\alpha}},\ a_1 = 0, & b_1 = \lambda\ \mu = \lambda
\end{array} \quad (22)$$

where λ is any arbitrary constant.

In view of (22) we obtain the kinks solutions for $\frac{c-1}{\alpha} > 0$:

$$\begin{aligned} u_1(x,y,t) &= \sqrt{\tfrac{c-1}{\alpha}}\tanh\left[\sqrt{\tfrac{c-1}{\alpha}}(x+y-ct)\right], \\ u_2(x,y,t) &= \sqrt{\tfrac{c-1}{\alpha}}\tanh\left[\tfrac{1}{2}\sqrt{\tfrac{c-1}{\alpha}}(x+y-ct)\right], \\ u_3(x,y,t) &= \sqrt{\tfrac{c-\alpha\lambda^2-1}{3\alpha}} + \lambda\tanh\left[\lambda(x+y-ct)\right], \end{aligned} \tag{23}$$

and the solutions

$$\begin{aligned} u_4(x,y,t) &= \sqrt{\tfrac{c-1}{\alpha}}\coth\left[\sqrt{\tfrac{c-1}{\alpha}}(x+y-ct)\right], \\ u_5(x,y,t) &= \sqrt{\tfrac{c-1}{\alpha}}\coth\left[\tfrac{1}{2}\sqrt{\tfrac{c-1}{\alpha}}(x+y-ct)\right], \\ u_6(x,y,t) &= \sqrt{\tfrac{c-\alpha\lambda^2-1}{3\alpha}} + \lambda\coth\left[\lambda(x+y-ct)\right], \end{aligned} \tag{24}$$

However, for $\frac{c-1}{\alpha} < 0$, we obtain complex solutions.

4. Using the Hirota's Bilinear Method

Substituting

$$u(x,y,t) = e^{\theta_i}, \theta_i = k_i x + k_i y - c_i t, \tag{25}$$

into the linear terms of the STO-KP equation

$$\left[u_t + \alpha(u^3)_x + \frac{3}{2}\alpha(u^2)_{xx} + \alpha u_{xxx}\right]_x + u_{yy} = 0, \tag{26}$$

to obtain the dispersion relation

$$c_i = k_i + \alpha k_i^3, i = 1, 2, \cdots N, \tag{27}$$

and hence θ_i becomes

$$\theta_i = k_i(x+y) - (k_i + \alpha k_i^3)t. \tag{28}$$

To determine R, we substitute

$$u(x,y,t) = R\frac{\partial \ln f(x,y,t)}{\partial x} = R\frac{f_x}{f}, \tag{29}$$

where $f(x,y,t) = 1 + e^{k_1(x+y)-(k_1+\alpha k_1^3)t}$ into the STO-KP equation (26) and solve to find that $R = 1$. This means that the single soliton solution is given by

$$u(x,y,t) = \frac{\partial \ln f(x,t)}{\partial x} = \frac{k_1 e^{k_1(x+y)-(k_1+\alpha k_1^3)t}}{(1 + e^{k_1(x+y)-(k_1+\alpha k_1^3)t})}. \tag{30}$$

For two-soliton solutions, we substitute

$$u(x,y,t) = \frac{\partial \ln f(x,y,t)}{\partial x}, \tag{31}$$

where

$$f = 1 + e^{\theta_1} + e^{\theta_2} + a_{12}e^{\theta_1+\theta_2}, \tag{32}$$

into the STO-KP equation (26), where θ_1 and θ_2 are given in (28), to obtain that

$$a_{12} = 0, \tag{33}$$

and hence

$$a_{ij} = 0, 1 \le i < j \le N. \tag{34}$$

This in turn gives

$$f(x, y, t) = 1 + e^{k_1(x+y)-(k_1+\alpha k_1^3)t} + e^{k_2(x+y)-(k_2+\alpha k_2^3)t}. \tag{35}$$

To determine the two-soliton solutions explicitly, we substitute the last result for $f(x, y, t)$ into (29) to obtain the two-soliton solutions

$$u(x, y, t) = \frac{k_1e^{k_1(x+y)-(k_1+\alpha k_1^3)t} + k_2e^{k_2(x+y)-(k_2+\alpha k_2^3)t}}{1 + e^{k_1(x+y)-(k_1+\alpha k_1^3)t} + k_2e^{k_2(x+y)-(k_2+\alpha k_2^3)t}}. \tag{36}$$

Following the discussion presented before, we can determine the three-soliton solutions, therefore we set

$$f(x, y, t) = 1 + \exp(\theta_1) + \exp(\theta_2) + \exp(\theta_3) + b_{123}\exp(\theta_1 + \theta_2 + \theta_3), \tag{37}$$

into (29) and substitute it in the STO-KP equation to find that

$$b_{123} = a_{12}a_{13}a_{23} = 0. \tag{38}$$

In view of the last result we find

$$f(x, y, t) = 1 + \exp(\theta_1) + \exp(\theta_2) + \exp(\theta_3). \tag{39}$$

To determine the three-soliton solutions explicitly, we substitute the last result for $f(x, y, t)$ into (29) to obtain the three-front solutions given by

$$u(x, y, t) = \frac{k_1e^{k_1(x+y)-(k_1+\alpha k_1^3)t} + k_2e^{k_2(x+y)-(k_2+\alpha k_2^3)t} + k_3e^{k_3(x+y)-(k_3+\alpha k_3^3)t}}{1 + e^{k_1(x+y)-(k_1+\alpha k_1^3)t} + k_2e^{k_2(x+y)-(k_2+\alpha k_2^3)t} + k_3e^{k_3(x+y)-(k_1+\alpha k_1^3)t}}. \tag{40}$$

The N-soliton solution is therefore given by

$$u(x, y, t) = \frac{\sum_{i=1}^{N} k_ie^{k_i(x+y)-(k_i+\alpha k_i^3)t}}{1 + \sum_{i=1}^{N} k_ie^{k_i(x+y)-(k_i+\alpha k_i^3)t}}. \tag{41}$$

This confirms that the STO-KP equation is completely integrable and possesses multiple-soliton solutions of any order.

5. Conclusion

The tanh-coth method is used to develop a single kink solution for the completely integrable Sharma-Tasso-Olver-Kadomtsev-Petviashvili equation. The Hirota's direct method and the Hereman's simplified version of the direct method are effectively used to establish the multiple kink solutions for this equation. The effectiveness of the simplified version is confirmed.

Two facts can be confirmed here:
(i) the first is that soliton solutions are just polynomials of exponentials as emphasized by Hirota,
and
(ii) every solitonic equation that has generic $N = 3$ soliton solutions, then it has also soliton solutions for any $N \geq 4$ [14–20].

In closing, it is not always true that the extension of a completely integrable equation in the KP sense will give a completely integrable equation. This will be examined in a forthcoming work.

References

[1] A. M. Wazwaz, New solitons and kinks solutions to the Sharma-Tasso-Olver equation, *Appl. Math. Comput.*, **188** (2007) 1205-1213.

[2] Z. Lian and S.Y.Lou, Symmetries and exact solutions of the Sharma-Tasso-Olver equation, *Nonlinear Analysis,* **63**(2005) 1167–1177.

[3] S. Wang, X.Tang, S.Y.Lou, Soliton fission and fusion: Burgers equation and Sharma-Tasso-Olver equation, *Chaos, Solitons and Fractals,* **21** (2004) 231–239.

[4] W. Malfliet, Solitary wave solutions of nonlinear wave equations, *Am. J. Phys.*, **60**(7) 650–654 (1992).

[5] W. Malfliet, The tanh method: a tool for solving certain classes of nonlinear evolution and wave equations, *J. Comput. Appl. Math.,* (2004) 529–541.

[6] W. Malfliet and Willy Hereman, The tanh method:I. Exact solutions of nonlinear evolution and wave equations, *Physica Scripta,* **54** (1996) 563–568.

[7] W. Malfliet and Willy Hereman, The Tanh Method: II. Perturbation technique for conservative systems, Physica Scripta,, *Physica Scripta,* **54** (1996) 569–575.

[8] A.M.Wazwaz, Multiple-front solutions for the Burgers equation and the coupled Burgers equations, *Applied Mathematics and Computation,* **190** (2007) 1198–1206.

[9] A.M.Wazwaz, The tanh method for travelling wave solutions of nonlinear equations, *Appl. Math. Comput.*, **154** (3) (2004) 713-723.

[10] A.M.Wazwaz, *Partial Differential Equations:Methods and Applications,* Balkema Publishers, The Netherlands, 2002.

[11] A.M. Wazwaz, Compactons in a class of nonlinear dispersive equations, *Mathematical and Computer Modelling,* **37**(3/4) (2003) 333-341.

[12] A.M.Wazwaz, The extended tanh method for new solitons solutions for many forms of the fifth-order KdV equations, *Appl. Math. Comput.,* **184** (2007) 1002–1014.

[13] A.M.Wazwaz, The tanh-coth method for solitons and kink solutions for nonlinear parabolic equations, *Appl. Math. Comput.*, **188** (2007) 1467–1475.

[14] R. Hirota, *The Direct Method in Soliton Theory,* Cambridge University Press, Cambridge (2004).

[15] R. Hirota, Exact solutions of the Korteweg-de Vries equation for multiple collisions of solitons, *Physical Review Letters,* **27**(18) (1971) 1192–1194.

[16] R. Hirota, Exact solutions of the modified Korteweg-de Vries equation for multiple collisions of solitons, *J. Physical Society of Japan,* **33**(5) (1972) 1456–1458.

[17] R. Hirota, Exact solutions of the Sine-Gordon equation for multiple collisions of solitons, *J. Physical Society of Japan,* **33**(5) (1972) 1459–1463.

[18] R. Hirota, Exact N-soliton solutions of a nonlinear wave equation, *J. Math Phys.,* **14**(7) (1973) 805–809.

[19] W.Hereman and W. Zhaung, Symbolic software for soliton theory, Acta Applicandae Mathematicae, *Phys. Letters A*, **76** (1980), 95–96.

[20] W. Hereman and A. Nuseir, Symbolic methods to construct exact solutions of nonlinear partial differential equations, *Mathematics and Computers in Simulation,* **43** (1997), 13–27.

In: Numerical Simulation Research Progress
Editors: Simone P. Colombo et al, pp. 257-287 ISBN: 978-1-60456-783-0

Chapter 7

ADVANCES IN NUMERICAL SIMULATION OF GRANULAR MATERIAL

Mohammad Hadi Bordbar*
Lappeenranta University of Technology, P.O. Box 20,
FIN-53851 Lappeenranta, Finland

Abstract

A granular flow is a mixture of grains and a fluid phase. Granular materials deform plastically like a solid under weak shear and they flow like a fluid under high shear. These materials exhibit other unusual behaviors, including pattern formation in the shaking of granular materials. The Discrete Element Method (DEM) and continuum type simulation are the most common methods used for the simulation of granular material in different flow regimes. In this chapter, some new improvements in the numerical simulation of this complex material are reviewed. In the first part, the collision between monosize and multisize spherical particles is reviewed and new activities in presenting a better contact force model for describing the collision between particles that has a main role in the accuracy of DEM models are presented. A detailed overview of a simplified model for collisions between particles of granular material is presented to be of use in molecular dynamic simulations. Detailed reported parameters of this new contact force model for two viscoelastic materials, glass and ice, are presented.

In the second part, the detailed overview of collective processes in gas-particle flows are used in developing a new set of hydrodynamic continuum equations to describe the deformation and flow of dense gas–particle mixtures. The constitutive equation used for the stress tensor provides an effective viscosity with a liquid-like character at low shear rates and a gaseous-like behavior at high shear rates. A review of the results of using this method in simulating a physical gas-particle system is done.

In the last part, a simplified discrete element method for the simulation of dense gas-particle flows is developed using the detailed overview of collective processes in gas-particle flows. The large eddy simulation technique is used for solving the governing equation of the continuous phase, and a Lagrangian method is used to predict the particle motion. In order to show the results of this simplified model, a simulation of a pentagonal prism containing glass balls with its base subjected to a sinusoidal vibration has been done.

* E-mail address: bordbar@lut.fi

1. Granular Material in General

Granular material is an accumulation of discrete solid grains, which can interact with each other. Usually, there is a fluid phase (e.g. gas, liquid) as the continuum phase between these macroscopic particles. The discrete particles are called the dispersed phase, while the fluid phase is the carrier phase supporting the particles. Energy and momentum is lost when the particles interact with each other. This kind of material can be found widely in nature and industry, for example sugar, sand, cereals, plastics, coal, etc. Some examples of granular materials are shown in Figure 1. The particles in the granular material must be large enough so that they are not subject to thermal motion fluctuations. Therefore, the lower size limit for particles in granular material is about 1 μm. On the upper size limit, the physics of granular materials may be applied to ice floes, where the individual grains are icebergs [1].

Figure 1. Some examples of granular materials.

Granular materials are commercially important in many applications, such as pharmaceutical industry, agriculture and energy production. The history of research into granular materials goes back to at least Coulomb, who presented the famous law of friction for granular materials.

2. Numerical Simulation of Granular Material

In order to design efficient industrial handling and processing systems, a good understanding of the physics of granular material and accurate computational tools for simulating granular systems are needed. The importance of this kind of numerical methods and tools is apparent when one considers the following statistics; in chemical industry, approximately one half of the products and at least three quarters of the raw materials are in granular form [2].

One of the not well understood topics in the simulation of granular material is collisions between particles, which have a key role in transferring energy and momentum in granular systems. A lot of research has been done in this field since Newton's time and there is still a lot of activity in this field. In most studies, theoretical considerations, the finite element method and experimental results are employed to develop more accurate models to describe the collision between particles. A comprehensive review of these activities and some new findings are presented in section three of this review article. The result of a collision study leads to more accurate models for describing the collision between single particles and can be used in presenting numerical methods in the simulation of whole gas particle systems, which is the subject of sections four and five.

To develop a continuum type simulation for granular material, a detailed overview of the interaction between gas flow and particles is needed. This result is used in developing a new set of hydrodynamic continuum equations for the simulation of dense granular material in section four and a simplified discrete element method in section five.

3. Collision Modeling

Profound understanding of energy loss and the amount of energy dissipation when two particles collide into each other, is one of the main targets of this section. One of the simplest collision situations is the binary collision between two monosized spherical grains that involves repulsive rigid elastic interactions as well as dissipative frictional contacts. A binary collision represents the simplest of the multibody interactions in a dense granular flow. Frictional interaction represents the most fundamental difference between a granular system and a molecular system. Note that even a highly polished grain has surface roughness in many different length scales [3].

In granular flow studies, The most widely applied approach is the linear viscoelastic contact model in combination with Mohr friction law [4-6]. In this linear model, the normal and tangential contact forces are a linear function of the displacement and relative velocity, and the Mohr friction law controls the maximum tangential force. This linear contact model simplifies the computational procedure. Many types of granular flow behavior have been studied with this simple contact force model. In fact, for most granular material the contact force has a nonlinear relationship with relative displacement and velocity, which has been proved by experimental data and analytical solutions [7, 8]. More than a century ago, the famous Hertzain law for a non-linear, normal contact between elastic spheres was established [9]. The original non-linear, tangential contact model was constructed by Mindlin[10]. These two theories have been combined to construct the basic Hertzain-Mindlin contact theory for elastic granular materials. This theory has been applied widely. Walton et al. [11] adopted a simple plastic model to handle collision dissipation. This model has been applied in several other subsequent studies [12]. Based on the Hertzain-Mindlin theory, but considering also the plastic behavior, a finite element method has been used to investigate the details of a sphere colliding with a plane [3, 13-14].

Gugan et al. [15] report that in viscoelastic collision between two spheres, the spheres lost about 40% of their kinetic energy over the range of studied speeds. Gugan et al. [15] also did some research about the area and duration of contact, and their result was consistent with the Hertz theory.

Based on Newton's Second Law, the particle momentum equation is

$$m\ddot{\xi} + c_n \xi^{\alpha} \dot{\xi} + k_n \xi^{\beta} = 0, \tag{1}$$

where m is the sphere mass, and α and β are the index for linear and non-linear contact models. In the simple linear model, the elastic force and viscous force are the linear function of displacement and relative velocity respectively. We have $F_e = k_e \xi$ and $F_V = c_n \dot{\xi}$ when $\alpha = 0$ and $\beta = 1$. Therefore, the momentum equation (Eq. 1) can be written as

$$m\ddot{\xi} + c_n \dot{\xi} + k_e \xi = 0. \tag{2}$$

This linear model has been widely applied to study the basic physical process of complex granular systems [5, 6, 16 and 17].

When two spheres collide, the contact area increases with the increase of the contact force. In this case, a non-linear model can describe the interaction process of the two spheres more accurately. In the non-linear model, the elastic normal force was established by Hertz [9]. In his theory there is no damping, and the elastic index in Eq. 1 is β=3/2. This model has been proven with physical experimentation and numerical solutions [7, 18-21].

In previous research on calculating viscous force, several different values for α have been determined on the bases of experimental data and analytical solution. Brilliantov et al. [22] found that α=1/2 for sphere-to-sphere contact. The experiments of Falcon et al. [19] showed that it was better to set α =1/4. Some analytical comparisons between a linear model and non-linear model with α =1/4 have been carried out [7, 20]. It was shown that the stress, displacement and relative velocity of the two colliding spheres were simulated more reasonably with the non-linear model. In earlier studies α =3/2 was also used to study the contact process [8] and it was determined that α =4/5 and $\beta = 8/5$ fit the experimental results.

Brilliantov et al. [22] used the nonlinear model for modeling the collision, and presented the following equations with the initial conditions $\dot{\xi}(0) = g^n, \xi(0) = 0$ (where g^n represents the relative velocity of particles) for their model

$$\ddot{\xi} + \frac{2E\sqrt{R_{eff}}}{3m_{eff}(1-\nu^2)}(\xi^{3/2} + \frac{3}{2}A\sqrt{\xi}\dot{\xi}) = 0, \tag{3}$$

$$\textit{in the case } A\dot{\xi} \ll \xi \ : \ddot{\xi} + \frac{2E\sqrt{R_{eff}}}{3m_{eff}(1-\nu^2)}(\xi + A\dot{\xi})^{3/2} = 0, \tag{4}$$

where ξ represents the overlapping distance, E is the Young modulus, ν is the Poisson ratio of the particle material , m_{eff} is the effective mass of the particles, and R_{eff} is the effective radius of the particles. Coefficient A was considered to be a fit parameter. The effective mass and radius of the particles were defined as follows:

$$m_{eff} = m_1 m_2 / (m_1 + m_2) \tag{5}$$

$$R_{eff} = R_1 R_2 / (R_1 + R_2) \tag{6}$$

Brilliantov et al. [22] confirmed their model by comparing the result of the model with the experimental result reported by Bridges et al. [23] of binary collision between ice spheres in the temperature of 150-175K.

Zamankhan and Bordbar [3] have presented a new contact force model for binary collision between viscoelastic monosize and multisize particles. Their model is characterized by a time dependent relation between stress and strain. Note that neither the Maxwell nor the Kelvin model represent the behavior of most viscoelastic materials, including the glass balls or ice balls. For example, the Maxwell model [24] predicts that the stress asymptotically approaches zero when the strain is kept constant. On the other hand, the Kelvin model does not describe a permanent strain after unloading. In the model presented by Zamankhan and Bordbar [3], a second spring is placed in series with the Kelvin model used by Brilliantov et al. [22], to develop a model of linear viscoelastic material equivalent to a Maxwell model with a spring in series. The schematic of the three-parameter model used in this model is shown in Figure 2.

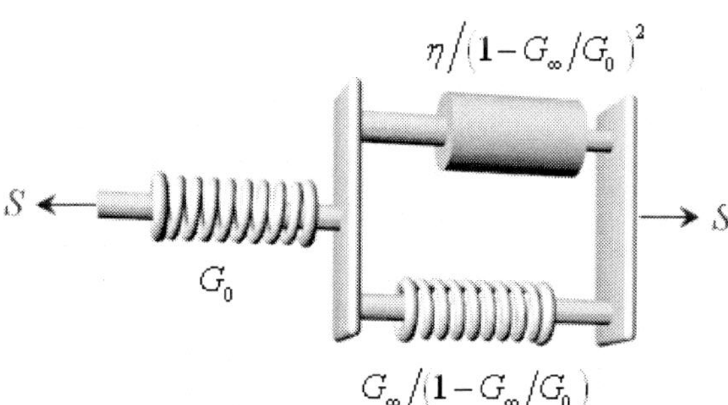

Figure 2. Binary collision model consisting of an auxiliary spring in series with the Kelvin model, presented by Zamankhan and Bordbar [3].

The following equation is suggested for the explanation of binary collision between two identical spherical viscoelastic particles in this new model:

$$\frac{d^2\xi}{dt^2} = -(2E/3m(1-\nu^2))\sigma_p \xi^{3/2} - K\tau(G_0^2/m(G_0^2 - G_\infty^2))\sigma_p^{1-\eta}\xi^{\eta}\, d\xi/dt \tag{7}$$

where σ_p, τ, G_0 and G_∞ represent the diameter of the particles, the relaxation time, instantaneous shear modulus and long time shear modulus of the material, respectively. K is the coefficient characterizing the inelastic behavior of the particles.

By using this modified model and fitting the result of the model with experimental data from Newton Cradle device measurements for coefficient of restitution of identical glass balls, the exponent(η) gets the value of one [3]. Figures 3 and 4 show the result reported by Zamankhan and Bordbar [3]. Figure 4(a) illustrates two sets of numerical results of the coefficient of restitution in a binary collision of glass balls. They were obtained using different values for the exponent η, where deltas and squares represent the coefficient of restitution as a function of impact velocity for $\eta = 1$ and $\eta = 1/2$, respectively. The circles are the experimental data from Newton cradle measurements. The results presented in Figure 4(a) show that the dependence of η of contact force on viscous dissipation appears to provide a better representation of the measurements. The new model can also predict the behavior of glass balls during a collision at the low velocity limit accurately. However, the Kelvin-type model (for which $\zeta^{1/2}$ dependence of contact force on viscous dissipation was obtained [22]) can not predict the coefficient of restitution either at the low or at the high impact velocity accurately. This observation implies that a Kelvin type model may not be quite appropriate to predict the glassy-type viscoelastic behavior of glass balls during a collision [3].

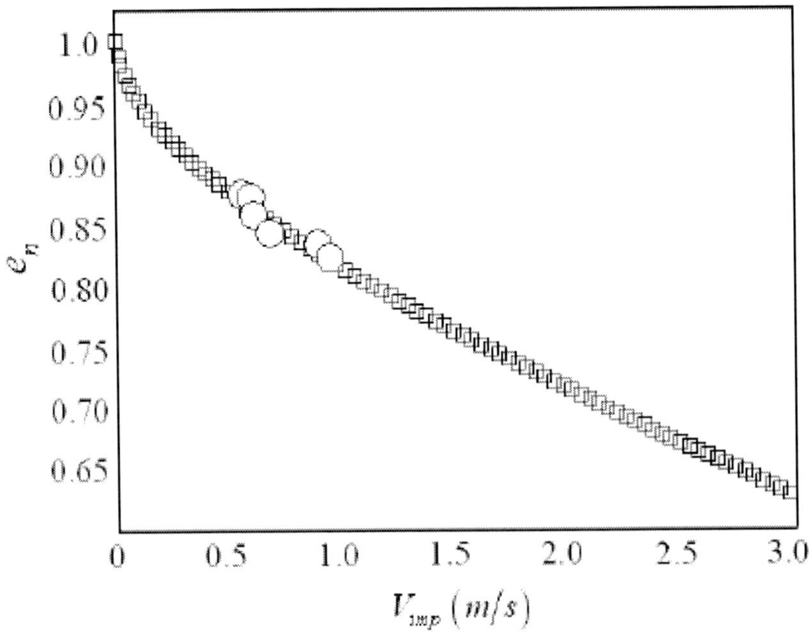

Figure 3. The coefficient of restitution versus impact velocity for viscoelastic glass spheres. Squares indicate the numerical results obtained using the finite element approach and circles represent experimental data obtained using Newton's cradle. The distribution of the normal coefficient of restitution is seen to follow the power law distribution (modified from Zamankhan and Bordbar [3], Fig.10).

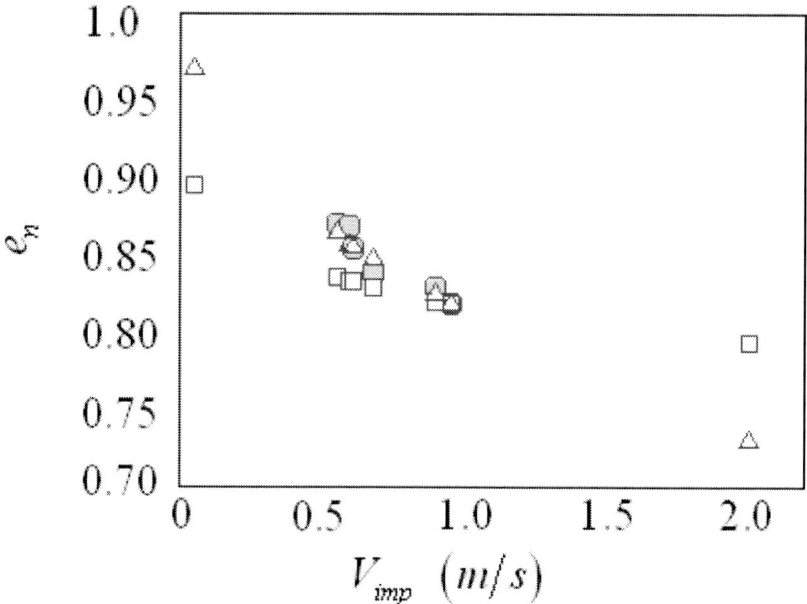

Figure 4. The coefficient of restitution versus impact velocity for a binary collision of viscoelastic glass balls. Deltas and squares indicate the numerical results obtained using the simplified model (7) for η =1, and η =1/2, respectively. Gray circles represent experimental data obtained using the Newton's cradle device. The values obtained for the parameter K for η =1, and η =1/2 are 143 and 1.67, respectively. The Kelvin-type model for η =1/2 does not seem to predict the behavior of glass balls during a collision at the low velocity limit accurately (from Zamankhan and Bordbar [3], Fig.11).

Currently, Bordbar and Hyppänen [25] have adapted this new model for modeling the collision between multisize viscoelastic particles. The following equation is reported for the force between two multisize spherical particles [25]:

$$F = \frac{4E_{eff}\sqrt{R_{eff}}}{3m_{eff}}\xi^{\frac{3}{2}} + 2K\frac{E_{eff}R_{eff}^{(1-\eta)}}{m_{eff}}\xi^{\eta}\dot{\xi}\ , \tag{8}$$

where E_{eff} represents the effective young modulus of two particles and is defined as follows:

$$\frac{1}{E_{eff}} = \frac{(1-\nu_1^2)}{E_1} + \frac{(1-\nu_2^2)}{E_2}, \tag{9}$$

By comparing the result of the model and fitting them for ice balls with the experimental result reported by Bridges et al.[23], Bordbar and Hyppänen [25] found that for ice balls, the value of 0.65 provides better conformity with experimental data than the Kelvin model($\eta = 1/2$). Figure 5 shows the result of the modeling collision of ice balls by using this new model in comparison with experimental data [23] and the result of the Kelvin model [22].

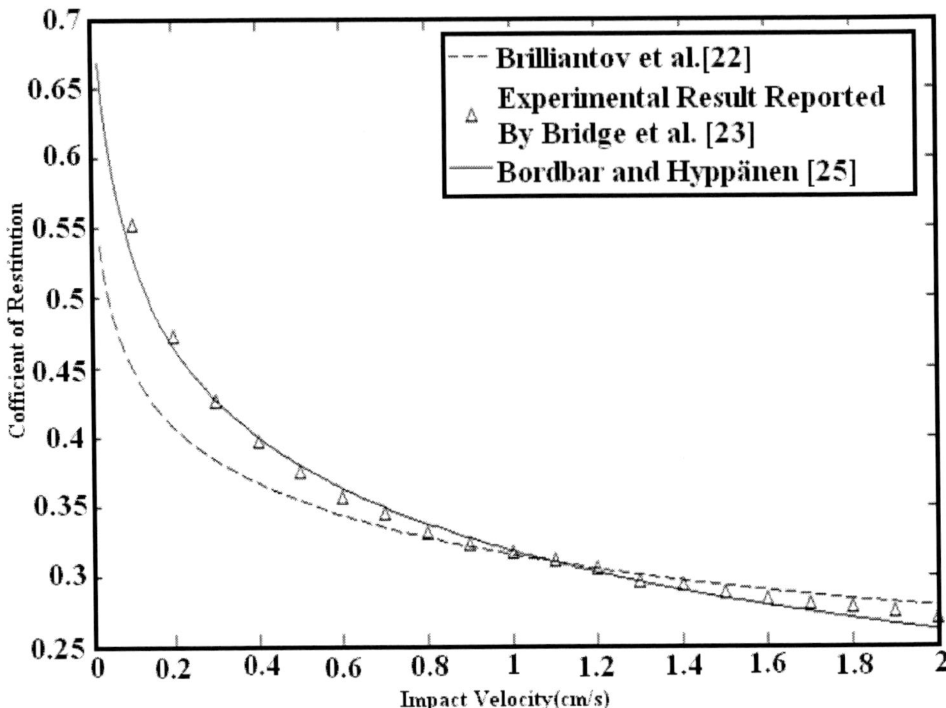

Figure 5. Coefficient of restitution versus impact velocity measured in (cm/s) for the spherical ice particles in the diameter of $1cm$ for the model presented in [25] compared with the experimental result reported in [23] and the model reported in [22] (from Bordbar and Hyppänen [25], Fig.6).

4. Continuum Type Approach

Modeling the granular material by using the DEM method explained in section five, is the most accurate method if the accurate equations exist for defining the forces between different phases. At the same time, this kind of method is computationally expensive. Considering more active forces in the system and more complex equations for defining them will lead to processor demands and very long run time. The abovementioned problems practically limit the time length of the simulation and the number of particles in the system. These restrictions have forced the researchers to find faster methods. An alternative way to simulate granular systems is to average the physics across many particles and thereby treat the material as a continuum. To find the property of the equivalent continuum material, the result of simulating small granular systems with a limited number of particles by using DEM would be very useful. In other words, the result of DEM for a limited system of particles can provide useful information of the bulk and cell viscosity of the equivalent continuum material. The gas and solid phase is coupled by using one-way or two way coupling. During the last two decades, a lot of research have been conducted to develop a continuum approach for modeling gas particle flows [26-30, 39]. Currently, Bordbar and Zamankhan [32] do a new research to develop a new set of hydrodynamic continuum equations for describing the deformation and flow of dense gas–particle mixtures. To develop this set of continuum equation, they present a detailed overview of the collective processes in gas particle flows in [31]. The details of this set of hydrodynamic continuum equation are reviewed in this section.

In most granular flow systems, the flow dynamics fall into an intermediate regime where both collisional and frictional interactions between particles should be taken into account. In addition, in some examples (such as granular materials under vertical vibration) gas trapped in the granular material has been found to be a major source of heaping [33]. Computational experiments may lead to scale-dependent closures for quantities such as the drag, stresses and effective dispersion [31]. These quantities represent important frontiers in this class of problems.

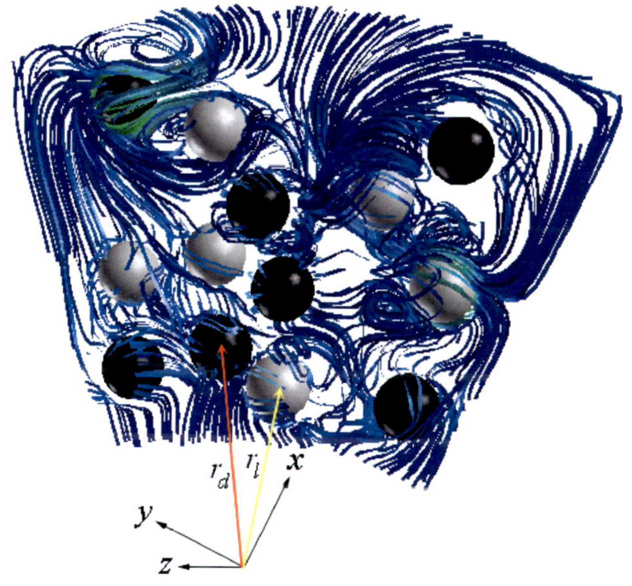

Figure 6. Schematic of a granular bed under shear composed of a binary mixture of differently colored, monosized, spherical particles (from Bordbar [36], Fig.4.38).

A granular material can be considered as a binary mixture of differently colored, monosized, spherical particles as illustrated in Figure 6. For this system, light and dark colors are represented by subscripts l and d, respectively. Zamankhan et al. [34] report the following equation as the mass balance of particles of kind l in a curvilinear coordinate system:

$$\frac{\partial \rho^l}{\partial t} + V_i \frac{\partial \rho^l}{\partial x_i} + \rho^l \frac{\partial V_i}{\partial x_i} + \frac{\partial}{\partial x_i}\left(\rho^l u_i^l\right) = 0, \qquad (10)$$

where ρ^l is the apparent density of particles of kind l, defined as $\rho_s \left\langle \phi_s^l \right\rangle$, ρ_s represents the particle material density, $\left\langle \phi_s^l \right\rangle$ is the average solid fraction of particles of kind l, x_i represents curvilinear coordinates, V_i is the slowly varying mean velocity and u_i^l is a rapidly fluctuating velocity of particles of kind l.

In Eq. 10, subscript i is a free index to which the summation convention is applied.

4.1. Constitutive Equations in Rate-Independent Quasi-static Regime

The third term on the left hand side of Eq.10 includes the divergence of the velocity field, which is known as the phase space compressibility of the granular assembly. In this section we are going to develop an approximate theory for predicting both the dilatancy and contractancy of the granular material, as illustrated in Figure 7. To obtain a representation of behavior for the slow flow dynamics of granular assembly at high solid fractions that can be merged, the kinetic theory results in moderately high shear rates for the assembly shown in figure 7, by following Savage [35], it is useful to define a yield function, *Y*, as follows:

$$Y \equiv \frac{\sigma_1^2+\sigma_2^2+\sigma_3^2-\sigma_1\sigma_2-\sigma_2\sigma_3-\sigma_3\sigma_1}{3}+\sin^2\theta_f\left[\left(\frac{\sigma_1+\sigma_2+\sigma_3}{3}+P\right)^2-P^2\right]=0. \quad (11)$$

where θ_f corresponds to the value of the angle of internal friction, σ_i, $i=1,2,3$ are the principal stresses (which are the eigenvalues of the stress tensor P_{ij}), and $P(\phi_s,T)$ is the pressure at the solids fraction ϕ_s and granular temperature T .

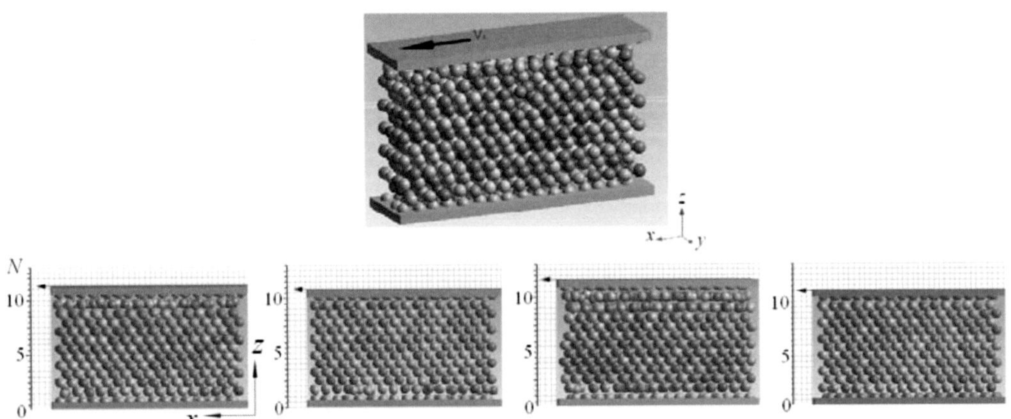

Figure 7. A layer of granular material under shearing (from Bordbar [36], Fig.4.39).

Note that the yield surface as described in Eq. 11 is convex whose size grows with increasing ϕ_s and T . Using the principle of coaxially, the rate of strains defined as $\varepsilon_{ij}=\frac{1}{2}\left(\partial V_i/\partial x_j+\partial V_j/\partial x_i\right)$ may be derived from a plastic potential function. In the simplest case when the yield and plastic potential functions are identical, the rate of strains may be given by:

$$\varepsilon_{ij} = \lambda \frac{\partial Y}{\partial \sigma_{ij}} , \tag{12}$$

where λ is a scalar . By solving Eq. 11 for stress components in terms of the rate of strains and then substituting the results in the yield function, the relations between stress and the rate of strain tensors may be found as [32]:

$$\sigma_{ij} = \frac{P}{\left[\left(\frac{4}{3}\sin^2\theta_f + 1\right)\mathrm{I}_1^2 - 4\sin^2\theta_f \mathrm{I}_2\right]^{\frac{1}{2}}} \times \left[2\sin^2\theta_f \varepsilon_{ij} + \left(1 - \frac{2}{3}\sin^2\theta_f\right)\varepsilon_{kk}\delta_{ij}\right] - P\delta_{ij}, \tag{13}$$

where I_1 and I_2 are the first and second invariants of the rate of strain tensor, respectively. Equation 13 represents the stress tensor for a rate-independent, plastic behavior of the granular assembly.

Making use of the expressions for the rate of strains, the divergence of the velocity may be expressed as

$$\frac{\partial V_i}{\partial x_i} = -\frac{\left[\left(\frac{4}{3}\sin^2\theta_f + 1\right)\mathrm{I}_1^2 - 4\sin^2\theta_f \mathrm{I}_2\right]^{\frac{1}{2}}}{P}\left(\frac{1}{3}\sigma_{ii} + P\right), \tag{14}$$

where $\frac{1}{3}\sigma_{ii}$ represents the average normal stress.

4.2. Constitutive Equations in Transitional Regime

The particle diffusive motion, which can be observed even at the high solids fraction [38, 39], is accounted for in Eq. 10. If the color of the particles is ignored, Eq. 10 becomes similar to the one used by Dasgupta et al. [40], in which the particle diffusive motion is regarded as correlated fluctuations in the solids fraction and solids velocity. In this case, the time-smoothing of volume-averaged continuity equation for the gas phase may be given as in [40];

$$\frac{\partial \langle \phi_g \rangle}{\partial t} + U_i \frac{\partial \langle \phi_g \rangle}{\partial x_i} + \langle \phi_g \rangle \frac{\partial U_i}{\partial x_i} + \frac{\partial}{\partial x_i}\left(\langle \phi_g' U_g' \rangle\right) = 0 , \tag{15}$$

where $\langle \phi_g \rangle$ and U_i represent the averaged values of volume fraction and velocity of the gas phase for which $\phi_g = 1 - \phi_s$. The primes designate the fluctuation parts about the averaged values.

Polashenski et al. [43] have shown that the net diffusive drift of particles in a collision-dominated regime in which fluctuations are driven by the collective motion of the particle assembly may be expressed as

$$\rho u_i = \rho D_{ij}^T \frac{\partial \ln T}{\partial x_j} + B_{ij} \frac{\partial \rho \Pi}{\partial x_j} + D_{ij} \frac{\partial \rho}{\partial x_j}, \tag{16}$$

where ρ is particle apparent density, defined as $\rho_s \langle \phi_s \rangle$, D_{ij}^T, B_{ij} and D_{ij} are thermal diffusion, particle mobility and ordinary diffusion tensors, respectively, and Π is a potential whose expression has been given by Polashenski et al.[39]. Here, the suffix "l" is dropped assuming that the color label plays no role in particle dynamics.

By analyzing a perturbation solution of the Boltzmann equation through the first order in the gradient of solids fraction, Garzo [41] has derived the elements of D_{ij} for granular shear flows. Note that in the absence of shearing motion diffusion tensor D_{ij}, may be reduced to $D_0\delta_{ij}$, where D_0 represents the diffusion coefficient [42] and δ_{ij} is the Kronecker delta.

The significance of the first and second terms on the right side of Eq. (16) far from the solid walls is not quite clear. Moreover, no analytical expressions exist for the diffusion coefficient in dense granular flows for which frictional contact between the particles plays a key role in flow dynamics. In this light, more rigorous theoretical studies are required to examine the significance of diffusion processes in systems such like those studied in vertically vibrated granular layers. In this case, the results of previous molecular dynamic type simulations [43] can be used as a predictive method to characterize particle diffusive motion.

4.3. Viscous-like Behavior

Here, the somewhat standard form of the linear momentum equation is used for the particle phase that in a curvilinear coordinate system is given as [34]:

$$\rho_s \left(\frac{\partial (\phi_s V_i)}{\partial t} + \frac{\partial (\phi_s V_i V_j)}{\partial x_j} \right) = \frac{\partial \sigma_{ij}}{\partial x_j} + F_i + \rho_s \phi_s g_i, \tag{17}$$

where σ_{ij} represents the stress tensor for the particle phase. Here, F_i is the total force of the interphase interaction per unit volume, which includes the drag force as well as the buoyancy contribution to the overall hydrodynamic force.

By averaging Eq. (17) over the ensemble of the spheres, an equation for balance of momentum may be obtained similar to that of the time-smoothing of volume averaged momentum equation suggested by Dasgupta et al. [40] ,which is:

$$\rho_s\left(\frac{\partial(\phi_s V_i)}{\partial t}+\frac{\partial}{\partial x_j}(\phi_s V_i V_j)\right)+\left(\frac{\partial \rho u_i}{\partial t}+\frac{\partial}{\partial x_j}\left(\rho u_j V_i+\rho u_i V_j\right)\right)=\frac{\partial\langle\sigma_{ij}\rangle}{\partial x_j}+\langle F_i\rangle+\rho_s\phi_s g_i, \tag{18}$$

where the third correlation $\langle\phi_s' V_i' V_j'\rangle$ has been neglected and the fluctuations are assumed to be relatively small, as suggested by the molecular dynamics type simulations presented in [31].

Note that for very low shear rates where the strain-rate fluctuations are very small, the constitutive behavior of the granular assembly may be predicted by using Eq. 13 for $\langle\sigma_{ij}^s\rangle$, which has a rate-independent form. However, as suggested by Savage [35], at higher shear rates fluctuations exist in σ_{ij} due to fluctuations in the strain rate ε_{ij}, which may be even larger than the mean strain rates $\langle\varepsilon_{ij}\rangle$. Therefore, it is not permissible to neglect the effect of viscosity associated with the pseudo-thermal motion of grains. By assuming that the strain rate fluctuations are isotropic and they are distributed according to a Gaussian distribution with a standard variation of ε, namely $f\left(\varepsilon_{ij}\right)=1/\varepsilon(2\pi)^{\frac{1}{2}}e^{-\left(\varepsilon_{ij}-\langle\varepsilon_{ij}\rangle\right)^2/2\varepsilon^2}$, a stress-strain rate relation can be developed similar to that of a Newtonian viscous fluid. Note that the variance of the strain rate fluctuations, ε^2, characterizes the pseudo-thermal motion of grain. Therefore, it is analogous to the granular temperature *T*, which is a term that measures the energy of the random motion of the grains.

The aim in this section is to provide a concise framework within which the seemingly complex behavior (such as the existence of solid-like and fluid-like qualities side-by-side in the continuous flow of granular material) involving very large strains can be described. To proceed, it may be conjectured that the mean value of stress can be calculated from the fluctuating strain rate by multiplying Eq. 13 for σ_{ij} by $f\left(\varepsilon_{ij}\right)$ and integrating over the strain rate space. It is convenient to evaluate the resulting integral over e_1, e_2, e_3, which represent principal strain rates. That is

$$\langle\sigma_i\rangle=1/\varepsilon^3(2\pi)^{\frac{3}{2}}\times\int e^{-\left[\left(\varepsilon_1-\langle\varepsilon_1\rangle\right)^2+\left(\varepsilon_2-\langle\varepsilon_2\rangle\right)^2+\left(\varepsilon_3-\langle\varepsilon_3\rangle\right)^2\right]/2\varepsilon^2}\sigma_i d\varepsilon_1 d\varepsilon_2 d\varepsilon_3, \tag{19}$$

where σ_i (i=1,2,3) represents the principal stresses in a three-dimensional system whose expressions are given by

$$\sigma_i=\frac{P}{\left[\left(\frac{4}{3}\sin^2\theta_f+1\right)\mathrm{I}_1^2-4\sin^2\theta_f\mathrm{I}_2\right]^{\frac{1}{2}}}\times\left[2\sin^2\theta_f\varepsilon_i+\left(1-\frac{2}{3}\sin^2\theta_f\right)\mathrm{I}_1\right]-P. \tag{20}$$

Following the approach of Savage [35], which is valid only for a two-dimensional case, the right hand side of Eq. 19 can be expanded in powers of $\langle e_i \rangle / \varepsilon$ which is a small parameter when the values of ε are large compared to the mean strain rates. Transforming to spherical coordinates by introducing $e_1 = \varepsilon r \cos\varphi \sin\theta$, $e_2 = \varepsilon r \sin\varphi \sin\theta$, and $e_3 = \varepsilon r \cos\theta$ into Eq. 20 and integrating over $0 < r < \infty, 0 < \theta < \pi, 0 < \varphi < 2\pi$, the mean principal stresses as a linear function of the mean principal strain rates may be found as

$$\langle \sigma_i \rangle = 2\mu \langle \varepsilon_i \rangle + (\zeta - \mu)\langle \mathrm{I}_1 \rangle - P, \tag{21}$$

where μ and ζ are the coefficients of viscosity, with the form of $\mu = PA_1 / \varepsilon$, $\zeta = PB_1 / \varepsilon$. The coefficients A_1 and B_1 as well as the ratio of viscosities, ζ / μ, depend on the angle of repose θ_f.

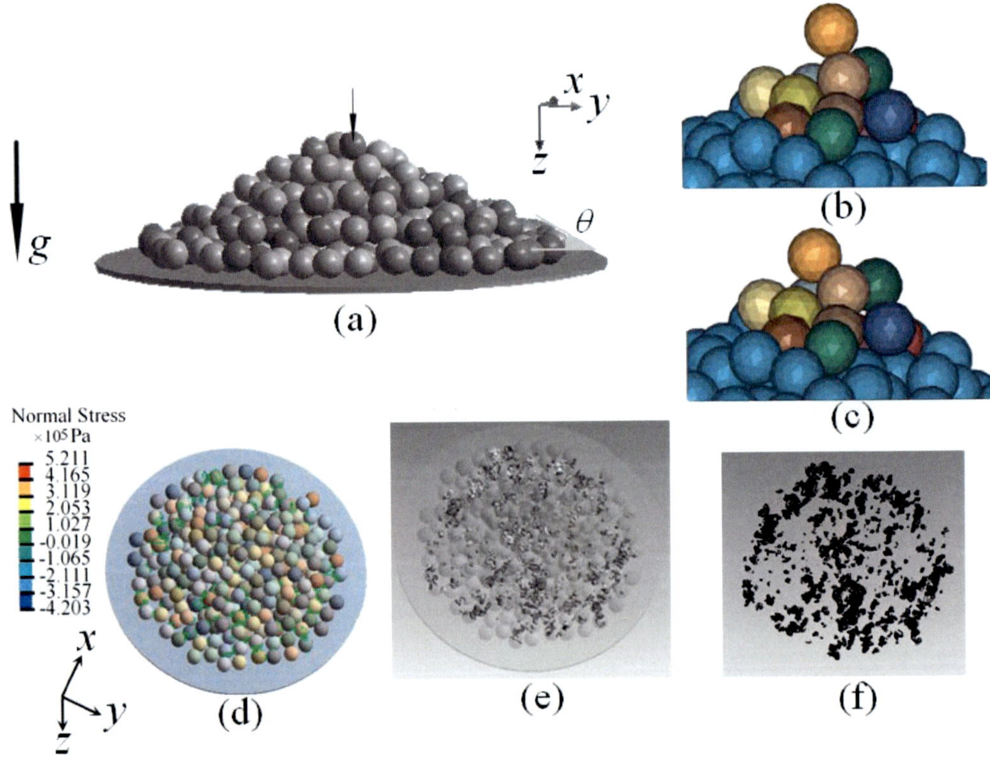

Figure 8. (a) Illustration of a conical spill using computer simulations. Note that a leak approaches the spill. The angle made between the surface and the outside of the cone is referred to as the angle of repose, θ_f. (b, c) Configuration of the particles in the spill at two different times after the collision. (d-f) Top view of the stress field developed in the spill $t = 10^{-4}$ sec after the collision (from Bordbar [36], Fig.4.41).

Note that the angle of repose, θ_f, is that made between the flat surface and the outside of the cone as illustrated in Figure 8. Figure 8 depicts a spill of particles onto a flat surface collected in a pile. The spill takes on the shape of a flat cone. In Figure 8 the numerical results have been obtained using the finite element approach. The resulting shape of the spill of particles could be affected by the distance between the leak and the flat surface where the leak collects. The roughness of the flat surface upon which the material collects is another important factor that affects the shape of the spill. In order to obtain an angle of repose of $\theta_f \approx 30^\circ$, the sliding coefficient of friction is set to 0.6 in the present simulations. Note that a theory of contact mechanics, valid for randomly rough surfaces, has been developed recently. This theory can be applied to viscoelastic solids in order to take into account more precisely the adhesive force between the glass balls in direct contact.

In a straightforward way the expressions for the mean stress components referring to a curvilinear coordinate system, x_i, may be given as

$$\langle\sigma_{ij}\rangle = 2\mu\langle\varepsilon_{ij}\rangle + (\zeta - \mu)\langle\varepsilon_{kk}\rangle\delta_{ij} - P\delta_{ij}. \tag{22}$$

Equation (22) resembles a viscous-like character for a liquid granular material in the sense that the viscosity decreases with the increase of the strain rate fluctuations. It may be reasonably straightforward to shown that the above averaged equations for a three-dimensional case can smoothly merge with those describing rapid granular flows at the limiting case of large fluctuations [35]. To proceed, an explicit form for P is required. Without a firm physical basis, Savage [35] suggests that P is composed of a pressure density contribution for quasi-static deformation, as well as a collisional stress contribution which depends on granular temperature. Hence

$$P(\phi_s, T) = P_0 \log\frac{\phi_\infty - \phi_0}{\phi_\infty - \phi_s} + \rho_s\phi_s\left[1 + 2(1+e)\phi_s g_c\right]T\ , \tag{23}$$

where P_0 is a reference value of P, ϕ_∞ is the solids fraction at close packing, ϕ_0 is the minimum solids fraction, e is the coefficient of restitution, g_c is the equilibrium value of the radial distribution function at contact, and T is the granular temperature defined as $T = \langle u^2\rangle/3$. A simple representation of the radial distribution function g_c follows from the Enskog approximate model of dense gases of identical rigid spheres, which is based on a heuristic concept of the free volume of each sphere to be smoothed over so as to present a regular sphere of a concentration dependent radius. It is suggested that [45]

$$g_c = \frac{1}{1 - (\phi_s/\phi_\infty)^{\frac{1}{3}}}, \tag{24}$$

The value of P_0 in Eq. 23 may be determined using the numerical method for predicting the stress distribution in a system. The bulk density ρ_b can easily be measured by filling a rectangular container with glass balls of a certain size. Note that the bulk density can be increased with the application of elevated stresses or when the material is vibrated. The bulk density is related to solid volume fraction by $\rho_b = (1-\phi_s)\rho_g + \phi_s\rho_s$ where ρ_g is the density of the gas. Apparently, an unequal transmission of the vertical force components occurs, which could imply a purely diffusive large-scale behavior of the force.

The data obtained using numerical simulations can be fit according to $P_0 \log(\phi_\infty - \phi_0/\phi_\infty - \phi_s)$ with $P_0 = 0.5\,kPa$, $\phi_\infty = 0.63$ and $\phi_0 = 0.487$. Here, ϕ_0 represents the solid volume fraction of the initially loose sample, ϕ_∞ is the maximum solid volume fraction, and P_0 is determined from a least-square fit of the numerical simulation data.

Equations 22 and 23 may not predict the anisotropy caused by an anisotropic distribution of collision directions due to presence of a chain-like granular structure. Moreover, the present approach regards the grains as possessing no rotational motion. However, these constitutive equations could be of use for interpretation of previously unexplained hydrodynamic phenomena in transitional regimes that are observed in three-dimensional granular flows.

The total force of the interphase interaction per unit volume in Eq. 16 may be approximated by [46, 47]:

$$F_i = \rho_s\phi_s\left[\beta_1 K_1(\phi_s) + \beta_2 K_2(\phi_s)\|W\|\right]W_i - \phi_s\left[\rho_g\phi_g + \rho_s\phi_s\right]g_i, \tag{25}$$

where W_i is the interstitial slip velocity of the gas, namely $U_i - V_i$, whose norm is denoted by $\|W\|$, and the expressions for the coefficients β_1, β_2, $K_1(\phi_s)$, and $K_2(\phi_s)$ are given as [46, 47]:

$$\beta_1 = \frac{18\nu_g}{\sigma^2},\ \beta_2 = \frac{3}{8\sigma},\ K_1 = \frac{1}{\phi_s^{5/2}},\ K_2 = \left(\frac{\phi_s}{1-1.17\phi_s^{2/3}}\right)^2, \tag{26}$$

In addition, an approximate expression for $\langle F_i\rangle$ is given by

$$\langle F_i\rangle = F_i + \rho_s\phi_s\left[\beta_1 \left.\frac{\partial K_1}{\partial \phi_s}\right|_{\phi_s}\langle\phi_s' W_i'\rangle + \beta_2 K_2(\phi_s)\langle W_j' W_j^0 W_i'\rangle + \beta_2\left.\frac{\partial K_2}{\partial \phi_s}\right|_{\phi_s}\left(\langle\phi_s' W_j^0 W_j'\rangle W_i + \langle\phi_s' W_i'\rangle\|\langle W\rangle\|\right)\right] \tag{27}$$

where $W_i' = U_i' - V_i'$, $W_j^0 = \langle W_j\rangle / \|\langle W\rangle\|$.

4.4. Fluctuation Energy

Note that by introducing the link between ε and T in the preceding section, an additional equation is required for the calculation of the granular temperature. The conservation of fluctuation energy for the particle phase [34] includes the work performed by $\langle\sigma_{ij}\rangle$, the divergence of a pseudo-energy flux owing to the fluctuating motion, as well as the production and dissipation of the fluctuation energy due to the interaction with the interstitial fluid, and to the energy dissipation due to strain-rate fluctuations. The rate of energy dissipation due to strain-rate fluctuations, γ, may be given as [35]:

$$\gamma = 1/\varepsilon^3 (2\pi)^{\frac{3}{2}} \times \int e^{-\left[(\varepsilon_1-\langle\varepsilon_1\rangle)^2+(\varepsilon_2-\langle\varepsilon_2\rangle)^2+(\varepsilon_3-\langle\varepsilon_3\rangle)^2\right]/2\varepsilon^2} \sigma_i'\varepsilon_i' d\varepsilon_1 d\varepsilon_2 d\varepsilon_3, \quad (28)$$

where $\sigma_i' = \sigma_i - \langle\sigma_i\rangle$, $\varepsilon_i' = \varepsilon_i - \langle\varepsilon_i\rangle$.

The flux of the fluctuation energy is expressed using the kinetic theory type approach [48]:

$$Q_i = -\lambda \frac{\partial T}{\partial x_i}, \quad (29)$$

λ is the coefficient of the fluctuation energy transfer which characterizes the processes of mean flow energy being converted into the energy of the fluctuations. This coefficient is defined as $\lambda = \alpha P A_1 / \varepsilon$, where the parameter α is a function of the average solid volume fractions given as

$$\alpha = \frac{30(-3+e)\left\{8(1+e)^2(-49+33e)g_c^2\phi_s^2 - \pi\left[5+6(1+e)g_c\phi_s\right]\left[5+3(1+e)^2(-1+2e)g_c\phi_s\right]\right\}}{(-49+33e)\left\{96(-3+e)(1+e)^2 g_c^2\phi_s^2 - \pi\left[5+4(1+e)g_c\phi_s\right]\left[5+2(1+e)^2(-1+3e)g_c\phi_s\right]\right\}} \quad (30)$$

The remaining mechanisms have relevance to the production and dissipation of the fluctuations due to interaction with the interstitial fluid. The former can be considered as energy inflow to the fluctuations due to the mean relative flow, and the latter characterizes the dissipated energy per unit volume and per unit time caused by hydrodynamic resistance to the fluctuations. Consideration of these mechanisms may be of interest, recalling that the presence of a complex motion has been reported by Pak et al. [49].

To obtain the rate of generation of the granular temperature due to hydrodynamic interactions, it would be necessary to determine the response of particles to fluid velocity fluctuations. Buyevich [47] has suggested reliable expressions for the energy supply caused by both hydrodynamic drag and buoyancy, and the dissipation power, which is due the drag alone:

$$\tilde{Q}_{+} = \rho_s\phi_s\left[\beta_1 K_1(\phi_s)\langle U_i' V_i'\rangle + \beta_1 \left.\frac{dK_1}{d\phi_s}\right|_{\phi_s}\langle\phi_s' V_i'\rangle W_i\right] +$$

$$\rho_s\phi_s\left\{\beta_2 K_2(\phi_s)\left[\langle U_i' V_i'\rangle + \langle(W_i^0 V_i')(W_j^0 U_j')\rangle\right]\|\langle W\rangle\| + \beta_2 \left.\frac{dK_2}{d\phi}\right|_{\phi_s}\langle\phi_s' V_i'\rangle W_i\|\langle W\rangle\|\right\} - \phi_s(\rho_s - \rho_g)\langle\phi_s' V_i'\rangle g_i, \tag{31}$$

and

$$\tilde{Q}_{-} = \rho_s\phi_s\left(3\beta_1 K_1(\phi_s) + 4\beta_2 K_2(\phi_s)\|\langle W\rangle\|\right)T. \tag{32}$$

These formulas appear to be sufficient to develop a conservation of "translational" fluctuation energy for the particle phase, which is given as

$$\frac{3}{2}\rho_s\left(\phi_s\frac{\partial T}{\partial t} + \phi_s V_i\frac{\partial T}{\partial x_j}\right) = \langle\sigma_{ij}\rangle\frac{\partial V_i}{\partial x_j} - \frac{\partial Q_i}{\partial x_i} + \tilde{Q}_{+} - \tilde{Q}_{-} - \gamma\ . \tag{33}$$

4.5. Continuous Phase

Turning attention to the flow of the interstitial fluid, on average the momentum conservation equation can be written as

$$\rho_g\left(\frac{\partial(\phi_g U_i)}{\partial t} + \frac{\partial(\phi_g U_i U_j)}{\partial x_j}\right) = \phi_g\frac{\partial E_{ij}^g}{\partial x_j} - F_i + \rho_g\phi_g g_i, \tag{34}$$

where E_{ij}^g is the stress tensor for the fluid phase. The time-smoothing of Eq. 34 can be approximately written as

$$\rho_g\left(\frac{\partial\langle\phi_g U_i\rangle}{\partial t} + \frac{\partial}{\partial x_j}(\phi_g U_i U_j) + \frac{\partial}{\partial x_j}\left(U_j\langle\phi_g' U_i'\rangle + U_i\langle\phi_g' U_j'\rangle\right)\right) =$$
$$\left(\left\langle\phi_g\frac{\partial E_{ij}^g}{\partial x_j}\right\rangle - \frac{\partial}{\partial x_j}\left(\phi_g\langle U_i' U_j'\rangle\right)\right) - \langle F_i\rangle + \rho_g\phi_g g_i. \tag{35}$$

The approximate expression for the time-smoothed total force of the interphase interaction per unit volume, $\langle F_i\rangle$, is given in Eq. 27.

A Newtonian stress-strain rate type constitutive equation is conjectured for the time-smoothed stress tensor for the gas phase, given as

$$\left\langle E_{ij}^{g}\right\rangle = -p_g\delta_{ij} + \mu_{eg}\left[\frac{\partial U_i}{\partial x_j} + \frac{\partial U_j}{\partial x_i}\right], \tag{36}$$

where the effective viscosity of the gas phase is given [50] as $\mu_{eg} = \mu_g\left(1+2.5\phi_s+7.6\phi_s^2\right)\left(1-\phi_s/\phi_m\right)$. Here, μ_g represents the molecular viscosity of the gas.

It may be speculated that Eq. 35 can be closed by a turbulence model, such as the $k-\omega$ model. Given the form of the conservation of momentum Eq. 35, the modeling of turbulence in gas particle flows is quite complex. Consideration of turbulence may be important in order to improve predictions of air heaping and bubbling. To capture the important features of turbulence in a vibrated granular material, Bordbar and Zamankhan [32] have used a highly simplified model which is integrable up to the wall proposed by Wilcox [50]. Hence, by neglecting the anisotropy of the fluid velocity fluctuations, the simplified transport equations of $k-\omega$ for the continuous phase whose density and velocity are defined as $\rho_m = \rho_g\left\langle\phi_g\right\rangle$ and $U_i^0 = \left\langle\phi_g\right\rangle\left\langle U_i\right\rangle$, respectively, are given as:

$$\begin{aligned}
&\frac{\partial}{\partial t}(\rho_m k)+\frac{\partial}{\partial x_i}\left[\rho_m U_i^0 k-\rho_m\left(v_g+\varsigma^* v_T\right)\frac{\partial k}{\partial x_i}\right]=E_{ij}^{g}\frac{\partial U_i}{\partial x_j}-\chi^* k\omega+\left(\tilde{Q}_- -\tilde{Q}_+\right),\\
&\frac{\partial}{\partial t}(\rho_m\omega)+\frac{\partial}{\partial x_i}\left[\rho_m U_i^0\omega-\rho_m\left(v_g+\varsigma v_T\right)\frac{\partial\omega}{\partial x_i}\right]=\Sigma\frac{\omega}{k}E_{ij}^{g}\frac{\partial U_i}{\partial x_j}-\chi\omega^2+C_\omega\chi^*\omega\left(\tilde{Q}_- -\tilde{Q}_+\right).
\end{aligned} \tag{37}$$

Here, $k=\left\langle U_iU_i\right\rangle/2$, the kinematic eddy viscosity is defined as $v_T = k/\omega$, and the closure coefficients and auxiliary relations are listed as

$$\begin{aligned}
&\Sigma=\frac{13}{25},\ \chi=\chi_0 f_\beta,\ \chi^*=\chi_0^* f_{\beta^*},\ \varsigma=\frac{1}{2},\ \varsigma^*=\frac{1}{2}\\
&\chi_0=\frac{9}{125},\ f_\beta=\frac{1+70\chi_\omega}{1+80\chi_\omega},\ \chi_\omega\equiv\left|\frac{\Omega_{ij}\Omega_{jk}S_{ki}}{\left(\chi_0^*\omega\right)^3}\right|,\\
&\chi_0^*=\frac{9}{100},\ C_\omega=1.2\\
&f_{\beta^*}=\begin{cases}1, & \chi_k\le 0\\ \dfrac{1+680\chi_k^2}{1+400\chi_k^2}, & \chi_k>0\end{cases},\ \chi_k\equiv\frac{1}{\omega^3}\frac{\partial k}{\partial x_j}\frac{\partial\omega}{\partial x_j}\\
&\varepsilon=\chi^*\omega k\ \text{ and }\ l=k^{\frac{1}{2}}/\omega\\
&\Omega_{ij}=\frac{1}{2}\left(\frac{\partial U_i}{\partial x_j}-\frac{\partial U_j}{\partial x_i}\right),\quad S_{ij}=\frac{1}{2}\left(\frac{\partial U_i}{\partial x_j}+\frac{\partial U_j}{\partial x_i}\right)
\end{aligned} \tag{38}$$

In formulating the turbulence model (37), a number of correlations appear in the balances of k and ω mainly due to the influence of the particle phase on the gas phase. These terms have been modeled on an ad hoc basis. However, the value of parameter C_ω has been estimated using the molecular dynamic-type simulations presented in [31].

4.6. Result of the Continuum Approach in Simulating the Flow of a Vibro-Granular Bed

By using this set of hydrodynamic continuum equations, Bordbar and Zamankhan [32] have simulated the deformation and flow of dense gas particle assembly in a vibrated sandbox. They were able to simulate the bubbly flow and spatial structure named Oscillon, which have been studied experimentally in previous research [33, 52-53]. Bordbar and Zamankhan [32] used the aforementioned hydrodynamic set of equations for simulating the flow of dense gas particle mixture in a vertical cylindrical container, whose base wall was subjected to sinusoidal oscillation in the vertical direction given as:

$$z = A\sin(k_v x)\sin(l_v y)\cos(\omega_v t), \quad \text{where } \omega_v = C\sqrt{l_v^2 + k_v^2} \text{ and } k_v = m_v \pi, l_v = n_v \pi , \tag{39}$$

where C is the speed of the wave in the base wall of the container, and A is a constant. Their result, shown in figure 9, predicted bubbling behavior analogous to those observed experimentally [33] for a heptagonal prism-shaped container under a certain vertical vibrations while for a pentagonal prism-shaped container under a certain vertical vibrations, their modeling reached an oscillon structure which was observed previously [51-52] in experimental research, shown in figure 10.

5. Discrete Elements Method

The discrete elements method is a numerical method for simulating a system consisting of distinct interacting general-shaped bodies or particles. The discrete elements can be rigid or deformable. The origin of this method goes back to 1971 when Cundall [53] developed this theory to be used in rock mechanics. Williams et al. [54] presented this method as a generalized finite elements method. Due to some similarities between this method and the molecular dynamics method, this method is called in some documents molecular dynamics. Moreover, this method can be used for modeling non-spherical particles that is in contrast with molecular dynamics. During the last decades, this Lagrangian solution has received increasing attention in modeling small-scaled granular systems. Especially in slow, frictional (which are usually called Coulomb) flow regimes where the particles are in contact for longer times and friction is the overwhelming interaction, this method can provide detailed information about the position and dynamics of every individual particle in each time of simulation.

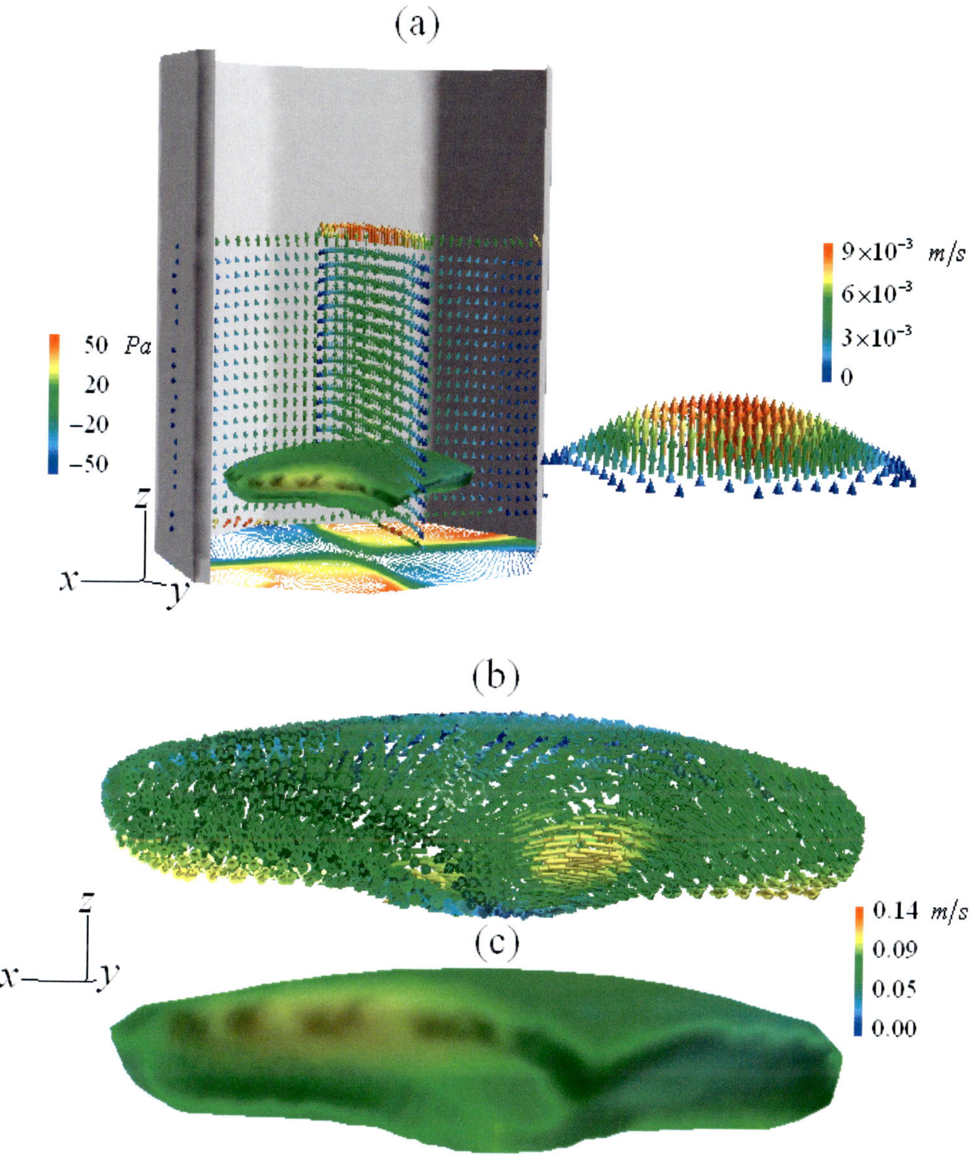

Figure 9. The result of simulation of bubble formation in vibrated dense granular material by using the continuum approach [32] (form Bordbar and Zamankhan [32], Fig.10). (a)A large bubble color-coded with the local gas pressure in the middle of a heptagonal vibro-granular media device. The snapshot is taken at $t = 0.228$ sec . The contours of pressure are shown at the base of the container. The vector velocity field of the glass balls slightly higher than the top side of the bubble is magnified and presented in the inset of (a). (b) The vector velocity field of the gas phase around the bubble. (c) To obtain a better visualization, a single, large bubble in (a) is magnified. The glass particle diameter is $1\,mm$, and initially $10\,cm$ of a heptagonal prism is poured with them.

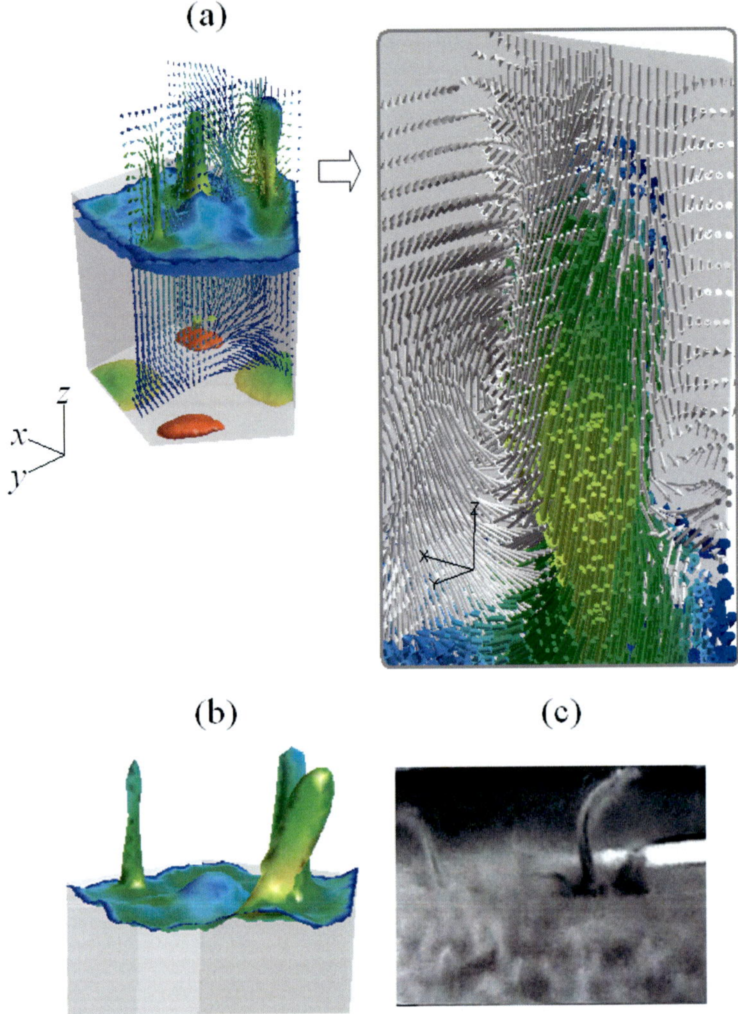

Figure 10. Result of simulation of oscillon structure by the continuum approach [32] (form Bordbar and Zamankhan [32], Fig.12). (a) A side-view of the instantaneous surface shape of glass balls in a pentagonal vibro-granular media device taken after 0.185 sec. (b) A view of (a) excluding the details of velocity vector fields. (c) Photograph of rouge oscillon[52], Note that the oscillon captured by camera is likely to bend due to gas-oscillon interaction.

For gas particle systems, DEM can be employed by considering a viscous additional drag force acting on the particles, and the effect of particles' motion should be taken into account in the gas flow field calculation. In the other words, simulating fluid flow filed requires coupling the particle dynamics with fluid dynamics calculations. In this regard, one-way coupling is used widely in flow simulation of gas particle systems due to its simplicity and short computational time. In one-way coupling, the effect of the carrier fluid flow field on the dynamics of the particles is taken into account by considering the drag force on the particles calculated on the bases of local fluid conditions in a converged fluid flow field, while the effect of the particles on the fluid flow field is ignored. Using this type of coupling has the advantage of simplicity and computational time. It is preferred when the momentum transfer from fluid to particles is small, relative to the local fluid momentum. A more complex and time consuming coupling method is full coupling [55-58], which takes into account momentum exchange and mass and energy exchange if necessary.

In the beginning of each DEM simulation, the particles are put in certain positions and an initial velocity is given to them. Based on this initial velocity and location, as well as the related physical laws, the amount of the force acting on each particle can be calculated. The acting forces on each particle are added up to find the total acting force. There are several integration methods, such as the Verlet algorithm, velocity Verlet, symplectic integrators and the leapfrog method, which are used to compute the change in the position and velocity of each particle during a predefined time step by using Newton's laws of motion. Then the new data (position and velocity) is used to compute the new forces acting on each particle. This iterated numerical procedure will be continued until the end of simulation. Hence, this method is processor intensive, considering that a complex particle-to-particle force model increases the computational cost. How ever, friction, drag, damping, gravity and recoil force are usually considered in DEM models. When two rough particles touch each other, friction should be considered as an acting force on both particles in opposite directions. Drag force is usually calculated by using the local carrier fluid flow condition. For viscoelastic particles, some part of energy is lost during the collision of the particle, so the damping force should be taken into account for this kind of cases. When using the DEM approach in modeling molecular systems, some other forces, such as Van Der Waals, Coulomb force, and Pauli repulsion force should be taken into account.

In the following section, a review of a new DEM simulation, which is done currently [36, 59] for modeling a granular bed subjected to vertical vibration, is presented.

5.1. Simulating a Vibro-Granular Bed Using the DEM Approach

Currently, a lot of research has been conducted to present more accurate models for calculating inter-particle forces useful for developing discrete element models. In this regard, a simplified model [36, 59] is proposed for providing detailed information of a virtual type, vibrofluidized, granular medium device, in a reasonable computational time. The system, as shown schematically in Figure 11, consists of roughly 20000 identical, slightly overlapping, spherical glass particles with a diameter of 600 micrometer used to fill a pentagonal prism container. More than half of the container is filled up by particles in the initial state, as illustrated in part (a) of Figure 1. The base wall of the container is subjected to sinusoidal oscillation in the vertical direction given by Eq. 39, where m_v and n_v are numerically set to 17 and 13 m^{-1}, respectively. More details about the system and its physical properties can be found in [36, 59].

The following equation is presented as the equation of motion for a particle with mass of m in the aforementioned system:

$$m\frac{dV_i^p}{dt} = F_i^f + F_i^g + F_i^i + F_i^b \quad (40)$$

where F_i^f is the drag force from the gas, F_i^g is the gravitational force, F_i^i is the force due to particle-particle interaction, and F_i^b is the Brownian force due to the thermal motion of the gas. For the vibro-device system, the Brownian force is negligible because the work done by the thermal force is very small in comparison with work done by drag force [60].

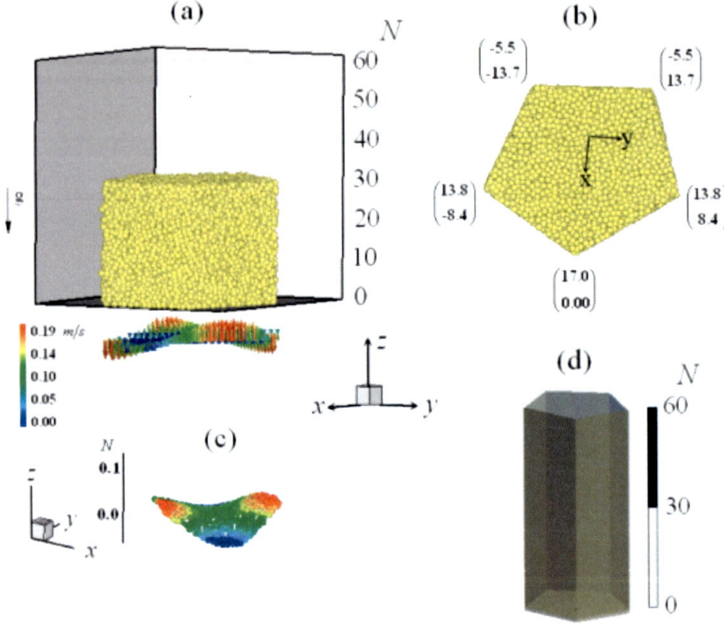

Figure 11. (from Bordbar [36], Fig.4-31and Zamankhan and Huang [59], Fig.1) ;(a) Initial configuration of spherical particles in a vibrofluidized granular matter device. Also shown is the typical instantaneous velocity vector field of particles located at the base of the container. (b) Top view of the initial configuration of spherical particles. All the sizes are normalized using particle diameter as a characteristic length. (c) Instantaneous configuration of particles located adjacent to the base of the container. To obtain a better visualization, the displacements of particles are rescaled. (d) Schematic of the container. The sidewalls of the container are neither moving nor deforming. The air flow in the container is resolved using 4×10^4 tetrahedral cells, nonuniformly distributed in the grid, where only one-seventh of the computational cells are used in $N \leq 30$.

The following dimensionless quantities, the Reynolds number (i.e. the ratio of inertial forces to viscous forces), Stokes number (i.e. the ratio of the stopping distance of a particle to a characteristic dimension of the obstacle) and Froude number (i.e. the ratio of total kinetic energy of the particle and gravitational energy), are defined as follows to simplify eq.40:

$$\mathrm{Re}_p = (1-\phi_s)\frac{\sigma\left|u_i - V_i^p\right|}{\nu_g},$$

$$\mathrm{St} = \frac{2}{9}(1-\phi_s)\left(\frac{\rho_g}{\rho_p}\right)^{-1}\frac{\sigma\left|u_i - V_i^p\right|}{\nu_g}, \tag{41}$$

$$\mathrm{Fr} = \frac{\left((1-\phi_s)\left|u_i - V_i^p\right|\right)^2}{gL},$$

where ρ_p is the material density of the particle, ρ_p is the fluid density, g is the gravitational acceleration, ϕ_s is the solid volume fraction set to 0.585, σ is the particle diameter, u_i is the velocity vector of gas, V_i^p is the particle velocity vector, ν_g is the kinematic viscosity of the gas, and L is the linear size of the vibro-device. By considering the special properties for the glass particle and gas, the order of magnitude of the above dimensionless quantities for the vibro-device are estimated as $\mathrm{Re}_p \sim 10, \mathrm{St} \sim 10^4$, and $\mathrm{Fr} > 1$. This order of the Stokes number of the particles highlights the role of the solid body in the particle velocity distribution, and the effect of subgrid scale eddies on particle motion is negligible. However, larger unsteady turbulent motions might affect the particles. Thus for the case $\mathrm{Re}_p \sim 10$, and $\mathrm{St} \gg 1$ with non-linear, visco-elastic, particle-particle interactions, the Langevin equation for translational motion of the pth particle in the vibro-device is modified to:

$$\frac{dV_i^p}{dt} = \frac{F_i^D}{m} - g\delta_{iz} + \sum_{p=1}^{N_j} F_{i,jp}^{(n)}, \tag{42}$$

where $F_i^D = 3\pi\sigma\rho_g\nu_g(1-\phi_s)^\beta\left(1+0.15\mathrm{Re}_p^{0.687}\right)\left(u_i - V_i^p\right)$ is the drag force, δ_{ij} represents Kronecker delta, and N_j represents the number of neighboring particles in contact with the jth particle at time t. The exponent β is not a constant and its value varies with the particle Reynolds number given as [59]:

$$\beta = \begin{cases} 3.7 & \mathrm{Re}_p < 5 \\ 3.689 + 2.7\times10^{-1}\,\mathrm{Re}_p - 5.66\times10^{-3}\,\mathrm{Re}_p^2 + 1.37\times10^{-4}\,\mathrm{Re}_p^3 & 5 \le \mathrm{Re}_p < 25 \end{cases}. \tag{43}$$

By using the result of viscoelastic collision modeling presented in section 3 and by employing the soft particle approach, the following expression can be derived for the total force per unit mass acting on the pth particle due to viscoelastic contacts between this particle and its neighboring j^{th} particles at time *t*, given as [3]:

$$F_{i,jp} = \left(\left(2E / \left(3\left(1 - v_p^2\right) m \right) \right) \sigma^{1/2} \xi_{jp}^{3/2} + \frac{K_n \tau}{m} \left(G_0^2 / (G_0 - G_\infty) \right) \xi_{jp} \frac{d\xi_{jp}}{dt} \right) k_{i,jp} - \left(k_t \chi_{ij}^{3/2} + c_t V_{ij}^{imp} . e_t \right) e_t \tag{44}$$

Which is the summation of the normal component, i.e. the first term on the right hand side and the tangential component (the second term on the right hand side) of the contact force per unit mass. In Eq. 44, σ, τ, E, v, m, G_0 and G_∞ represent the diameter of the particles, relaxation time, Young Modulus, Poisson's Ratio, mass of the particle, instantaneous shear modulus, and long time shear modulus of the material, respectively. As mentioned in section three, K is the coefficient characterizing the inelastic behavior of the particles, ξ_{jp} is the overlapping distance between the pth and j^{th} particles, and the exponent $\eta = 1$ for glass particles [3]. $k_{i,jp}$ is the unit vector directed from the center of particle j to that of particle p at the moment of impact, while χ is the tangential displacement and k_t is the tangential elastic constant, and c_t is the tangential damping coefficient.

e_t is the unit vector in tangential direction and is defined as $e_t = \varepsilon_{imq} k_{m,jp} (\varepsilon_{qst} e_{s,jp}^{V_{imp}} k_{t,jp})$. Coulomb friction law is used to describe the friction between two colliding particles with a surface friction coefficient, μ_p, when there is mutual slipping at the point of contact. The tangential displacement χ_{jp} is calculated as necessary to satisfy:

$$\left| F_{tjp} \right| = \mu_p \left| F_{i,njp} k_{i,jp} \right|. \tag{45}$$

Otherwise, the contact surfaces are considered as stuck, while $\left| F_{tjp} \right| < \mu_p \left| F_{njp} \right|$ where $\left| F_{tjp} \right|$ and $\left| F_{njp} \right|$ represent the magnitude of the normal and tangential components of the contact force, respectively. Clearly, it can be written for the rate of change of tangential displacement:

$$\frac{d\chi_{jp}}{dt} = V_{i,jp}^{imp} e_i. \tag{46}$$

The displacement, χ_{jp}, is set initially to zero when a new contact is occurred and once the contact is broken, all memory of the prior displacement will be lost.

The torque on particle *j* induced by the friction force is defined as [3, 61];

$$T_{i,j} = -\frac{1}{2}\sum_{p=1}^{N_j} \varepsilon_{iqr} m_j r_{jp} k_{q,jp} F_{r,jp} \,. \tag{47}$$

Therefore, in addition to Eq. 42, the following torque equation for the rotational motion of particle j should be considered:

$$I_j \frac{d\omega_{i,j}}{dt} = T_{i,j} \tag{48}$$

where I_j and $\omega_{i,j}$ $(i = x, y, z)$ represent the moment of inertia and the angular velocity vector of the jth particle, respectively.

To model the contact between the walls of the system and particles, σ in Eq. 42 should be changed to 2σ.

In [36, 59], for predicting the gas flow field in the vibro-device, a generalized form of the Navier-Stokes equations for a gas interacting with a solid phase is used as follows:

$$\left(\rho_g(1-\phi_s)u_i\right)_{,i} - \left(\rho_g(1-\phi_s)\right)_{,t} = 0, \tag{49}$$

$$\left(\rho_g(1-\phi_s)u_i\right)_{,t} + \left(\rho_g(1-\phi_s)u_i u_j\right)_{,j} = -\left((1-\phi_s)p\right)_{,i} + \left((1-\phi_s)\sigma_{ij}\right)_{,j} - (1-\phi_s)\rho_g g\delta_{iz} + f_i. \tag{50}$$

The effect of particle motion on the gas is considered in the above equation by using the term f_i on the right hand side of Eq. 50. This term is given as the sum of all hydrodynamic forces on the particles in a computational cell:

$$f_i = -1/\left(V_c(1-\phi_s)\right)\sum_{s=1}^{N_c} F_i^D \,, \tag{51}$$

where V_c is the volume of a computational cell, and N_c is the number of particles in the cell. The large eddy simulation technique is employed for solving the governing equations of the gas phase. In this regard, by adding the gradient of the subgrid scale stress tensor $\tau_{ij,j}$ to the left hand side of equations 49 and 50, the filtered equations for the gas phase will be achieved. This scale stress tensor is given as:

$$\tau_{ij} = \overline{\rho u_i u_j} - \overline{\rho u_i}\,\overline{\rho u_j}/\overline{\rho} \,, \tag{52}$$

where $\rho = \rho_g\left(1-\phi_s\right)$.

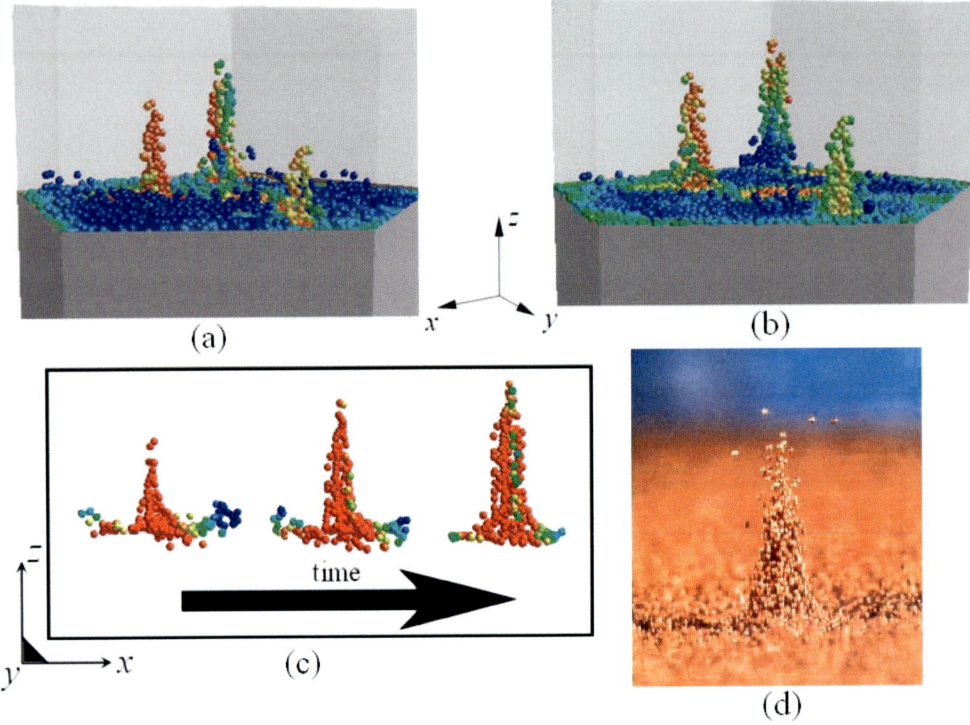

Figure 12. (a-b) Observed oscillon structure that has been formed in the discrete elements modeling of the system shown in figure 11; shown at two different times. The particles have been colored by local granular temperature. The next oscillon will form in the central region of part (b) (red hole) that has a high granular temperature. (c) Changing the shape of an oscillon structure with time (d) The oscillon structure observed experimentally in a brass ball bed subjected to vibration at a critical amplitude, (Modified from Bordbar [36] and Umbanhowar et al.[51]).

By using the aforementioned simplified discrete elements model as illustrated in Figure 12, an oscillon-type structure is observed in the free surface of the vibro-granular bed illustrated in Figure 11. The oscillon-type structure observed in this simulation is analogous with those previously observed experimentally by Umbanhowar et al. [51].

6. Summary

A comprehensive review of current activities in the field of simulation of granular material has been done. Due to the importance of this kind of material in industry and nature, providing new numerical methods for simulating industrial systems containing this complex material with reasonable computational cost is highly appreciated.

As collision between particles has the main role in transferring energy and momentum transfer within these materials, the new achievements in modeling binary collision between viscoelastic particles have been reviewed in section three of this chapter. A new model has been made by considering an additional spring in series with the Kelwin model. The result of modeling the collision of glass balls and ice balls using this method has a better conformity with experimental data.

In the next section, a new set of hydrodynamics continuum equation for modeling a gas particle system has been presented. Since the discrete elements method is computationally expensive, developing this kind of faster models for simulating a granular system is highly appreciated. Even though the continuum approach cannot provide detailed information of each individual particle during the simulation, these methods are computationally fast. The result of using the new set of continuum equations in modeling a complex granular system has been reviewed in section four of this chapter. In contrast with the continuum approach, the discrete elements method, which is the topic of section five, can provide detailed information of each individual particle during the simulation. A simplified DEM model has been reviewed and the result of using this model in simulating a granular bed which base was subjected to a complex sinusoidal vibration is presented.

References

[1] Duran, J. *Sands, Powders, and Grains: An Introduction to the Physics of Granular Materials*; Springer-Verlag: New York, NY, 2000.

[2] Nedderman, R. M. *Statics and Kinematics of Granular Materials*; Cambridge University Press: Cambridge, UK, 1992.

[3] Zamankhan, P.; Bordbar, M. H. *J. Appl. Mech.-T ASME* 2006, 73, 648-657.

[4] Cundall, P. ; Strack, O. A. *Geotechnique* 1979,29, 47-65.

[5] Babic, M.; Shen, H. H. ; Shen, H. T. *J. Fluid Mech.* 1990, 219, 81-118.

[6] Campbell, C. S. *J. Fluid Mech.* 2002,465, 261-291.

[7] Zhang, D. Z.; Whiten, W. *Powder Technol.* 1996, 88, 59-64.

[8] Mishra, B. K.; Murty, C. V. R. *Powder Technol.* 2001,115, 290-297.

[9] Hertz, H.; *J.reine und angewandte Mathematik* 1882, 92, 156-171.

[10] Mindlin, T. D. *J. Appl. Mech.-T ASME* 1949,16, 259-268.

[11] Walton, O. R.; Braun, R. L. *J. Rheology* 1986, 30, 949-980.

[12] Thornton, C.; Ning, Z. *Powder Technol.* 1998, 99, 154-162.

[13] Vu-Quoc, L.; Zhang, X.; Lesburg, L. *Int. J. Solids Struct.*2001, 38 ,6455-6490.

[14] Sellgren, U.; Björklund, S.; Andersson, S. *Wear* 2003, 254, 1180-1188.

[15] Gugan, D. *Am. J. Phys.* 2000, 68, 920-924.

[16] Zhang, D. Z.; Rauenzahn, R. M. *J. Rheology* 2000, 44, 1019-1041.

[17] Zhang, D. Z.; Rauenzahn, R. M. *J. Rheology* 1997,41, 1275-1298.

[18] Kuo, H. P.; Knight, P. C.; Parker, D. J.; Tsuji, Y.; Adams, M. J.; Seville J. P. K. *Chem. Eng. Sci.* 2002, 57, 3621-3638.

[19] Falcon, E.; Laroche, C.; Fauve, S.; Coste, C. *Eur. Phys. J. A* 1998, 3, 45-57.

[20] Tsuji, Y.; Tanaka, T.; Ishida, T. *Powder Technol.* 1992, 71, 239-250.

[21] Tsuji, Y.; Kawanuchi, T.; Tanaka, T. *Powder Technol.* 1993, 77, 79-87.

[22] Brilliantov, N. V.; Spahn, F.; Hertzsch, J. M.; Pöschel, T. *Phys. Rev. E* 1996, 53, 5382-5392.

[23] Bridges, F. G.; Hatzes, A.; Lin, D. N. C. *Nature* 1984, 309, 333-335.

[24] Findley, W. N.; Lai, J. S.; Onaran, K. *Creep and Relaxation of Nonlinear Viscoelastic Materials (with an Introduction to Linear Viscoelasticity)*; Dover Publications Inc.: New York, NY, 1989.

[25] Bordbar, M. H.; Hyppänen, T. *J. Numer. Anal. Indust. Appl. Math.*2007, 2, 115-128.

[26] Mathiesen, V.; Solberg, T.; Hjertager, B. H. *Powder Technol.* 2000, 112, 34-45.

[27] Eggers, J.; Riecke, H. *Phys. Rev. E* 1999, 59, 4476-4483.

[28] Arastoopour, H.; Gidaspow, D. *Powder Technol.* 1979, 22, 77-87.

[29] Tsuo, Y. P.; Gidaspow, W. *AIChE J.* 1990, 36, 885- 896.

[30] Mathiesen, V.; Solberg, T.; Arastoopour, H.; Hjertager, B. H. *AIChE J.* 2004, 45,2503-2518.

[31] Bordbar, M. H.; Zamankhan, P. *Commun. Nonlinear Sci. and Numer. Simul.* 2007, 12, 254-272.

[32] Bordbar, M. H.; Zamankhan, P., *Commun. Nonlinear Sci. and Numer. Simul.* 2007, 12, 273-299.

[33] Pak, H. K.; Behringer R. P. *Nature* 1994, 371, 231-233.

[34] Zamankhan, P.; Vahedi Tafreshi, H.; Chen, J. C. *Phys. Rev. E* 1997, 56, 2972-2980.

[35] Savage, S. B. *J. Fluid Mech.* 1998, 377, 1-26.

[36] Bordbar, M. H.; *Theoretical Analysis and Simulation of Vertically Vibrated Granular Materials, Ph.D dissertation*; Acta Universitatis: Lappeenranta, 2005.

[37] Zamankhan, P.; Huang, J. *J. Appl. Mech.-T ASME* 2007, 74, 691-702.

[38] Hsiau, S. S.; Hunt, M. L. *J. Fluid Mech.* 1993, 251, 299-313.

[39] Gidaspow, D. *Multiphase Flow and Fluidization. Continuum and Kinetic Theory Descriptions*; Academic Press: Boston, 1994.

[40] Dasgupta, S.; Jackson, R.; Sundaresan, S. *AICHE J.* 1994 , 40, 215-228.

[41] Garzo, V. *Phys.Rev.E* 2002, 66, 021308.

[42] Savage, S. B.; Dai, R. *Mech.Mater.* 1993, 16, 225-238.

[43] Polashenski, W. Jr.; Zamankhan, P.; Mäkiharju, S.; Zamankhan, P. *Phys.Rev.E* 2002, 66, 021303.

[44] Persson, B. N. J. *J. Chem.Phys.* 2001, 115, 3840-3861.

[45] Hirschfelder, J. O.; Curtiss, Ch. F.; Brid, R. B. *Molecular Theory of Gases and Liquids*, Wiley: New York, NY, 1954.

[46] Buyevich, Yu. A.; Kapbasov, Sh. K. *Chem. Eng. Sci.* 1994, 49,1229-1243.

[47] Buyevich, Yu. A. *Chem. Eng. Sci.* 1994, 49, 1217-1228.

[48] Jenkins, J. T.; Richman, M. W. *Arch. Ral. Mech. Anal.*1985, 87, 355-377.

[49] Pak, H. K.; Behringer, R. P. *Nature* 1994, 371, 231-233.

[50] Wilcox, D. C. *Turbulence Modeling for CFD*; DCW Industries Inc.: a La Canada, CA, 2000.

[51] Umbanhowar, P.; Melo, F.; Swinney, H. *Nature* 1996, 382, 793-796.

[52] Thrasher M. E. (1997). Still playing in a sand box. http://oas.ucok.edu/OJAS/97/T97/MTHRA.HTM

[53] Cundall, P.A.; Strack, O.D.L. *Geotechnique* 1979, 29, 47–65.

[54] Williams, J.R., Hocking, G., and Mustoe, G.G.W., *Proceeding of International Conference on Numerical Methods of Engineering, Theory and Applications* 1985, 897-906.

[55] Davidson, M. R. *Appl. Math. Model.* 1998, 22, 39-55.

[56] Druzhinin, O. A. *Phys. Fluids* 2001, 13, 3738-3755.

[57] Ferrante, A.; Elghobashi, S. *Phys. Fluids* 2003, 15, 315-329.

[58] Kitagawa, A.; Murai, Y.; Yamamoto, F. *Int. J. Multiphase Flow* 2001, 27, 2129-2153.

[59] Zamankhan, P.; Huang, J. *J. Fluid Eng.-T ASME* 2007, 129, 236-244.
[60] Phan-Thien, N. Understanding Viscoelasticity; Springer: Heidelberg, 2002.
[61] Zamankhan, P.; Soleymani, A.; Polashenski, W. Jr.; Zamankhan, P. *Chem. Eng. Sci.* 2004, 59, 235-246.

INDEX

A

absorption, 104
accelerator, 66, 67
accuracy, x, 4, 8, 57, 61, 67, 120, 147, 192, 199, 257
acoustic, 6, 27, 33, 35, 50, 90, 91, 92, 93, 94, 139, 154
acoustic waves, 154
ad hoc, 276
Adams, 285
adaptation, 228, 237
adhesive force, 271
adiabatic, 5, 17, 25, 32, 93
adjustment, 2, 24
agriculture, 258
air, 185, 275, 280
algorithm, 103, 105, 107, 108, 119, 122, 211, 215, 229, 230, 236, 238, 239, 241, 244, 246, 247, 248, 279
alternative, ix, 66, 112, 161, 264
amplitude, vii, 1, 4, 18, 58, 66, 72, 73, 76, 77, 80, 101, 103, 116, 120, 121, 122, 125, 126, 129, 284
angular velocity, 283
anisotropy, 132, 133, 134, 139, 272, 275
annealing, 210, 247
anode, 24, 25
aorta, 90, 94
aortic valve, 90
application, vii, viii, 1, 2, 90, 93, 100, 104, 112, 119, 125, 136, 152, 272
argument, 193, 230, 233
arteries, 90
aspect ratio, 126
assessment, 113
assumptions, 104, 162, 210, 212, 225, 230
astrophysics, viii, 99
asymptotic, 217, 250
asymptotically, 261
atmosphere, viii, 100, 141
averaging, 142, 161, 162, 268

B

beams, 71, 75, 76, 77, 94
beating, 90
behavior, ix, x, 100, 126, 135, 209, 211, 217, 224, 257, 259, 260, 261, 262, 263, 266, 267, 269, 272, 276, 282
Beijing, 141
benefits, 118
blood, 90, 94
blood flow, 90
Bose-Einstein, viii, 141
Boston, 247, 286
boundary conditions, viii, 10, 18, 24, 99, 101, 102, 111, 112, 113, 117, 118, 119, 129, 135, 138, 153, 168, 170, 171, 210, 212, 219, 228
boundary value problem, ix, 209, 229, 238
brass, 284
bubble, 277
buildings, 205

C

calculus, 158
Canada, 1, 286
carrier, 258, 278, 279
cathode, 25, 27, 31
Cauchy problem, 216, 233
cell, 2, 106, 264, 283
cereals, 258
chaos, 101
charge density, 14, 58, 64, 122
charged particle, viii, 99, 100, 121
chemical industry, 259
China, 141, 157
classes, 255
classical, vii, ix, 1, 4, 49, 109, 141, 142, 161, 162, 178, 211, 228, 229, 237, 246
closure, 163, 275
coal, 258

codes, vii, viii, 1, 2, 3, 5, 23, 66, 67, 79, 93, 94, 100, 101, 103, 122, 135, 136
collisions, x, 24, 120, 122, 151, 256, 257, 259
colors, 265
combined effect, 49, 66, 93
communication, 146
communication systems, 146
compatibility, ix, 117, 184, 188, 209, 211, 219
compensation, 213
complexity, 115
components, 129, 174, 176, 267, 271, 272, 282
composition, 107, 144, 164
compressibility, 266
computation, viii, 2, 112, 113, 115, 141, 153, 156
Computational Fluid Dynamics, 206, 286
computational mathematics, 141
computer simulations, 270
computing, 103, 104, 117, 173
concentrates, 134
concentration, 271
concrete, 162, 246
condensation, viii, 141
conductivity, 109
configuration, viii, 66, 99, 101, 126, 129, 137, 280
conformity, 183, 263, 284
conjecture, 223
conservation, ix, 100, 101, 103, 105, 107, 120, 135, 143, 145, 146, 149, 153, 156, 161, 162, 180, 205, 250, 273, 274, 275
constraints, 210, 214, 237
construction, 158, 162, 194, 205, 206, 228, 237
consumption, 210, 217, 224, 241
continuity, 91, 100, 103, 221, 222, 227, 232, 267
contracts, 90
control, viii, ix, 4, 17, 23, 99, 101, 120, 209, 211, 246, 247
convection, 200, 206
convective, 90, 199, 200
convergence, 71, 210, 211, 229, 238, 241, 244
convex, 211, 212, 225, 226, 266
cooling, 11, 30, 45, 48
coordination, ix, 209, 210, 213, 214, 215, 218, 222, 227, 228, 229, 235, 236, 238, 242, 244, 245, 247
correlation, 269
correlations, 276
costs, 213
Coulomb, 68, 124, 258, 276, 279, 282
Coulomb gauge, 68
couples, 119
coupling, 4, 93, 151, 264, 278, 279
CPU, 111, 242, 244
CRC, 205
critical value, 93
CRM, 247
CSF, 179
cyclotron, 32, 51, 109

D

damping, 114, 116, 120, 121, 122, 138, 260, 279, 282
decay, 73, 121, 126
decomposition, 102, 114, 117, 118, 130
definition, 55, 106, 107, 112, 127, 181, 194, 230, 242
deformation, x, 6, 20, 24, 73, 84, 257, 264, 271, 276
demand, 210, 211, 212, 240, 242, 243
density, 2, 4, 5, 6, 7, 13, 14, 15, 17, 18, 20, 21, 23, 30, 31, 33, 34, 35, 40, 42, 47, 48, 49, 51, 53, 54, 57, 58, 59, 61, 63, 64, 66, 71, 72, 73, 75, 76, 77, 78, 79, 80, 84, 85, 86, 87, 90, 92, 93, 94, 102, 103, 108, 109, 110, 118, 119, 127, 171, 265, 268, 271, 272, 275, 281
derivatives, 93, 102, 105, 106, 111, 112, 113, 117, 129, 130, 174, 175, 189, 212, 214, 232, 250, 251
detachment, 32, 43
deviation, 40
diastole, 90
differential equations, 1, 2, 91, 93, 99, 115, 157, 163, 164, 204, 205, 206, 211, 219, 228
diffusion, 79, 93, 129, 200, 268
diffusion process, 268
diffusivity, 170
dimensionality, 103, 108
directives, 131
Dirichlet boundary conditions, 117
discharges, 4, 17, 24
discontinuity, 214, 229
discretization, 111, 113, 136, 144, 145, 192, 194, 210, 240, 242
dispersion, 115, 120, 121, 151, 205, 251, 252, 253, 265
displacement, 109, 259, 260, 282
distribution, viii, 4, 6, 7, 8, 11, 12, 13, 14, 17, 18, 19, 20, 23, 24, 25, 26, 27, 28, 29, 30, 31, 33, 34, 35, 36, 37, 39, 40, 44, 45, 46, 48, 49, 50, 51, 61, 64, 66, 67, 68, 70, 73, 74, 75, 76, 77, 78, 85, 86, 87, 88, 89, 93, 94, 99, 101, 102, 103, 104, 105, 106, 107, 108, 118, 119, 121, 122, 123, 125, 142, 262, 269, 271, 272, 281
distribution function, viii, 4, 6, 7, 8, 11, 12, 13, 14, 17, 18, 19, 20, 23, 24, 25, 26, 27, 28, 29, 31, 33, 36, 37, 39, 44, 45, 46, 48, 49, 50, 51, 67, 68, 70, 74, 75, 76, 77, 78, 85, 86, 87, 88, 89, 93, 102, 103, 104, 105, 106, 107, 108, 118, 119, 121, 122, 123, 125
divergence, 110, 127, 131, 266, 267, 273
divertor, 4, 32
division, 40, 48, 57, 58, 59, 61, 64
duality, 124
duration, 260
dynamic viscosity, 109
dynamical system, 158
dynamical systems, 158

E

eddies, 281
elasticity, 142
electric current, 4
electric field, vii, 1, 2, 4, 5, 6, 8, 9, 14, 16, 18, 19, 21, 22, 25, 30, 33, 35, 36, 38, 40, 42, 43, 49, 50, 51, 53, 54, 55, 56, 57, 61, 62, 64, 66, 73, 75, 76, 78, 79, 80, 83, 84, 87, 93, 94, 109, 114, 116, 119, 121, 122, 123, 125, 126, 129
electric potential, 24, 124
electric power, 210
electricity, 246
electrodes, 24
electromagnetic, viii, 67, 69, 73, 99, 100, 108, 109
electromagnetic fields, viii, 69, 99, 100, 109
electromagnetic wave, 100
electromagnetic waves, 100
electron, 4, 6, 7, 9, 12, 13, 14, 16, 20, 21, 23, 29, 30, 34, 35, 40, 42, 48, 51, 54, 66, 67, 72, 73, 74, 75, 76, 77, 78, 79, 80, 84, 85, 86, 87, 90, 118, 119, 122, 123, 125, 139
electron density, 13, 35, 72, 73, 76, 77, 78, 84
electrons, vii, 1, 5, 6, 7, 8, 11, 12, 14, 16, 17, 18, 19, 20, 22, 23, 24, 25, 29, 30, 32, 33, 34, 35, 37, 39, 40, 42, 43, 44, 48, 49, 51, 54, 58, 64, 66, 67, 69, 71, 73, 79, 80, 84, 85, 93, 94, 104
energy, viii, 4, 16, 17, 23, 27, 43, 44, 48, 49, 93, 100, 101, 103, 109, 116, 120, 121, 122, 126, 128, 131, 132, 133, 134, 135, 141, 143, 145, 153, 154, 155, 158, 173, 183, 184, 210, 241, 250, 258, 259, 260, 269, 273, 274, 279, 280, 284
energy density, 145
energy supply, 273
energy transfer, 126, 273
entropy, 100
environment, vii
equality, 8
equilibrium, viii, 5, 17, 54, 99, 101, 119, 126, 129, 131, 132, 135, 212, 271
erosion, 4
etching, 3
Eulerian, vii, 1, 2, 3, 5, 17, 18, 31, 33, 50, 54, 67, 79, 93, 94, 102, 103, 116, 121, 122
evolution, vii, viii, 4, 20, 40, 71, 87, 100, 101, 103, 104, 105, 107, 108, 109, 110, 114, 116, 119, 121, 122, 123, 124, 125, 126, 128, 132, 137, 138, 141, 142, 143, 151, 153, 154, 156, 250, 255
excitation, 24, 110
execution, 103
expansions, 105, 107
extraction, vii, 1, 4, 24, 25, 31, 93

F

family, 221
Fermi, 138
FFT, 102, 112, 119, 135
fiber, 146
fibers, 146, 157
finite differences, 113
finite element method, viii, 99, 101, 135, 260
finite volume, viii, 99, 101, 102, 135
finite volume method, viii, 99, 135
Finland, 257
fission, 255
floating, 7, 16, 35, 54, 55, 61, 66
flow, ix, x, 32, 49, 90, 91, 94, 136, 144, 161, 162, 168, 169, 170, 176, 177, 178, 179, 181, 185, 188, 206, 235, 257, 259, 264, 265, 266, 268, 269, 273, 274, 276, 278, 279, 280, 283
flow field, 278, 283
flow rate, 235
fluctuations, viii, 99, 101, 120, 129, 137, 258, 267, 268, 269, 271, 273, 275
fluid, vii, viii, ix, 1, 66, 90, 94, 99, 100, 101, 102, 103, 104, 108, 109, 110, 111, 112, 118, 124, 125, 135, 136, 141, 142, 161, 171, 176, 185, 204, 257, 258, 269, 273, 274, 275, 278, 279, 281
fluid mechanics, 118, 135, 141
Fourier, 102, 112, 115, 131, 132, 154, 157
France, 98, 99, 100, 102, 104, 106, 108, 110, 112, 114, 116, 118, 120, 122, 124, 126, 128, 130, 132, 134, 136, 138, 161
free energy, 101
free volume, 271
friction, 258, 259, 266, 271, 276, 279, 282
fuel, 210, 211, 213, 217, 239, 241, 242, 243, 246
fulfillment, 219
fusion, 94, 100, 255

G

galactic, 102
gas, viii, x, 99, 100, 257, 258, 259, 264, 265, 267, 272, 274, 275, 276, 277, 278, 280, 281, 283, 285
gas phase, 267, 274, 275, 276, 277, 283
gases, 271
gauge, 68
Gaussian, 167, 180, 269
generalization, 108, 144, 194
generation, 49, 66, 93, 144, 210, 212, 213, 214, 217, 218, 223, 230, 240, 242, 244, 245, 246, 273
generators, 177
genetic algorithms, 210
Germany, 207
glass, x, 257, 261, 262, 263, 271, 272, 277, 278, 279, 281, 282, 284
grain, 259, 269
grains, ix, 257, 258, 259, 269, 272
granular flow, ix, 257, 259, 265, 268, 271, 272
graph, 122
gravitational force, 280
gravity, 170, 279
grazing, vii, 1, 4, 32, 43, 44, 49, 93

grids, 3, 94, 130
groups, 175, 176, 183, 195, 206
growth, 124, 126, 131, 132, 136
growth rate, 124, 126, 131, 136
guidance, 24

H

Hamiltonian, 68, 104, 124, 142, 143, 144, 145, 146, 147, 148, 155, 156, 157, 158, 159, 162
handling, 23, 259
harmonics, 115, 132
heart, vii, 1, 3, 90
heat, 4, 32, 93, 163, 166, 167, 175, 187, 193, 206
heating, 17, 66, 79, 94
height, 185, 186, 224, 242
helium, 18, 19, 20
heuristic, 210, 271
high resolution, viii, 99, 111
high-speed, 146
horizon, 210
human, 90
hydro, ix, 224, 240
hydrodynamic, x, 124, 257, 259, 264, 265, 268, 272, 273, 276, 283
hydrodynamics, 285
hydrogen, 32
Hydrothermal, ix, 209, 210, 211, 213, 214, 215, 217, 219, 221, 223, 225, 227, 229, 231, 233, 235, 237, 239, 241, 243, 245, 246, 247, 248
hydrothermal system, 209, 211, 239, 246, 248
hyperbolic, vii, 1, 2, 90, 91, 93, 100, 105, 107, 110, 250
hypothesis, 112, 173, 184, 185, 191, 216, 217, 224, 231, 232, 233, 234, 238, 245

I

ice, x, 257, 258, 261, 263, 264, 284
identity, 147, 178
IMA, 247
implementation, 102, 104, 135, 136, 210
in transition, 272
inactive, 229
incidence, vii, 1, 4, 32, 40, 43, 44, 49, 93
inclusion, 24, 212, 219, 220
incompatibility, 181
incompressible, 2, 110, 114, 117, 118, 124, 127, 135, 136, 167, 207
independent variable, 145, 251
indication, 24
industrial, vii, 1, 3, 4, 24, 93, 161, 259, 284
industrial application, 4, 24, 161
industrial revolution, vii, 1, 3
industry, 258, 259, 284
inelastic, 24, 262, 282
inertia, 283
infinite, 108, 142, 210, 250
inhomogeneities, 101
inhomogeneity, 111, 124, 129, 135
initial state, 142, 154, 157, 279
injection, 27, 31
instabilities, viii, 99, 100, 101, 102, 103, 126, 131, 132, 135, 136, 137
instability, viii, 99, 101, 124, 125, 126, 132, 136, 137, 139, 154
integration, 4, 36, 49, 66, 102, 103, 104, 107, 108, 119, 137, 154, 157, 158, 162, 167, 210, 224, 250, 279
intensity, vii, 1, 79, 135
interaction, vii, 1, 2, 3, 4, 5, 67, 73, 79, 89, 93, 94, 103, 108, 120, 121, 151, 152, 259, 260, 268, 272, 273, 274, 276, 278, 280
interaction process, 260
interactions, 3, 66, 79, 100, 101, 104, 126, 132, 134, 142, 155, 156, 259, 265, 273, 281
interphase, 268, 272, 274
interpretation, 93, 229, 272
interstitial, 272, 273, 274
interval, ix, 44, 105, 106, 119, 149, 156, 209, 210, 211, 212, 215, 218, 220, 227, 228, 229, 232, 235, 237, 240, 242
intervention, 231
invariants, 91, 92, 182, 184, 190, 267
inversion, 10, 102, 164
ion beam, 25
ionization, 100
ions, vii, 1, 5, 6, 7, 8, 11, 12, 14, 16, 17, 18, 19, 20, 21, 22, 23, 24, 25, 29, 30, 32, 33, 34, 35, 37, 38, 39, 40, 42, 43, 44, 45, 48, 49, 50, 51, 54, 55, 61, 66, 67, 69, 71, 73, 77, 79, 80, 84, 85, 86, 87, 89, 90, 93, 94, 108, 109, 118, 119, 177
irradiation, 79
island, 132
isothermal, ix, 161, 162, 171, 176, 185, 187, 188, 189, 190, 191
isotropic, 269
isotropy, 187
Italy, 99
iteration, 71

J

Jacobian, 143
Japan, 96, 256
Jung, 97

K

kernel, 180
kinetic effects, 66, 67, 79, 94, 103
kinetic energy, 48, 120, 132, 134, 260, 280
kinetic equations, vii, 1, 3, 4, 5, 94
kinks, 253, 255

Klein-Gordon, 142, 152, 153, 154, 156
Kolmogorov, 162, 173
Korteweg-de Vries, 256

L

Lagrangian, x, 3, 102, 138, 162, 210, 247, 257, 276
Lagrangian approach, 102
lamina, 167
laminar, 167
Laminar, 167, 170
Landau damping, 114, 116, 121, 122, 138
Landau theory, 120
Langmuir, 30, 114, 116, 120, 121, 138
Langmuir wave, 120, 121
language, 151
large-scale, 124, 210, 272
laser, vii, 1, 4, 5, 66, 67, 71, 72, 73, 76, 79, 83, 84, 89, 94
law, 5, 17, 25, 30, 32, 93, 105, 107, 145, 153, 162, 183, 184, 185, 258, 259, 262, 282
laws, vii, ix, 143, 161, 162, 173, 180, 181, 250, 279
lead, 66, 67, 79, 94, 100, 101, 173, 264, 265
left ventricle, 90
liberty, 205
Lie group, 157, 164, 205, 206
limitation, 2, 108, 142, 162
limitations, 103, 115, 142
linear, 9, 66, 77, 100, 103, 104, 113, 114, 115, 116, 117, 119, 120, 121, 122, 126, 127, 131, 136, 141, 151, 158, 163, 167, 176, 178, 185, 192, 210, 249, 251, 253, 259, 260, 261, 268, 270, 281
linear function, 259, 260
linear model, 259, 260
linear programming, 210
links, 109
location, 201, 279
London, 96
long-term, 162, 246
losses, 217, 218, 224, 240, 242

M

machinery, ix, 209, 211
machines, 102
magnetic, vii, 1, 4, 5, 32, 33, 34, 35, 40, 43, 44, 45, 49, 50, 51, 54, 58, 64, 66, 93, 100, 102, 104, 109, 110, 111, 117, 126, 127, 128, 129, 131, 132, 133, 134, 137, 139
magnetic field, vii, 1, 4, 5, 32, 33, 34, 35, 40, 43, 44, 45, 49, 50, 51, 54, 58, 64, 66, 93, 100, 102, 104, 109, 111, 117, 126, 127, 128, 129, 131, 132
manifold, 158, 194
manifolds, 158
mapping, 163, 219, 220
market, 246
Marx, 161, 162, 164, 166, 168, 170, 172, 174, 176, 178, 180, 182, 184, 186, 188, 190, 192, 194, 196, 198, 200, 202, 204, 206
Massachusetts, 205
material surface, 3, 4
mathematical methods, 137
mathematics, 159, 206
matrix, 9, 10, 36, 114, 117, 118, 145, 147, 178, 181, 188
Maxwell equations, 102
measures, 269
media, 137, 277, 278
memory, 108, 111, 122, 282
MHD, 2, 100, 109, 110, 111, 117, 127, 131, 136, 137
micrometer, 279
mixing, 137
mobility, 268
modeling, 4, 24, 94, 162, 179, 206, 247, 260, 263, 264, 275, 276, 279, 281, 284, 285
models, ix, x, 93, 157, 161, 162, 176, 179, 180, 181, 182, 183, 185, 188, 189, 190, 191, 205, 206, 212, 242, 245, 257, 259, 260, 279, 285
modulation, 75, 77, 121, 122, 125
modulus, viii, 92, 130, 141, 149, 150, 151, 261, 262, 263, 282
molecular dynamics, 269, 276
momentum, 67, 68, 71, 75, 79, 94, 108, 145, 169, 258, 259, 260, 268, 274, 275, 279, 284
Moscow, 138
motion, x, 2, 6, 20, 32, 35, 38, 48, 80, 91, 100, 101, 102, 104, 168, 171, 179, 257, 258, 267, 268, 269, 272, 273, 278, 279, 280, 281, 283
MPI, 131
multidimensional, 104, 108
multiplication, 166
myocardium, 90

N

nanotechnology, 17
natural, 142, 144, 212, 224, 240, 241, 242
Navier-Stokes, ix, 100, 161, 205, 206, 283
Navier-Stokes equation, 205, 206, 283
NBC, 117, 118
negativity, 240
neglect, 269
Netherlands, 248, 255
network, 192
New Jersey, 136
New York, 95, 96, 157, 246, 247, 248, 285, 286
Newton, 259, 260, 262, 263, 279
Newtonian, 171, 176, 269, 274
Nielsen, 139, 185
nodes, 238
noise, 2, 5, 18, 31, 50, 93, 94, 103, 116, 121, 122
nonlinear, viii, 66, 67, 90, 99, 100, 101, 103, 104, 111, 119, 121, 122, 124, 126, 131, 132, 134, 136, 137, 138, 139, 141, 142, 143, 146, 151, 152, 153,

154, 156, 157, 158, 210, 249, 250, 251, 255, 256, 259, 260, 281
nonlinear dynamics, 101, 104
nonlinear optics, viii, 141, 152
nonlinear wave equations, 255
nonlinearities, 121, 122
normal, 19, 23, 24, 33, 34, 42, 43, 44, 49, 50, 69, 100, 259, 260, 262, 267, 282
normalization, 6, 18, 25
numerical tool, 90

O

observations, 131, 139, 157
OCT, 210
one dimension, 68
operator, 107, 111, 115, 144, 147, 156, 163, 174, 180, 182
opposition, 163, 167
optical, 146, 157
optical fiber, 146, 157
optical solitons, 146
optics, viii, 141, 142, 152
optimization, ix, 209, 210, 211, 212, 214, 215, 225, 236, 239, 240, 242, 246, 248
orbit, 49
oscillation, 8, 21, 40, 42, 44, 45, 48, 58, 64, 66, 73, 78, 121, 122, 201, 276, 279
oscillations, 4, 20, 22, 32, 43, 49, 57, 59, 61, 73, 79, 93, 122, 124, 138, 139

P

paper, 120, 206, 212, 214, 229, 236
parabolic, 2, 79, 100, 110, 256
parameter, 2, 27, 32, 67, 129, 163, 164, 171, 177, 195, 199, 240, 251, 252, 261, 263, 270, 276
Paris, 207
partial differential equations, vii, viii, 2, 99, 109, 135, 142, 144, 145, 153, 155, 156, 157, 158, 256
particle density, 108, 119
particles, vii, viii, x, 1, 2, 4, 7, 35, 42, 68, 76, 78, 99, 100, 102, 103, 104, 108, 120, 121, 122, 124, 257, 258, 259, 260, 261, 262, 263, 264, 265, 267, 268, 270, 271, 273, 276, 278, 279, 280, 281, 282, 283, 284
performance, ix, 104, 201, 209, 211, 241, 242
periodic, viii, ix, 52, 99, 101, 102, 111, 112, 119, 124, 129, 130, 154, 249, 250
periodicity, 112, 114
permit, 164
personal, 242, 244
perturbation, 103, 121, 122, 123, 125, 126, 268
perturbations, 103, 120, 121, 122, 123, 126, 130, 137
pharmaceutical, 258
pharmaceutical industry, 258
phase space, viii, 99, 102, 103, 104, 107, 108, 118, 119, 122, 124, 125, 126, 143, 266
phenomenology, 123
Philadelphia, 248
physical properties, ix, 161, 162, 205, 279
physical sciences, 205
physics, vii, viii, 1, 2, 3, 5, 24, 32, 49, 66, 79, 84, 93, 94, 100, 102, 103, 124, 141, 142, 152, 162, 179, 201, 205, 258, 259, 264
planning, 246, 247
plants, ix, 209, 210, 211, 213, 214, 224, 229, 230
plasma, vii, viii, 1, 2, 3, 4, 5, 6, 8, 14, 16, 17, 18, 19, 23, 24, 25, 30, 31, 32, 33, 34, 35, 40, 43, 49, 50, 54, 55, 57, 58, 59, 61, 64, 66, 67, 73, 77, 78, 79, 80, 83, 84, 87, 90, 93, 94, 99, 100, 101, 102, 103, 104, 108, 109, 110, 112, 119, 120, 122, 123, 124, 125, 126, 129, 135, 136, 138, 139, 141, 142, 154, 157
plasma physics, vii, viii, 1, 2, 3, 79, 93, 100, 102, 103, 124, 141, 142
plastic, 259, 260, 266, 267
plastics, 258
play, 49, 100, 104, 213
Poisson, 5, 6, 8, 9, 18, 33, 36, 50, 51, 55, 69, 70, 102, 110, 119, 124, 125, 135, 137, 143, 261, 282
Poisson equation, 33, 102, 110, 119, 124, 135
Poisson ratio, 261
polarization, 40, 71, 77, 146
polynomial, 3, 175
polynomials, 3, 102, 112, 251, 255
pond, 13
poor, 115, 201
poor performance, 201
population, 44, 75, 121
power, ix, 54, 66, 67, 103, 112, 162, 210, 211, 212, 213, 214, 217, 224, 229, 230, 231, 240, 241, 242, 243, 247, 249, 262, 273
power generation, 214, 242
power plant, 211, 213, 217, 231
power plants, 211, 213, 217
powers, 270
Prandtl, 189, 191
prediction, 114, 116, 121, 122
present value, 8
pressure, 3, 4, 40, 50, 55, 56, 61, 62, 66, 90, 91, 93, 108, 109, 110, 117, 118, 127, 129, 135, 167, 171, 177, 178, 181, 182, 187, 266, 271, 277
probability, viii, 99
probability distribution, 99
probe, 4, 32
production, 4, 94, 126, 258, 273
program, 239
programming, 210, 246, 247
propagation, 32, 94, 106, 119, 121, 124, 146
property, viii, 141, 145, 149, 151, 154, 190, 201, 226, 264
proposition, 238
protons, 104
pseudo, 102

pulse, vii, 1, 67, 71, 72, 73, 76, 77, 79, 90, 94, 137, 146
pulses, 4, 5, 66, 67, 79, 94
pumping, 213, 214, 224, 228, 229, 240, 241

Q

quantum, 141
quantum fields, 141

R

radial distribution, 271
radiation, 71
radical, 189
radius, 51, 55, 261, 271
random, 269
range, viii, 99, 104, 108, 109, 111, 161, 205, 229, 260
raw material, 259
raw materials, 259
real numbers, 154
real time, 4
reality, 130
recall, 167, 224, 226, 235
recalling, 273
recovery, 213
recurrence, 142, 154, 157, 229, 237
reduction, 3, 167, 250
reflection, 181, 188
regular, viii, 48, 99, 102, 135, 193, 195, 196, 200, 271
relationship, 259
relaxation, 131, 210, 247, 262, 282
relaxation time, 262, 282
relevance, 273
reliability, 210
research, vii, 211, 258, 259, 260, 264, 276, 279
researchers, 264
reservoir, 101, 240, 242
resistance, 273
resistive, 126
resistivity, 111, 126, 127
resolution, ix, 102, 108, 111, 113, 132, 138, 164, 173, 192, 204, 209
resources, 103, 104, 111
restitution, 262, 263, 264, 271
Reynolds, 127, 185, 280, 281
Reynolds number, 127, 185, 280, 281
rotations, 163, 172, 174, 177, 181, 182, 187, 189
roughness, 259, 271
routines, 102
Russian, 205

S

safety, 210
sample, 102, 272
sand, 258, 286
saturation, 73, 121, 122, 137
scalar, 68, 108, 182, 183, 185, 190, 191, 267
scaling, 162, 163, 166, 167, 168, 170, 173, 178, 181, 182, 183, 188, 189, 190, 191, 194, 195
scaling law, 162, 173, 181
scheduling, 246, 247
scientists, 137
search, 228, 229, 235, 238
Second Law of Thermodynamics, 184
separation, vii, 1, 18, 32, 40, 43, 50, 54, 58, 64, 66, 79, 93, 94, 162, 165
series, 215, 261, 284
shape, 271, 278, 284
shares, 100
shear, ix, x, 66, 167, 168, 169, 171, 172, 204, 206, 257, 262, 265, 266, 268, 269, 282
shear rates, x, 257, 266, 269
shock, 2, 91
shock waves, 91
short run, 75
short-term, ix, 209, 210, 246, 247
sign, 8, 67, 73, 91, 105, 106, 189
similarity, 124, 169, 180, 181, 190
simulation, vii, ix, x, 1, 4, 7, 8, 11, 18, 25, 51, 55, 61, 66, 67, 71, 73, 75, 76, 79, 84, 89, 94, 103, 105, 111, 113, 119, 120, 121, 122, 123, 125, 126, 127, 129, 131, 132, 134, 136, 137, 158, 161, 162, 205, 206, 207, 257, 259, 264, 272, 276, 277, 278, 279, 283, 284, 285
simulations, viii, x, 8, 54, 66, 79, 94, 99, 101, 121, 124, 125, 131, 137, 141, 142, 149, 150, 152, 154, 155, 257, 268, 269, 270, 271, 272, 276
Singapore, 136
singular, 118
SLAC, 4
smoothing, 6, 34, 267
SO2, 247
software, 256
solar, 101, 110, 137
solar wind, 110, 137
solid phase, 264, 283
soliton, viii, 141, 142, 144, 146, 149, 151, 153, 156, 157, 158, 250, 251, 253, 254, 255, 256
solitons, 141, 142, 152, 157, 250, 255, 256
solutions, viii, ix, 4, 17, 36, 37, 90, 99, 105, 107, 121, 135, 145, 146, 154, 157, 162, 164, 166, 167, 168, 169, 170, 171, 172, 173, 192, 201, 204, 209, 210, 211, 227, 231, 232, 246, 249, 250, 251, 252, 253, 254, 255, 256, 259, 260
sound speed, 6, 18, 33, 45, 110
space-time, 101
Spain, 209

spatial, viii, 9, 33, 67, 71, 94, 99, 102, 104, 106, 108, 111, 112, 113, 122, 123, 124, 130, 131, 132, 135, 142, 145, 147, 178, 197, 250, 276
spectral component, 114, 116, 119, 121, 122, 123, 125, 126
spectral techniques, viii, 99
spectrum, 4, 103, 111, 116, 126, 130, 132, 133, 134
speed, 2, 27, 35, 67, 90, 104, 109, 146, 276
speed of light, 67, 104, 109
spheres, 259, 260, 261, 262, 268, 271
stability, 105, 111, 123, 124, 125, 135, 183
stars, 100
statistics, 259
steady state, 4, 20, 29, 35, 57, 93
STO, ix, 249, 250
storage, 230
strain, 109, 171, 182, 261, 267, 269, 270, 271
strains, 266, 267, 269
streams, 124
stress, x, 70, 109, 179, 257, 260, 261, 266, 267, 268, 269, 270, 271, 272, 274, 283
subsonic, 101
substitution, 10
sugar, 258
summation rule, 61
Sun, 142, 144, 146, 148, 150, 152, 154, 156, 158
supercomputers, 103
surface friction, 282
surface roughness, 259
surface treatment, 4
switching, 157
symmetry, ix, 161, 162, 163, 164, 165, 166, 168, 171, 174, 175, 176, 177, 179, 180, 181, 183, 185, 187, 188, 191, 192, 193, 194, 195, 201, 205, 206, 250
symplectic, viii, 100, 135, 141, 142, 143, 144, 148, 149, 151, 152, 156, 157, 158, 162, 279
systems, 4, 5, 100, 101, 104, 108, 110, 124, 125, 139, 144, 146, 158, 162, 210, 213, 246, 247, 255, 259, 260, 264, 265, 268, 276, 278, 279, 284

T

targets, 5, 94, 259
Taylor expansion, 9, 70, 106, 107, 113, 114, 118, 200
technology, 146
telecommunication, 146
temperature, 4, 6, 13, 14, 23, 30, 31, 35, 40, 45, 48, 49, 51, 58, 64, 73, 100, 109, 118, 170, 171, 176, 189, 192, 261, 266, 269, 271, 273, 284
temporal, viii, 99
theory, ix, 102, 103, 121, 123, 126, 136, 141, 161, 162, 164, 166, 171, 201, 205, 206, 207, 209, 247, 250, 256, 259, 260, 266, 271, 273, 276
thermal expansion, 170
thermodynamic, 110
thermodynamic properties, 110
thermodynamics, 108, 183, 184, 185
thermonuclear, 100
third order, 111
three-dimensional, 126, 127, 129, 132, 135, 139, 269, 272
threshold, 54
time, viii, 3, 4, 8, 11, 18, 21, 23, 29, 36, 37, 38, 40, 48, 50, 52, 55, 56, 57, 58, 60, 61, 62, 63, 64, 65, 67, 69, 77, 79, 80, 81, 82, 87, 95, 99, 100, 101, 102, 103, 104, 105, 106, 107, 108, 109, 111, 112, 113, 117, 118, 119, 120, 121, 122, 123, 124, 125, 126, 127, 131, 133, 135, 137, 139, 142, 144, 146, 151, 153, 154, 161, 162, 171, 177, 181, 182, 185, 187, 189, 193, 195, 196, 197, 199, 201, 210, 211, 217, 229, 242, 243, 244, 246, 259, 261, 262, 264, 267, 273, 274, 276, 278, 279, 281, 282, 284
time consuming, 279
time increment, 197
time use, 95, 242, 244
tokamak, 4, 49, 61, 94
tolerance, 210
topology, 126
torque, 282, 283
total energy, 30, 120, 132, 133, 134
trajectory, 211
trans, 36
transfer, 101, 126, 132, 213, 273, 279, 284
transformation, 143, 163, 164, 165, 166, 168, 174, 177, 178, 180, 181, 182, 183, 193, 195, 196, 232
transformations, 143, 163, 164, 166, 172, 173, 175, 176, 177, 178, 181, 183, 187, 188, 189, 190, 192, 195, 205, 207, 250
transition, vii, 1, 4, 5, 6, 17, 32, 34, 54, 71, 94, 100, 144, 229
transitions, 3, 101, 144
translation, 105, 107, 108, 165, 171, 181, 190, 195, 197, 201
translational, 2, 274, 281
transmission, 16, 146, 240, 242, 272
transparency, 89, 90
transport, 32, 40, 43, 57, 275
transportation, 205
travel, 40, 151
turbulence, ix, 32, 49, 61, 66, 101, 126, 132, 137, 139, 161, 162, 171, 176, 179, 181, 182, 185, 188, 191, 205, 206, 207, 275, 276
turbulent, 100, 101, 108, 135, 161, 179, 183, 190, 206, 281
two-dimensional (2D), 4, 5, 66, 67, 68, 71, 124, 127, 131, 132, 205, 270

U

uniform, 9, 10, 11, 35, 40, 102, 127, 130, 231
USSR, 205

V

vacuum, 69, 71, 73, 79, 80, 86
validity, 71
values, vii, 3, 7, 8, 13, 16, 17, 20, 21, 22, 40, 50, 54, 82, 103, 111, 112, 113, 114, 117, 118, 126, 134, 135, 168, 193, 196, 198, 210, 211, 213, 214, 220, 230, 237, 240, 242, 260, 262, 263, 267, 270
variability, 135
variable, 9, 103, 111, 130, 145, 151, 165, 167, 169, 171, 176, 192, 242, 250, 251, 252
variables, 111, 117, 153, 164, 165, 168, 224
variance, 103, 269
variation, 3, 5, 21, 42, 53, 55, 59, 61, 73, 79, 101, 158, 173, 174, 206, 217, 269
vector, 67, 68, 71, 73, 77, 111, 114, 117, 126, 130, 149, 150, 173, 174, 175, 176, 177, 178, 187, 277, 278, 280, 281, 282, 283
velocity, 3, 4, 5, 6, 8, 9, 13, 16, 19, 20, 25, 30, 31, 32, 33, 34, 35, 45, 48, 50, 52, 54, 66, 67, 77, 90, 91, 103, 104, 105, 107, 108, 109, 110, 114, 117, 118, 119, 120, 121, 122, 123, 124, 125, 127, 129, 131, 151, 168, 171, 176, 178, 185, 186, 199, 259, 260, 262, 263, 264, 265, 266, 267, 272, 273, 275, 277, 278, 279, 280, 281, 283
vibration, x, 257, 265, 279, 284, 285
viscoelastic materials, x, 257, 261
viscosity, x, 109, 110, 111, 127, 129, 168, 171, 183, 257, 264, 269, 270, 271, 275, 281
visible, 14, 76, 85, 121, 122, 123, 132
visualization, 277, 280
vortex, 85, 86, 87, 120, 123, 124, 171, 173
vortices, 124, 125, 126

W

Warsaw, 97
water, 2, 3, 90, 94, 124, 210, 212, 213, 217, 220, 224, 228, 229, 230, 231, 232, 240, 241, 242, 243, 245
wave equations, 156, 157, 255
wave number, 121, 123, 130
wave packet, 151, 157
wave packets, 151, 157
wave propagation, 83, 90, 92, 93, 137, 157
wave vector, 126, 130, 133, 134
wavelengths, 120, 121, 132
wealth, 5
wind, 110, 137

Y

yield, 173, 266, 267